Numerical Modelling in
Geomechanics

Numerical Modelling in
Geomechanics

edited by
Manuel Pastor &
Claudio Tamagnini

First published in 2002 by Lavoisier and Hermes Science Publications
First published in Great Britain and the United States in 2004 by Kogan Page
Science, an imprint of Kogan Page Limited

Kogan Page Limited
120 Pentonville Road
London N1 9JN
UK
www.koganpagescience.com

Kogan Page India
4737/23 Ansari Road
New Delhi- 110002

First South Asian Edition 2007

ISBN 1-9039-9642-2

British Library Cataloguing-in-Publication Data
A CIP record for this book is available from the British Library.

Typeset by Kogan Page
Printed in Brijbasi Art Press Ltd., I-72, Sector-9, Noida, U.P. India.

Contents

Foreword

The alliance of Laboratories in Europe for Research and Technology (ALERT) "Geomaterials" was created in 1989 as a pioneering attempt to develop a European school of thought in the field of mechanics of geomaterials. The generic name "Geomaterials" is viewed as the gathering together of materials whose mechanical behaviour depends on the pressure level, which can be dilatant under shearing and are multiphase because of their porous structure. So the "geomaterials" label brings together mainly concrete, soils and rocks.

ALERT now includes 20 European universities or organisations that are most active in the field of numerical modelling of geomaterials and geostructures.

Its main areas of interest are:

- micromechanics and constitutive modelling for geomaterials engineering;

- failure, strain localisation and instabilities;

- large scale computations for geomaterials and geostructures;

- integrity of geostructures and inverse analysis in geomechanics;

- environmental geomechanics and durability of geomaterials.

Numerical modelling has originated a deep revolution in the available tools and methods in geomechanics and the present situation is certainly just the beginning. Among the various numerical methods available (finite elements, discrete elements, finite differences, boundary integrals, etc) this school focuses only on the first ones, which are certainly the most utilised today. In this wide domain, many aspects are considered: brittle and ductile materials, water saturated and partially saturated geomaterials, small and large strains, steady state and transient problems, soil dynamics, strain localisation, and some applications related to natural hazards.

This school is organised and coordinated by Manuel Pastor and Claudio Tamagnini. On behalf of the ALERT Board of Directors and of all the members of ALERT, I would like to warmly acknowledge Manuel and Claudio for all the work they have done on this publication.

Félix Darve
Director, ALERT Geomaterials
INPG, Grenoble, France

Chapter 1

Steady State Problems of Elliptic Type

Manuel Pastor and Pablo Mira

Department of Computational Engineering, Centro de Estudios y Experimentacion de Obras Publicas, Madrid, Spain, and Department of Applied Mathematics, ETS de Ingenieros de Caminos, Madrid, Spain

1. Introduction

Many engineering problems involve complex phenomena. Just to mention an example, let us consider the case of the fast flowslide of Santa Tecla, El Salvador, which was caused by the earthquake of January 13th, 2001. Santa Tecla is a small residential village located on the outskirts of San Salvador. The slope which failed was 100m high approximately, with an inclination of 45°. The material was a loose, collapsible, non-saturated soil of volcanic origin. While collapsing, the tendency to compact made pore air pressures increase, arriving at a state which could be described as "dry liquefaction". The mass of soil started to move like a fluid. Basal friction probably increased temperature with the subsequent increment of pore pressure. Suspension of fine particles, and air entrapment on the surface could have also contributed to a higher mobility. The fluidized soil moved at high speed towards the village, burying the houses along its path until it came to rest, and causing numerous victims.

To understand such complex phenomena we need to make some simplifying assumptions which allow us to formulate a mathematical model. Quite frequently, the model consists of a set of partial differential equations (PDEs) for which there is no analytical solution. In these cases it is necessary either to further simplify the model or to obtain an approximate solution using numerical methods.

There exists today a wide variety of numerical methods which can be used to approximate either the solution of a PDE or the equation itself. Finite differences (FD), finite elements (FEM), boundary integral or element methods (BEM) and finite volumes (FV) are worthy of mention.

This course is devoted to FEM, as it is perhaps the most widespread today.

Partial differential equations can be classified into three main types: elliptic, parabolic and hyperbolic. As examples of elliptic PDEs we can mention:

(i) Structural engineering problems where loads are very slowly applied (There are exceptions: in some cases, the problem changes from elliptic to hyperbolic and becomes ill-posed. This is the case of localized failure in elastoplastic softening materials)

(ii) Steady state seepage and heat transfer problems.

Parabolic and hyperbolic problems correspond to transient problems in general, such as transient heat transfer and seepage, convective transport of pollutants and solid and fluid dynamics problems. We will see later that one main difference between both types is that parabolic problems introduce damping or diffusion while hyperbolic problems do not.

Numerical methods for problems of a different type present important particularities, and this is why we will present them separately.

This chapter is devoted to presenting an overview of finite element techniques applied to elliptic problems.

2. Alternative Formulations and Mathematical Models

2.1. *Introduction to strong, weak and variational formulations for 1D elliptic problems*

There exist in mechanics different alternative ways to formulate a problem. For a moving particle under a given force, it is possible to: (i) integrate the equations of motion (ii) use the balance of linear momentum equations, and (iii) use conservation of energy principles. The choice depends on the particular problem being solved. In the case of solid and structural mechanics, it is also possible to integrate the equilibrium equations, or apply energy or conservation principles.

The situation is different when seeking approximate solutions of PDEs. We will see that there are approximate solutions which are continuous on the whole domain, but their derivatives are not continuous. Therefore, they cannot be used in formulations involving second order derivatives.

We will consider first two simple one dimensional cases, heat conduction in a bar, and deformation of an elastic bar under axial loading.

The former can be stated as: find the temperature $\phi(x)$ that satisfies

$$-\frac{d}{dx}\left(k\frac{d\phi}{dx}\right) = s \qquad x \in (0, L) \tag{1}$$

$$\phi(0) = 0 \qquad -k\frac{d\phi}{dx}\,|_{x=L} = q_L \tag{2}$$

where k is the thermal conductivity, s the heat sources and q_L the prescribed flux at $x = L$, while the latter would be: find the displacement field $u(x)$ that satisfies

$$-\frac{d}{dx}\left(EA\frac{du}{dx}\right) = b \qquad x \in (0, L) \tag{3}$$

$$u(0) = 0 \qquad EA\frac{du}{dx}\,|_{x=L} = F_L \tag{4}$$

where E is the Young modulus, A the cross sectional area, b the distributed forces and F_L the force acting at $x = L$. Both problems are two-point boundary value problems, which can be written in a non-dimensional general form as:

$$-\frac{d^2\phi}{dx^2} = f \qquad x \in (0, 1) \tag{5}$$

$$\phi(0) = \phi_0 \qquad -\frac{d\phi}{dx}\,|_{x=1} = q \tag{6}$$

The solution of the problem is a smooth function belonging to the class $C^2([0, 1])$ of continuous functions with continuous first and second order derivatives in the interval $[0, 1]$. The open domain can be noted as Ω. Its boundary is usually denoted by $\partial\Omega$ or Γ, and consists here in the two points $x = 0$ and $x = 1$. We will use $\overline{\Omega}$ to refer to the union of the domain and its boundary, i.e., its closure $\overline{\Omega} = \Omega \cup \partial\Omega$.

If we multiply the PDE by a smooth function $\psi(x)$ satisfying $\psi(0) = 0$ and integrate by parts we obtain

$$\int_0^1 \frac{d\phi}{dx}\frac{d\psi}{dx}dx - \frac{d\phi}{dx}\psi\bigg|_0^1 = \int_0^1 \psi f\, dx \qquad (7)$$

and, taking into account that $-\frac{d\phi}{dx} = q$ at $x = 1$, and $\psi(0) = 0$

$$\int_0^1 \frac{d\phi}{dx}\frac{d\psi}{dx}dx + q\psi(1) = \int_0^1 \psi f\, dx \qquad (8)$$

The above equation can be written in a more compact way if we introduce the notation:

$$a(u, v) \quad = \quad \int_0^1 \frac{du}{dx}\frac{dv}{dx}dx \qquad (9)$$

$$(f, v) \quad = \quad \int_0^1 vf\, dx \qquad (10)$$

resulting in:

$$a(\phi, \psi) = (f, \psi) - q\psi(1) \qquad (11)$$

In the above, $a(.\,,.)$ is a bilinear form, and $(.\,,.)$ an inner product. It can be observed that both the solution ϕ and the function ψ do not need to belong to the class C^2 [0,1]. Indeed, we do not need to require even continuity of first derivatives of ϕ and ψ. The space V in which we seek the solution consists of functions v which are square integrable, i.e., $v \in L^2$ with first derivatives which are also square integrable. This space is a Sobolev space which is denoted by H^1 [0,1] (the superindex indicates the order of the derivatives which will be squarely integrable). The space is then defined as:

$$H^1[0, 1] = \left\{ v : [0, 1] \to \mathbb{R} \;\bigg|\; \int_0^1 v^2 dx < \infty\,, \int_0^1 \left(\frac{dv}{dx}\right)^2 dx < \infty \right\} \qquad (12)$$

Sometimes, we add the extra condition $v(0) = 0$, and we write $H_0^1[0, 1]$

The problem (11) can then be stated as:

Find $\phi \in U = \big\{ u : [0, 1] \to \mathbb{R} \mid u \in H^1[0,1] \text{ and } u(0) = \phi_0 \big\}$ such that

$$a(\phi, \psi) = (f, \psi) - q\psi(1) \quad \forall \psi \in V \qquad (13)$$

where the space V is defined by $V = \big\{ v : [0, 1] \to \mathbb{R} \mid u \in H^1[0, 1] \text{ and } v(0) = 0 \big\}$

The space U is usually referred to as solution space or space of trial functions, the space V being the space of variations or the space of test functions. This alternative formulation is referred to as "weak formulation" or "variational formulation" of the problem (5). It can be proved that if ϕ is a solution of the strong problem then it is the solution of the weak problem and viceversa.

Note 1 Boundary conditions of the type imposed at $x = 0$ i.e. $\phi(0) = \phi_0$ are called "essential". Boundary condition at $x = 1$ consists of prescribing the flux, and they will not enter the formulation explicitly if the prescribed flux is zero. Sometimes the function ψ is denoted $\delta\psi$ and it can be thought as the admissible variation of a function ψ belonging to U fulfilling the essential boundary condition so that the variation $\delta\psi$ is zero at $x = 0$. However, both spaces U and V can be different, as in the case of Petrov-Galerkin methods which have been used for convection dominated problems.

Note 2 In the case of the elastic bar, we can write:

$$a(u, \delta v) = \int_0^L EA\frac{du}{dx}\frac{d}{dx}(\delta v)\,dx \tag{14}$$

$$(f, \delta v) = \int_0^L f\,\delta v\,dx \tag{15}$$

The weak formulation is precisely the principle of virtual work:

$$\int_0^L EA\frac{du}{dx}\frac{d}{dx}(\delta v)\,dx = \int_0^L f\,\delta v\,dx + F_L\delta v\,|_{x=L} \tag{16}$$

$$\int_0^L \sigma A\,\delta\varepsilon\,dx = \int_0^L f\,\delta v\,dx + F_L\delta v\,|_{x=L}$$

i.e., under a virtual displacement field compatible with essential boundary conditions the virtual work done by the internal and by the external forces is the same. An alternative statement of the problem consists of finding the function u which makes stationary the functional

$$\Pi = \frac{1}{2}\int_0^L \sigma A\,\varepsilon\,dx - \int_0^L f\,v\,dx - F_L v\,|_{x=L} \tag{17}$$

Note 3 The spaces U and V are infinite dimensional. We can find approximate solutions in finite dimensional spaces $U^h \subset U$ and $V^h \subset V$. The problem of finding $u_h \in U$ satisfying $a\,(u_h, v_h) = (f, v_h) - qv_h(1)\ \forall v_h \in V^h$ is the basis for Galerkin methods.

We will consider here two classical problems of elliptic type, seepage of water through soil and deformation of an elastic body, which will be treated later in more detail.

2.2. Two and three dimensional problems

2.2.1. Flow in Porous Media

Let us define the average velocity of water relative to soil skeleton w which is obtained dividing the real velocity in the pores by the factor $n.S_w$ where n is the porosity and S_r the degree of saturation. The balance of momentum equation is:

$$-\nabla p_w + \rho_w \mathbf{g} = \mathbf{k}_w^{-1}.\mathbf{w} \tag{18}$$

where ∇ is the gradient operator $(\partial_x, \partial_y, \partial_z)^T$ in 3D, and $(\partial_x, \partial_y)^T$ in 2D situations; p_w and ρ_w the pore water pressure and density, g the vector of body forces and $\mathbf{k}_w$ is the tensor of permeability. Combining it with the balance of mass equation

$$\nabla^T.\mathbf{w} + \left(C_s + \frac{1}{Q}\right)\frac{\partial p_w}{\partial t} = 0 \tag{19}$$

where ∇^T is the divergence operator, C_s is the specific storage coefficient and $\frac{1}{Q}$ a factor accounting for the compressibility of water and soil, we obtain

$$\nabla^T\left(\mathbf{k}_w\left(\nabla\phi\right)\right) = \left(C_s + \frac{1}{Q}\right)\frac{\partial p_w}{\partial t} \tag{20}$$

where we have introduced the potential ϕ such that $\nabla\phi = \nabla p_w + \rho_w \mathbf{g}$

The steady state condition is:

$$\nabla^T\left(\mathbf{k}_w \nabla\phi\right) = 0 \quad \text{in } \Omega \tag{21}$$

Essential or Dirichlet boundary conditions are imposed at $\partial\Omega_\phi$

$$\phi - \bar{\phi} = 0 \tag{22}$$

and the fluxes are prescribed at $\partial\Omega_q$ (Neumann BCs) as:

$$\left(\mathbf{k}_w \nabla\phi\right)^T.\mathbf{n} + \bar{q} = 0 \tag{23}$$

where n is is the unit normal vector to the boundary.

The weak formulation can be obtained by multiplying the PDE by a function ψ, integrating it over the domain and then applying Gauss theorem:

$$\int_\Omega \nabla\psi^T \mathbf{k}_w \nabla\phi \, d\Omega = \int_{\partial\Omega_q} \psi\bar{q} \, d\Gamma \tag{24}$$

Introducing $a(\phi, \psi) = \int_\Omega \nabla\psi^T \mathbf{k}_w \nabla\phi \, d\Omega$, the problem can be stated as:

Find $\phi \in U = \left\{\phi : \bar{\Omega} \to \mathbb{R} \mid \phi \in H^1(\bar{\Omega}) \text{ and } \phi - \bar{\phi} = 0 \text{ on } \partial\Omega_\phi\right\}$ such that $a(\phi, \psi) = \int_{\partial\Omega_q} \psi\bar{q} \, d\Gamma \; \forall\psi \in V = \left\{\psi : \bar{\Omega} \to \mathbb{R} \mid \psi \in H^1(\bar{\Omega}) \text{ and } \psi = 0 \text{ on } \partial\Omega_\phi\right\}$

Note 4 We have applied Gauss's theorem to $\int_\Omega \psi \, \nabla^T \left(\mathbf{k}_w \left(\nabla \phi\right)\right) d\Omega = 0$. If we define the residual $R_\Omega = \nabla^T \left(\mathbf{k}_w \left(\nabla \phi\right)\right)$, the equation can be rewritten as $\int_\Omega \psi R_\Omega \, d\Omega = 0$.

The variational formulation of this problem would consist of finding $\phi \in U$ which makes stationary the functional

$$\Pi = \frac{1}{2} \int_\Omega \nabla \phi^T \mathbf{k}_w \nabla \phi \, d\Omega - \int_{\partial \Omega_q} \phi \bar{q} \, d\Gamma \tag{25}$$

with respect to variations $\delta\phi \in V$.

2.2.2. The Linear Elastic Problem

The balance of momentum equation can be written as:

$$\mathbf{S}^T.\boldsymbol{\sigma} + \mathbf{b} = 0 \tag{26}$$

where $\boldsymbol{\sigma}$ is the vector representation of Cauchy stress tensor and $\mathbf{S}^T$ is the discrete divergence operator, which in 3D are given by

$$\mathbf{S}^T = \begin{pmatrix} \partial_x & 0 & 0 & \partial_y & 0 & \partial_z \\ 0 & \partial_y & 0 & \partial_x & \partial_z & 0 \\ 0 & 0 & \partial_z & 0 & \partial_y & \partial_x \end{pmatrix} \text{ and } \boldsymbol{\sigma} = \begin{pmatrix} \sigma_{xx} \\ \sigma_{yy} \\ \sigma_{zz} \\ \sigma_{xy} \\ \sigma_{yz} \\ \sigma_{zx} \end{pmatrix} \tag{27}$$

In the case of plane strain, we will have

$$\mathbf{S}^T = \begin{pmatrix} \partial_x & 0 & \partial_y \\ 0 & \partial_y & \partial_x \end{pmatrix} \text{ and } \boldsymbol{\sigma} = \begin{pmatrix} \sigma_{xx} \\ \sigma_{yy} \\ \sigma_{xy} \end{pmatrix} \tag{28}$$

The stress is related to the infinitesimal strain ε by the elastic constitutive relation

$$\boldsymbol{\sigma} = \mathbf{D}^e.\boldsymbol{\varepsilon} \tag{29}$$

and, finally, the strain can be written in terms of the displacement field $\mathbf{u}$ by

$$\boldsymbol{\varepsilon} = \mathbf{S}.\mathbf{u} \tag{30}$$

Concerning the boundary conditions, we will assume Dirichlet boundary conditions on $\partial \Omega_u$ (prescribed displacements)

$$\mathbf{u} - \bar{\mathbf{u}} = 0 \tag{31}$$

and we will assume that on $\partial \Omega_t$ we know the forces per unit area $\mathbf{t}$ acting on the surface

$$\sigma_{ij} n_j - t_i = 0 \tag{32}$$

where we have used the tensor form. Of course, $\partial\Omega = \partial\Omega_u \cup \partial\Omega_t$ with $\partial\Omega_u \cap \partial\Omega_t = \{\varnothing\}$.

It can be seen that the problem involves stresses, deformation and displacements as variables, but it can written in terms of displacements as:

$$\mathbf{S}^T.\mathbf{D}^e\mathbf{S}.\mathbf{u} + \mathbf{b} = 0 \qquad (33)$$

There exist other alternatives, and the problem can be formulated in terms of (i) displacements and hydrostatic confining pressures $p = -\frac{1}{3}\mathrm{tr}\,(\sigma)$, (ii) stresses and displacements, (iii) stresses, deformations and displacements. These formulations are referred to as mixed formulations, and present important advantages. The main dissadvantage is that, in general, the number of unknowns increases and solving the discretized problem involves solving a much larger system of equations.

Concerning the displacement formulation, it involves second derivatives of displacements. If the solution is being approximated in a space U, it will be necessary to use functions of class C^2.

The alternative weak formulation is:

Find $\mathbf{u} \in U = \left\{ H^1\,(\bar{\Omega}) \rightarrow \mathbb{R}^{\mathrm{ndim}}\,|\mathbf{u} - \bar{\mathbf{u}} = 0 \text{ on } \partial\Omega_t \right\}$ such that

$$\int_\Omega (\mathbf{S}.\mathbf{v})^T.\sigma d\Omega = \int_\Omega \mathbf{v}^T.\mathbf{b}\,d\Omega + \int_{\partial\Omega_t} \mathbf{v}^T.\mathbf{t}\,d\Omega \qquad (34)$$

for all $\mathbf{v} \in V = \left\{ H^1\,(\bar{\Omega}) \rightarrow \mathbb{R}^{\mathrm{ndim}}\,|\mathbf{v} = 0 \text{ on } \partial\Omega_t \right\}$.

This weak formulation is indeed the expression of the Principle of Virtual Work (We can think that $\mathbf{v}$ represents an admissible variation of the displacement field $\mathbf{u}$, so we can write it as $\delta\mathbf{v}$. Then, $\mathbf{S}.\delta\mathbf{v} = \delta\varepsilon$ and the PVW follows).

Finally, we can make stationary the functional Π :

$$\Pi = \int_\Omega \left(W(\mathbf{u}) - \mathbf{v}^T.\mathbf{b}\,\right) d\Omega\ - \int_{\partial\Omega_t} \mathbf{u}^T.\mathbf{t}\,d\Omega \qquad (35)$$

with respect to variations $\delta\mathbf{v} \in V$. In above equation, $W(\mathbf{u})$ is the density of elastic energy.

3. Finite Element Spaces

3.1. *Introduction: elements, nodes and shape functions*

In the preceding section we have seen how a given problem can be cast in different ways: strong, weak, and variational formulations. Although they are equivalent, they present differences which will become apparent when obtaining approximate solutions.

The main purpose of this section is to show how to obtain approximate solutions using simpler finite dimensional spaces U^h and V^h. Here, h will refer to a characteristic discretization size inversely proportional to the dimension of the spaces.

A general function u^h belonging to U^h can be expressed in terms of a base $\{\phi_i, i = 1..n\}$ of the space:

$$u^h = \sum_{i=1}^{n} a_i \phi_i \tag{36}$$

Finite element spaces are a particular case of finite dimensional function spaces obtained using a decomposition of the domain Ω into a set of $nelem$ disjoint domains called **elements**, of simpler geommetrical form such as triangles and quadrilaterals in 2D or tetrahedra or hexahedra in 3D. The domain $\bar{\Omega}$ is approximated by $\bar{\Omega}^h$

$$\bar{\Omega} \approx \bar{\Omega}^h = \cup_{e=1}^{nelem} \bar{\Omega}^e \tag{37}$$

The idea is to approximate the solution u within each element $u^{h,e}$ using simple functions. The global solution can be expressed as

$$u^h = \sum_{e=1}^{nelem} u^{h(e)} \chi_e \tag{38}$$

where χ_e is a function which is zero outside the element e and one in it. Therefore the approximation has been built using "piecewise defined" functions.

The function is defined in each element as $u^{h(e)} = \sum_{i=1}^{ne} a_i \phi_i^{(e)}$ where ne is the number of functions used in each element to build the approximation. An important class of piecewise defined functions are interpolation polynomials defined over a set of $nnode$ points in each element, which are referred to as **nodes**. If we build the polynomials $N_{inode}^{(e)}(x)$, $x \in \bar{\Omega}^e$ in such a way that the $inode$ poyinomial is made zero at all nodes except at node $inode$, where its value is unity, the approximation within element $ielem$ can be expressed as:

$$u^{h(e)} = \sum_{inode=1}^{nnode} \hat{u}_{inode}^{(e)} N_{inode}^{(e)} \tag{39}$$

This approximating trial functions are commonly referred to as "**shape functions**".

We will describe next some cases of interest.

3.2. *Finite elements in one dimension*

We will consider the 1D domain $\Omega = \{x \in [0, L]\}$ and we will divide into n finite elements. We will also define the $n + 1$ nodes which are shown in Figure 1. We will use these nodes as support for linear interpolation polynomials within each element, as we show in Figure 2.

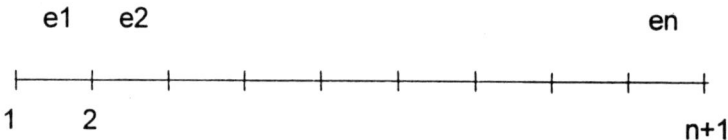

Figure 1. *1D finite element mesh*

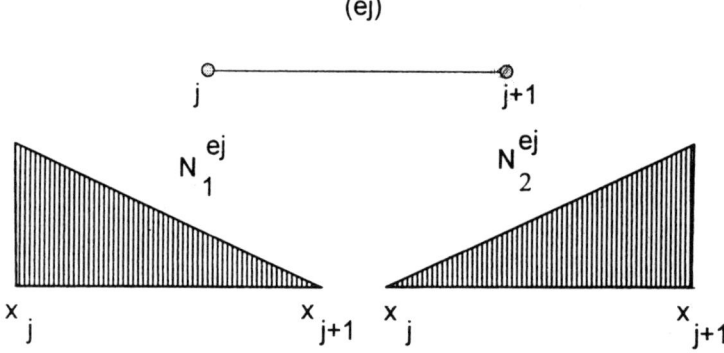

Figure 2. *Shape functions of element (ej)*

The function $u(x)$ is approximated within each element by

$$u^{h(ej)} = \sum_{inode=1}^{nnode} \hat{u}_{inode}^{(ej)} N_{inode}^{(ej)} \quad \text{where } nnode = 2, \text{ or in a more compact manner:}$$

$$u^{h(ej)} = \mathbf{N}^{(ej)}.\hat{\mathbf{u}}^{(ej)}$$

where

$$\mathbf{N}^{(ej)} = \left(\begin{array}{cc} N_1^{(ej)} & N_2^{(ej)} \end{array} \right)$$

and

$$\hat{\mathbf{u}}^{(ej)} = \left(\begin{array}{c} \hat{u}_1^{(ej)} \\ \hat{u}_2^{(ej)} \end{array} \right)$$

The shape functions are given by $N_1^{(ej)} = (x_{j+1} - x)/(x_{j+1} - x_j)$ and $N_2^{(ej)} = (x - x_j)/(x_{j+1} - x_j)$

The global approximation function u^h we have built belongs to the class C^0, has first derivatives which are discontinuous between elements, and the second derivatives are infinity between elements. However, it can be seen that $u^h \in H^1 (\bar{\Omega})$,i.e., both u^h and it first derivative are square integrable.

The global approximation can be also expressed as $u^h = \sum_{i=1}^{n+1} \hat{u}_i \, N_i^{(g)}$ where $N_i^{(g)}$ are **global shape functions** which are shown in Figure 3.

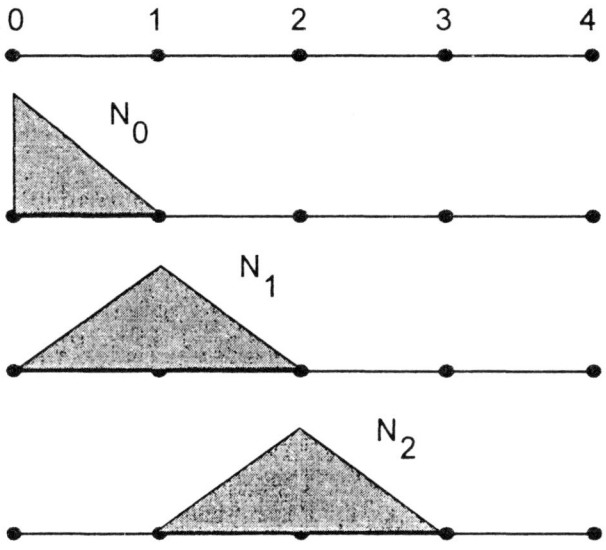

Figure 3. *Global shape functions*

We can also write
$$u^h = \mathbf{N}.\hat{\mathbf{u}}$$

where
$$\mathbf{N} = \begin{pmatrix} N_1 & N_2 & ... & N_{n+1} \end{pmatrix}$$

and
$$\hat{\mathbf{u}} = \begin{pmatrix} \hat{u}_1 \\ \hat{u}_2 \\ ... \\ \hat{u}_{n+1} \end{pmatrix}$$

Note 5 In order to make the functions u^h of the space U^h fulfill an essential boundary condition imposed at $x = 0$, $u - \bar{u} = 0$, the approximation could be written as $u^h = \sum\limits_{i=2}^{n+1} \hat{u}_i \, N_i^{(g)} + \bar{u} \, N_1^{(g)}$, and the space of trial functions U^h would consist of functions u^h that could be written in this way. The space of test functions V^h would consist of functions that could be written as $v^h = \sum\limits_{i=2}^{n+1} \hat{v}_i \, N_i^{(g)}$ which fulfill the condition of being zero at $\partial\Omega_u$, $i.e., x = 0$.

3.2.1. Mapping: isoparametric elements

Shape functions of element (ej) are given by $N_1^{(ej)} = (x_{j+1} - x) / (x_{j+1} - x_j)$ and $N_2^{(ej)} = (x - x_j)/(x_{j+1} - x_j)$, and vary from element to element. It would be convenient to define shape functions in a such way that they would be always the same. This is achieved by geometrical mappings which transform the elements into a standard one. In the case of the linear 1D elements we are considering, we introduce the abscissa $\xi \in [0, 1]$

$$\xi = \frac{x - x_j}{x_{j+1} - x_j}$$

Shape functions are now given by $N_1^{(ej)} = 1 - \xi$ and $N_2^{(ej)} = \xi$. The abscissa x is given by $x = x_j + \xi\,(x_{j+1} - x_j) = (1 - \xi)\,x_j + \xi x_{j+1}$ which can be written as

$$x = N_1^{(ej)} x_{j-1} + N_2^{(ej)} x_j$$

Therefore, the mapping can be defined by the shape functions which were used for building the approximation, and the element is referred to as **isoparametric** (Figure 4).

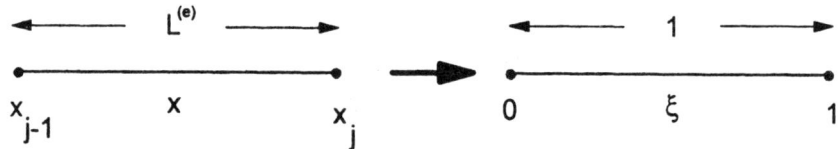

Figure 4. *Isoparametric Transformation*

Derivatives are obtained as:

$$\frac{\partial u^{h(ej)}}{\partial x} = \frac{\partial u^{h(ej)}}{\partial \xi} \cdot \frac{\partial \xi}{\partial x} = \frac{\partial \hat{u}^{(ej)}}{\partial \xi} \cdot \frac{1}{L^{(ej)}}$$

where $L^{(ej)}$ is the length of the element. If we now substitute $u^{h(ej)} = \mathbf{N}^{(ej)}.\hat{\mathbf{u}}^{(ej)} in$ above expression, we obtain

$$\frac{\partial u^{h(ej)}}{\partial x} = \frac{1}{L^{(e)}} \left[\begin{array}{cc} \frac{\partial N_1^{(ej)}}{\partial \xi} & \frac{\partial N_2^{(ej)}}{\partial \xi} \end{array} \right] \cdot \left(\begin{array}{c} \hat{u}_1^{(ej)} \\ \hat{u}_2^{(ej)} \end{array} \right)$$

$$= \frac{1}{L^{(e)}} \begin{bmatrix} -1 & 1 \end{bmatrix} \cdot \begin{pmatrix} \hat{u}_1^{(ej)} \\ \hat{u}_2^{(ej)} \end{pmatrix}$$

which is usually written as

$$\frac{\partial u^{h(ej)}}{\partial x} = \mathbf{B} . \hat{\mathbf{u}}^{(ej)}$$

where $\mathbf{B}^{(ej)}$ is the discrete first derivative operator. If we introduce $S = \frac{\partial}{\partial x}$, it follows that $\mathbf{B}^{(ej)} = S . \mathbf{N}^{(ej)}$

3.3. Finite elements in two and three dimensions

There exists today a wide variety of two dimensional elements, from the simplest linear triangles to much more complex elements such as, for instance, the 15 noded triangles which have become popular in geotechnical finite element codes because of their robustness. The choice of element depends, of course, on the element avail ability in the computer code we are using. First finite element programmers favoured simple triangle and quadrilaterals for the simplicity and speed of computations. However, it soon became apparent that such simple elements presented important inconveniences (poor convergence rate, poor behaviour in bending, locking in quasi-incompressible materials...). Because of this, and also because of availability of frontal solvers, during a second period, more complex elements were used. Today, further theoretical developments have allowed one to improve the simplest elements. This is the case of the "enhanced strain" quadrilaterals, with excellent behaviour in bending dominated situations and when the materials are close to imcompressibility, or some pressure-displacement mixed triangles.

In the domain of computational fluid dynamics, the situation is different, as tri-angles and tetrahedra are the favourite choices in large scale problems with more than 10^6 degrees of freedom.

The simplest element in 2D is the linear triangle. Figure 5 shows the domain Ω and its approximation Ω^h by a mesh of triangular finite elements. It can be seen how the boundary $\partial\Omega$ is also approximated by $\partial\Omega^h$. All the elements can be mapped into the normalized triangle which can be seen in the figure, using the same shape functions of the triangle:

$$N_1^{(e)} = 1 - \xi - \eta \tag{40}$$
$$N_2^{(e)} = \xi$$
$$N_3^{(e)} = \eta$$

$$\mathbf{x} = \begin{pmatrix} N_1^{(e)}(\xi,\eta) & N_2^{(e)}(\xi,\eta) & N_3^{(e)}(\xi,\eta) \end{pmatrix} \cdot \begin{pmatrix} x_1^{(e)} \\ x_2^{(e)} \\ x_3^{(e)} \end{pmatrix} \tag{41}$$

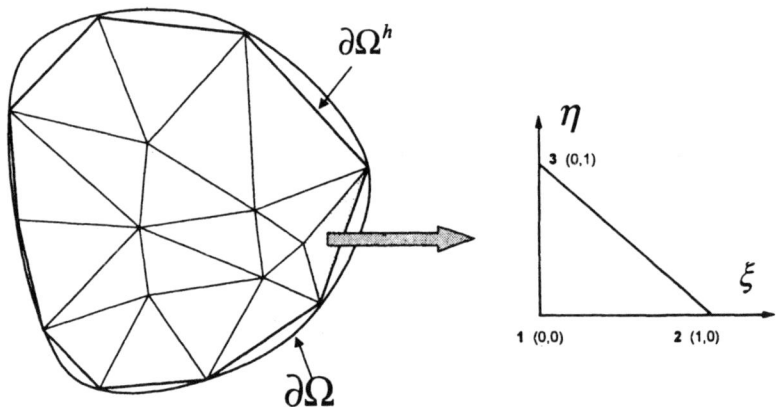

Figure 5. *Mesh of linear triangles*

Among triangles of higher order, we can mention the 6 noded quadratic triangle, which is sketched in Figure 6. This element provides a better approximation to curved boundaries, as it approximates them as segments of parabolic lines.

The bilinear quadrilateral is sketched in Figure 7, and the "serendiptic" 8 noded quadrilateral can be seen in Figure 8.

Concerning 3D problems, we could mention the linear and quadratic tetrahedra, with 4 and 10 nodes, and the 8 and 20 nodes hexahedral or "bricks". All of them are isoparametric elements.

An important point is the evaluation of derivatives inside elements, which are needed to obtain approximations of the gradient operator, fluxes, strain and stress, for instance.

The gradient of the approximation, $\nabla u^{h(e)}$ can be obtained substituing $u^{h(e)} = N^{(e)} . \hat{u}^{(e)}$, which results in

$$\nabla u^{h(e)} = \nabla N^{(e)} . \hat{u}^{(e)} = G^{(e)} . \hat{u}^{(e)} \tag{42}$$

where we have introduced the discrete gradient operator $G^{(e)} = \nabla N^{(e)}$. In the case of 3D, it is given by:

$$G^{(e)} = \begin{pmatrix} \partial_x N_1^{(e)} & \cdots & \partial_x N_{nnode}^{(e)} \\ \partial_y N_1^{(e)} & & \partial_y N_{nnode}^{(e)} \\ \partial_z N_1^{(e)} & & \partial_z N_{nnode}^{(e)} \end{pmatrix} \tag{43}$$

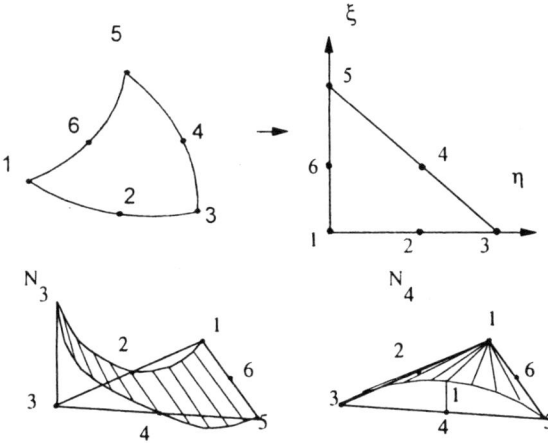

Figure 6. *Quadratic triangle*

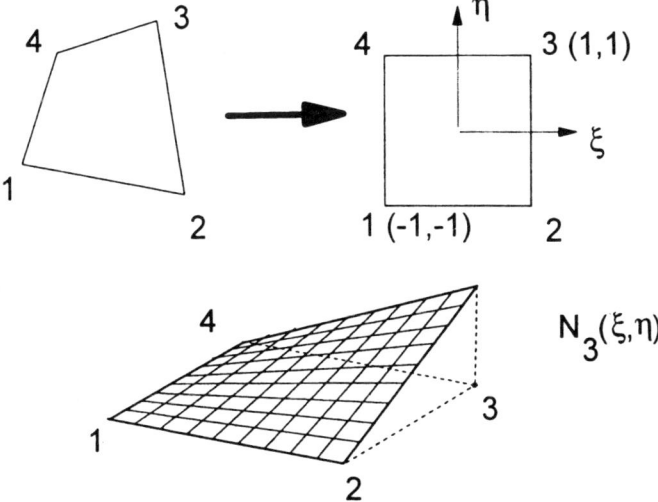

Figure 7. *Bilinear Quadrilateral*

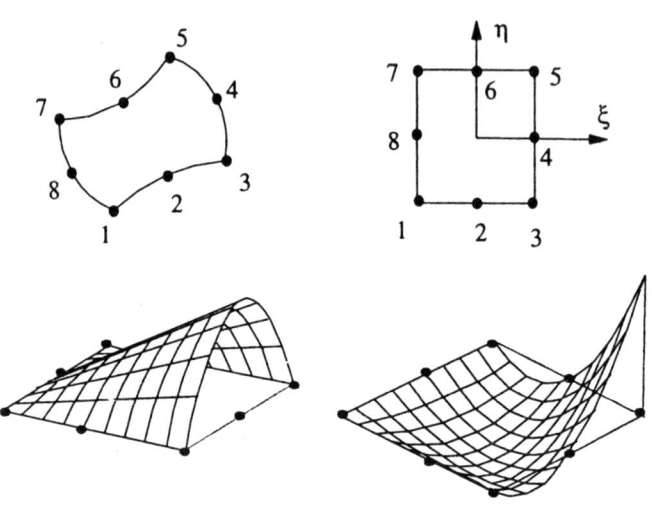

Figure 8. *8 noded quadrilateral*

We can follow a similar method to obtain the strain within a particular element. The strain is related to the displacements $\mathbf{u}^{h(e)}$ by $\varepsilon^{h(e)} = \mathbf{S}.\mathbf{u}^{h(e)}$ where $\mathbf{S}$ is the strain operator. If we write the displacement field in terms of the shape functions, we arrive at:

$$\varepsilon^{h(e)} = \mathbf{B}^{(e)}.\hat{\mathbf{u}}^{(e)} \tag{44}$$

where $\mathbf{B}^{(e)}$ is the discrete strain operator. In the case of a plane strain problem, it is given by

$$\mathbf{B}^{(e)} = \begin{pmatrix} \partial_x N_1^{(e)} & 0 & \cdots & \partial_x N_{nnode}^{(e)} & 0 \\ 0 & \partial_y N_1^{(e)} & \cdots & 0 & \partial_y N_{nnode}^{(e)} \\ \partial_y N_1^{(e)} & \partial_x N_1^{(e)} & \cdots & \partial_y N_{nnode}^{(e)} & \partial_x N_{nnode}^{(e)} \end{pmatrix} \tag{45}$$

The stress follows immediately as $\sigma^{(e)} = \mathbf{D}^e.\mathbf{B}^{(e)}.\hat{\mathbf{u}}^{(e)}$. It is important to notice that all entities related to derivatives will be discontinuous between elements.

3.4. A note on numerical integration techniques on finite element spaces

Quite often we will have to evaluate integrals over the domain Ω^h of functions which will involve products of shape functions and their derivatives. Of course, the integral will be decomposed into integral over the elements. In some rare cases the integrals can be easily obtained as it actually happens with linear triangles and tetrahedra. However, in the more general case, we will have to evaluate them using numerical integration techniques.

A numerical integration rule can be expressed as:

$$\int_{\Omega^{(e)}} f(\mathbf{x})d\Omega = \sum_{k=1}^{ngauss} W_k\, f(\mathbf{x}_k) = \sum_{k=1}^{ngauss} W_k\, f(\boldsymbol{\xi}_k)\det \mathbf{J} \qquad (46)$$

where W_k and $\mathbf{x}_k$ (with $k = 1..ngauss$) are the weights and the position of the integration points. If the integral is performed over the normalized element, we need to multiply by the determinant of the jacobian matrix of the mapping $\mathbf{J}$. The number of points in the integration rule n_{gauss} depends on the particular rule which has been chosen. A key point is the order of precision of the formula, which indicates the higher order of the polynomial which can be exactly integrated. Sometimes, the analysts use integration rules of lower degree of precission than required. This is done for several reasons, among which we can mention (i) speed up the computations, (ii) avoid volumetric locking, etc. However, the use of these so-called "reduced integration" formulae can cause spurious oscillations.

4. Finite Element Discretization

4.1. Galerkin and Weighted Residual Approximations

We will recall the weak formulation:

Given $U = \left\{ u \in H^1\left(\bar{\Omega}\right)\ |\ u - \bar{u} = 0\ x \in \partial\Omega_u \right\}$

and $V = \left\{ v \in H^1\left(\bar{\Omega}\right)\ |\ v = 0\ x \in \partial\Omega_u \right\}$

Find $u \in U$ such that $a(u, v) = \int_{\partial\Omega_q} vq\ d\Gamma\ \forall v \in V$

The basic idea of the method of Galerkin consists in building two finite dimensional spaces $U^h \subset U$ and $V^h \subset V$. We begin by choosing V^h and we build the members of U^h as $u^h = v^h + w^h$, where $v^h \in V^h$ and $w^h - \bar{u} = 0$ on $\partial\Omega_u$. If the dimension of V^h is nh, and we choose a basis $\left\{ \phi_j^h \right\}$ the approximate solution u^h can be expressed as

$$u^h = \sum_{j=1}^{nh} a_j \phi_j^h + w^h \qquad (47)$$

This version of Galerkin method in which the space of trial functions has been constructed from the space of tests functions is referred to as Boubnov-Galerkin, while more general formulations are referred to as Petrov-Galerkin.

The Galerkin weak formulation can be written as:

Given $V^h = \left\{ v^h \in H^1\left(\bar{\Omega}\right)\ |\ v^h = 0\ x \in \partial\Omega_u \right\}$

and $U^h = \left\{ u^h \in H^1\left(\bar{\Omega}\right) \mid u^h = v^h + w^h,\ v^h \in V^h \text{and } w^h - \bar{u} = 0 \text{ on } \partial\Omega_u \right\}$

Find $u^h \in U^h$ such that $a(u^h, v^h) = \int_{\partial\Omega_q} v^h q\, d\Gamma\ \forall v^h \in V$

The nh unknowns $\{a_j\}$ can be obtained applying the nh conditions:

$$a\left(\sum_{j=1}^{nh} a_j \phi_j^h + w^h, \phi_i^h\right) = \int_{\partial\Omega_q} \phi_i^h q\, d\Gamma \tag{48}$$

from which we obtain the following linear system:

$$\mathbf{K}.\mathbf{a} = \mathbf{f} \tag{49}$$

where

$$K_{ij} = a\left(\phi_i^h, \phi_j^h\right) \tag{50}$$

$$f_i = \int_{\partial\Omega_q} \phi_i^h q\, d\Gamma - a\left(\phi_i^h, w^h\right)$$

A finite element approximation can be obtained directly by choosing $\phi_i^h = N_i^{(g)}$:

$$K_{ij} = a\left(N_i^{(g)}, N_j^{(g)}\right) \tag{51}$$

$$f_i = \int_{\partial\Omega_q} N_i^{(g)} q\, d\Gamma - a\left(N_i^{(g)}, w^h\right)$$

the unknowns $\{a_i\}$ are now the nodal values $\{\hat{u}_i\}$. Concerning the function w^h, it can be approximated by the shape functions $N_i^{(g)}$ associated to nodes belonging to the boundary $\partial\Omega_u$. The matrix K_{ij} and the vector f_i are referred to as the **stiffness matrix** and the **force vector** in structural mechanics applications.

Note 6 An alternative formulation of the Galerkin method presented above starts with the original PDEs of the problem, which we will write in the general form:

$$\mathcal{L}u + s = 0 \text{ on } \Omega \tag{52}$$

$$u - \bar{u} = 0 \text{ on } \partial\Omega_u \tag{53}$$

$$Mu + q = 0 \text{ on } \partial\Omega_q \tag{54}$$

with $\partial\Omega = \partial\Omega_u \cup \partial\Omega_q$, $\partial\Omega_u \cap \partial\Omega_q = \{\varnothing\}$. We will build the approximation $u^h \in U^h$ as $u^h = \sum N_j^{(g)} \hat{u}_j$ and we will define the residual $R_\Omega = \mathcal{L}u^h + s \neq 0$ on Ω. The weak formulation will be obtained by applying the Gauss theorem to

$$\int_\Omega N_i^{(g)} R_\Omega\, d\Omega = 0 \tag{55}$$

We will next illustrate with some examples how to obtain the finite element equations in several alternative ways.

4.2. Example 1: 1D elastic bar in tension

We will consider first the bar in tension depicted in Figure 9. It consists of two parts, with sections A_1 and A_2, Young modulii E_1 and E_2, and lengths L_1 and L_2. We will assume that two forces F_2, F_3 are acting at points 2 and 3, and the bar is fixed at point 1. The bar will be discretized using two linear elements.

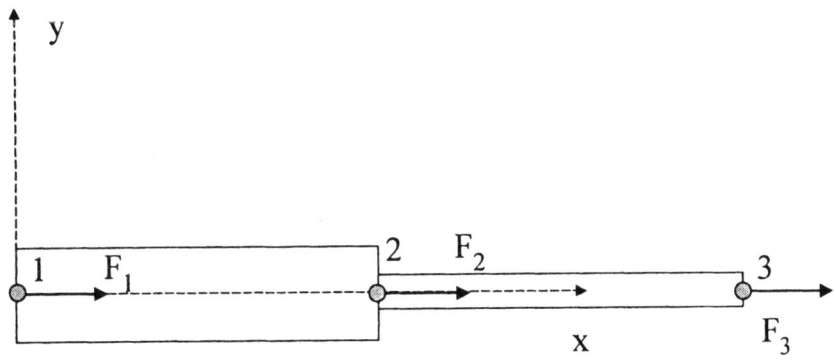

Figure 9. *One dimensional bar in tension*

The weak formulation of this problem is the following:

Given a finite dimensional space $V^h = \{ v^h \in H^1(\bar{\Omega}^h) \, | v^h = 0 \text{ at } x = 0 \}$, a base $\{ N_j^{(g)} \}$ with $j = 2, 3$, and the function $w^h = \bar{u} \, N_1^{(g)} = 0$, find $u^h = \hat{u}_2.N_2^{(g)} + \hat{u}_3.N_3^{(g)}$ such that

$$a(N_i^{(g)}, N_j^{(g)}) \, \hat{u}_j = F_3 N_i^{(g)} \mid_{x=L} \tag{56}$$

where $a(N_i^{(g)}, N_j^{(g)}) = \int_0^L \frac{\partial N_i^{(g)}}{\partial x} EA \frac{\partial N_j^{(g)}}{\partial x} dx$.

Figure 10 depicts $N_2^{(g)}$ and $N_3^{(g)}$ together with $N_1^{(g)}$, which is not needed in this case as the essential boundary condition is homogeneous, i.e., $\bar{u} = 0$.

The coefficients of the stiffness matrix are obtained easily as:

$$K_{22} = \int_{\Omega(e1)} \frac{\partial N_2^{(g)}}{\partial x} EA \frac{\partial N_2^{(g)}}{\partial x} dx + \int_{\Omega(e2)} \frac{\partial N_2^{(g)}}{\partial x} EA \frac{\partial N_2^{(g)}}{\partial x} dx = \frac{E_1 A_1}{L_1} + \frac{E_2 A_2}{L_2} \tag{57}$$

$$K_{23} = K_{32} = \int_{\Omega(e1)} \frac{\partial N_2^{(g)}}{\partial x} EA \frac{\partial N_3^{(g)}}{\partial x} dx + \int_{\Omega(e2)} \frac{\partial N_2^{(g)}}{\partial x} EA \frac{\partial N_3^{(g)}}{\partial x} dx = -\frac{E_2 A_2}{L_2} \tag{58}$$

$$K_{33} = \int_{\Omega(e1)} \frac{\partial N_3^{(g)}}{\partial x} EA \frac{\partial N_3^{(g)}}{\partial x} dx + \int_{\Omega(e2)} \frac{\partial N_3^{(g)}}{\partial x} EA \frac{\partial N_3^{(g)}}{\partial x} dx = \frac{E_2 A_2}{L_2} \tag{59}$$

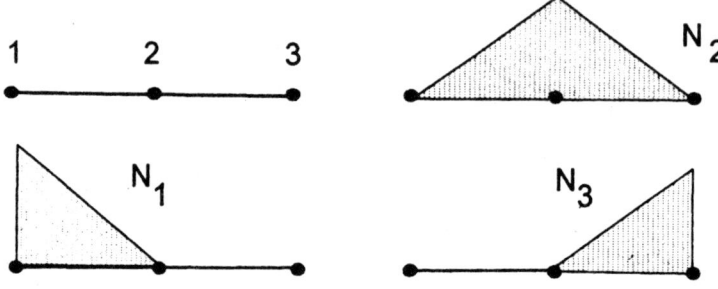

Figure 10. *Global Shape Functions for the Elastic Bar*

and the force vector is

$$f_2 = 0 \quad f_3 = F_3 \tag{60}$$

The system of equations is:

$$\begin{pmatrix} \frac{E_1 A_1}{L_1} + \frac{E_2 A_2}{L_2} & -\frac{E_2 A_2}{L_2} \\ -\frac{E_2 A_2}{L_2} & \frac{E_2 A_2}{L_2} \end{pmatrix} \cdot \begin{pmatrix} \hat{u}_2 \\ \hat{u}_3 \end{pmatrix} = \begin{pmatrix} 0 \\ F_3 \end{pmatrix} \tag{61}$$

which yields $\hat{u}_2 = F_3.L_1/(E_1 A_1)$ and $\hat{u}_3 = F_3 \left(\frac{L_1}{E_1 A_1} + \frac{L_2}{E_2 A_2} \right)$.

From a practical point of view, in order to authomatize the computation of **K**, it is more convenient to formulate the problem assuming that all BCs are of Neumann type, which is equivalent to assuming that we know the reactions acting on prescribed displacement nodes. The essential boundary conditions will be imposed on the discretized system of equations, which will have unknown reactions in the force term, and known prescribed displacements in the vector of unknowns.

4.3. Example 2: Flow in Porous Media

We will illustrate in this case the weighted residual version of the Boubnov-Galerkin Method. The strong formulation of the problem is:

Given a domain Ω with boundary $\partial\Omega = \partial\Omega_u \cup \partial\Omega_q$ with $\partial\Omega_u \cup \partial\Omega_q = \{\varnothing\}$ and smooth functions $k_w : \Omega \to \mathbb{R}^{ndim \times ndim}, \bar{\phi} : \partial\Omega_u \to \mathbb{R}$, and $\bar{q} : \partial\Omega_q \to \mathbb{R}^{ndim}$, obtain ϕ satisfying the following conditions:

$$R_\Omega := \nabla^T (k_w \nabla \phi) = 0 \quad \text{in } \Omega \tag{62}$$

together with essential or Dirichlet boundary conditions imposed at $\partial\Omega_\phi$

$$\phi - \bar{\phi} = 0 \qquad (63)$$

and fluxes prescribed at $\partial\Omega_q$ (Neumann BCs) :

$$(\mathbf{k}_w \nabla \phi)^T . \mathbf{n} + \bar{q} = 0 \qquad (64)$$

Multiplying by a test function $v \in V$ and integrating over the domain Ω, we will have $\int_\Omega v \, R_\Omega d\Omega = 0$. Applying Gauss's theorem to this integral, we arrive at:

$$-\int_\Omega (\nabla v)^T \mathbf{k}_w \nabla \phi \, d\Omega - \int_{\partial\Omega_u} v \, q \, d\Gamma - \int_{\partial\Omega_q} v \, \bar{q} \, d\Gamma = 0 \qquad (65)$$

We will assume known the unknown "reaction" fluxes q at the boundary $\partial\Omega_u$. We will discretize now the domain Ω with $nelem$ elements, and we will assume we have defined the approximation $u^h = \sum_{j=1}^{nh} \hat{\phi}_j N_j^{(g)} \in U^h = V^h$. The system of equations will be:

$$\left(\int_\Omega (\nabla N_i^{(g)})^T \mathbf{k}_w \nabla N_j^{(g)} \, d\Omega\right) \hat{\phi}_j = -\int_{\partial\Omega_u} N_i^{(g)} q \, d\Gamma - \int_{\partial\Omega_q} N_i^{(g)} \bar{q} \, d\Gamma \qquad (66)$$

which can be written as

$$\mathbf{K}.\hat{\phi} = \mathbf{f} + \mathbf{r} \qquad (67)$$

where we have introduced the vector of "reactions" $\mathbf{r} = -\int_{\partial\Omega_u} N_i^{(g)} q \, d\Gamma$. It is important to notice that, for a certain equation i, either we do not know the nodal value of the unknown $\hat{\phi}_i$ or the reaction r_i.

In this way the computation is systematized obtaining the global matrix $\mathbf{K}$ by assembling the contributions of all the elements, independently of whether a given element has nodes in $\partial\Omega_u$.

4.4. Example 3: Linear Elasticity

We will consider next the case of an elastic solid occupying the domain Ω with boundary $\partial\Omega$. We will derive the finite element equations from the principle of virtual work:

Find $\mathbf{u} \in U = \left\{ H^1(\bar{\Omega}) \to \mathbb{R}^{ndim} \,|\, \mathbf{u} - \bar{\mathbf{u}} = 0 \text{ on } \partial\Omega_t \right\}$ such that

$$\int_\Omega (\mathbf{S}.\mathbf{v})^T .\sigma d\Omega = \int_\Omega \mathbf{v}^T .\mathbf{b} \, d\Omega + \int_{\partial\Omega_t} \mathbf{v}^T .\mathbf{t} \, d\Omega \qquad (68)$$

for all $\mathbf{v} \in V = \left\{ H^1(\bar{\Omega}) \to \mathbb{R}^{ndim} \,|\, \mathbf{v} = 0 \text{ on } \partial\Omega_t \right\}$.

We will use a finite element discretization Ω^h of the domain and its boundary $\partial\Omega^h$, and define an approximation $\mathbf{u}^h = \mathbf{v}^h + \mathbf{w}^h$, where $\mathbf{w}^h$ satisfies the essential

BCs. Again, we will assume known the tractions on $\partial\Omega_u$, which means $\mathbf{w}^h = 0$. The approximation will be given by

$$\mathbf{u}^h = \mathbf{N}^{(g)}.\hat{\mathbf{u}} \tag{69}$$

The discrete form of the PVW is

$$\left(\int_\Omega \mathbf{B}^T \mathbf{D}\, \mathbf{B}\, d\Omega\right)\hat{\mathbf{u}} = \int_\Omega \mathbf{N}^T.\mathbf{b}\, d\Omega + \int_{\partial\Omega_t} \mathbf{N}^T.\mathbf{t}\, d\Omega + \int_{\partial\Omega_u} \mathbf{N}^T.\mathbf{t}\, d\Omega \tag{70}$$

or

$$\mathbf{K}.\hat{\mathbf{u}} = \mathbf{f} + \mathbf{r} \tag{71}$$

where $\mathbf{K} = \int_\Omega \mathbf{B}^T \mathbf{D}\, \mathbf{B}\, d\Omega$, $\mathbf{f} = \int_\Omega \mathbf{N}^T.\mathbf{b}\, d\Omega + \int_{\partial\Omega_t} \mathbf{N}^T.\mathbf{t}\, d\Omega$ and $\mathbf{r} = \int_{\partial\Omega_u} \mathbf{N}^T.\mathbf{t}\, d\Omega$

Note 7 It is possible to combine prescribed tractions and displacements at a particular part of the boundary. For instance, we can think of a vertical wall which horizontal movement is prescribed, but which can move freely along the vertical. The BC is essential along X direction (prescribed displacement equal to zero), and of Neumann type along Y (traction free). Once again, we will have components of the vector of unknowns which will be known (displacements at nodes belonging to $\partial\Omega_u$), and unknown reactions at these nodes.

5. References

[BRE 91] S.C.Brenner and L.R.Scott, The Mathematical Theory of Finite Element Methods, Springer Verlag, New York, 1991

[CAR 83] G.F.Carey and J.T.Oden, Finite Elements: A second Course, Prentice-Hall, Inc., Englewood Cliffs, New Jersey, 1983

[HUG 87] T.J.R.Hughes, The Finite Element Method: Linear Static and Dynamic Finite Element Analysis, Prentice-Hall, Inc., Englewood Cliffs, New Jersey, 1987

[ZIE 83] O.Z.Zienkiewicz and K.Morgan, Finite Element and Approximation, John Wiley and Sons, New York, 1983

[ZIE 2000] O.Z.Zienkiewicz and R.L.Taylor, The Finite Element Method 5th edition (3 Vols.), Butterword-Heinemann, Oxford, 2000

Chapter 2

Transient Problems of Parabolic Type

Manuel Pastor and José Antonio Fernandez Merodo

Department of Computational Engineering, Centro de Estudios y Experimentacion de Obras Publicas, Madrid, Spain, and Department of Applied Mathematics, ETS de Ingenieros de Caminos, Madrid, Spain

1. Introduction: Strong and weak formulations

Parabolic equations describe problems such as transient seepage, fickian diffusion of pollutants in the soil or heat conduction in radioactive waste disposal sites. The PDE equation is obtained by combining a first equation describing a flux which is proportional to the gradient of a variable (temperature, concentration, pressure...) with a balance equation relating variations in the flux field to sources.

This is the case of the seepage through porous media described in the chapter devoted to elliptic equations, where the flux equation was

$$-\nabla p_w + \rho_w \mathbf{g} = \mathbf{k}_w^{-1} . \mathbf{w} \qquad [1]$$

and the balance of mass was expressed as

$$\nabla^T . \mathbf{w} + \left(C_s + \frac{1}{Q} \right) \frac{\partial p_w}{\partial t} = 0 \qquad [2]$$

There, we defined the average velocity of water relative to soil skeleton $\mathbf{w}$ which is obtained by dividing the real velocity in the pores by the factor nS_w where n is the porosity and S_r the degree of saturation. ∇ is the gradient operator $(\partial_x, \partial_y, \partial_z)^T$ in 3D, and $(\partial_x, \partial_y)^T$ in 2D situations; p_w and ρ_w the pore water pressure and density, $\mathbf{g}$ the vector of body forces and $\mathbf{k}_w$ is the tensor of permeability. Combining both equations, we arrived at

$$\nabla^T \left(\mathbf{k}_w \left(\nabla \phi \right) \right) = \left(C_s + \frac{1}{Q} \right) \frac{\partial p_w}{\partial t} \qquad [3]$$

where ∇^T is the divergence operator, C_s is the specific storage coefficient and $\frac{1}{Q}$ a factor accounting for the compressibility of water and soil. We also introduced the potential ϕ defined from $\nabla \phi = \nabla p_w + \rho_w \mathbf{g}$.

The problem is non linear, as both permeability and the coefficients C_s and Q depend on saturation.

Essential or Dirichlet boundary conditions are imposed at $\partial\Omega_\phi$

$$\phi - \bar{\phi} = 0 \qquad [4]$$

and the fluxes are prescribed at $\partial\Omega_q$ (Neumann BCs) as:

$$(\mathbf{k}_w \nabla \phi)^T . \mathbf{n} + \bar{q} = 0 \qquad [5]$$

where $\mathbf{n}$ is is the unit normal vector to the boundary. Now, $\partial\Omega_\phi, \partial\Omega_q, \bar{\phi}$ and $\bar{q}$ depend on time.

Finally, the initial conditions are

$$\phi(x, 0) = \phi_0(x) \qquad [6]$$

Other parabolic problems of interest are:

(i) Heat conduction

$$\nabla^T \left(\mathbf{D} \left(\nabla \phi \right) \right) = \rho c \frac{\partial p_w}{\partial t}$$

where $\mathbf{D}$ is the conductivity tensor, ρ the density, and c the specific heat of the material.

(ii) One dimensional consolidation

$$\frac{\partial}{\partial z} \left(k_w \frac{\partial p_w}{\partial z} \right) = \frac{g \rho_w}{E_m} \frac{\partial p_w}{\partial t}$$

which describes the excess pore pressure p_w variation along the depth z of a soil layer with vertical permeability k_w and oedometric modulus of deformation E_m. In the above ρ_w is the density of water and g the acceleration of gravity.

A more rigorous description of the seepage problem is the following:

(SF) Given the domain Ω with boundary $\partial \Omega = \partial \Omega_u \cup \partial \Omega_q$ such that $\partial \Omega_u \cap \partial \Omega_q = \{\emptyset\}$, the time interval $I = (0, T)$, and the functions $\mathbf{k}_w : I \times \Omega \to \mathbb{R}^{ndim \times ndim}$, $C_s : I \times \Omega \to \mathbb{R}$ and $Q : I \times \Omega \to \mathbb{R}$ find $\phi : \bar{I} \times \bar{\Omega} \to \mathbb{R}$ such that

$$\nabla^T \left(\mathbf{k}_w \left(\nabla \phi \right) \right) = \left(C_s + \frac{1}{Q} \right) \frac{\partial \phi}{\partial t} \text{ on } \bar{I} \times \bar{\Omega}$$

$$\phi - \bar{\phi} = 0 \text{ on } \bar{I} \times \partial \Omega_\phi \tag{7}$$

$$\left(\mathbf{k}_w \nabla \phi \right)^T . \mathbf{n} + \bar{q} = 0 \text{ on } \bar{I} \times \partial \Omega_q$$

$$\phi(x, 0) = \phi_0(x) \text{ on } \Omega$$

The weak formulation is obtained by multiplying the PDE by a function ψ, integrating it over the domain and then applying Gauss theorem.

(WF) Given the domain Ω with boundary $\partial \Omega = \partial \Omega_u \cup \partial \Omega_q$ such that $\partial \Omega_u \cap \partial \Omega_q = \{\emptyset\}$, the time interval $I = (0, T)$, and the functions $\mathbf{k}_w : I \times \Omega \to \mathbb{R}^{ndim}$, $C_s : I \times \Omega \to \mathbb{R}$ and $Q : I \times \Omega \to \mathbb{R}$

Find $\phi(t) \in U = \left\{ \phi : I \times \bar{\Omega} \to \mathbb{R} \mid \phi \in H^1(\bar{\Omega}) \text{ and } \phi - \bar{\phi} = 0 \text{ on } \partial \Omega_\phi \right\}$ such that

$$\left((C_s + 1/Q) \frac{\partial \phi}{\partial t}, \psi \right) + a(\phi, \psi) = \int_{\partial \Omega_q} \psi \bar{q} \, d\Gamma \tag{8}$$

$$\forall \psi \in V = \left\{ \psi : \bar{\Omega} \to \mathbb{R} \mid \psi \in H^1(\bar{\Omega}) \text{ and } \psi = 0 \text{ on } \partial \Omega_\phi \right\}$$

In above, we have used the bilinear form $a(\phi, \psi) = \int_\Omega \nabla \psi^T \mathbf{k}_w \nabla \phi \, d\Omega$

If we try to find fundamental solutions $\phi(x, t) = A \exp(i \kappa x - i \omega t)$ where κ is the wave number and ω the angular frequency for the one dimensional equation

$$(C_s + 1/Q) \frac{\partial \phi}{\partial t} = \frac{\partial}{\partial x} \left(k_w \frac{\partial \phi}{\partial x} \right)$$

we obtain, after substitution in the PDE

$$(C_s + 1/Q)\,(-i\omega)A\exp(i\kappa x - i\omega t) = -k_w\kappa^2 A\exp(i\kappa x - i\omega t)$$

from where $(C_s + 1/Q)\,(-i\omega)A = -k_w\kappa^2 A$, and

$$-i\omega t = -\frac{k_w}{(C_s + 1/Q)}\kappa^2 t \qquad [9]$$

The solution is therefore

$$\phi = A.\exp(-\frac{k_w}{(C_s + 1/Q)}\kappa^2 t)\exp(i\kappa x) \qquad [10]$$

and we can obtain the following conclusions:

– The factor $\exp(-\frac{k_w}{(C_s+1/Q)}\kappa^2 t)$ will cause the smoothing of the solution with time.

– Damping depends on the wave number κ, and it is higher for shorter wave lengths.

– Any discontinuity in the initial conditions will be smoothed

2. Finite Difference Methods: An introduction to the study of stability

The finite difference method is based on constructing a grid in the domain $\Omega \times I$. For instance, if we consider the heat conduction in a one dimensional bar of length L at times $t_0 \le t \le t_f$,

$$\rho c\frac{\partial \phi}{\partial t} = D\frac{\partial^2 \phi}{\partial x^2} \qquad [11]$$

the mesh would be the one depicted in Figure 1. Any node $x = x_0 + j.\,\Delta x$, $t = t_0 + n$ Δt, can be identified by (j, n).

Partial derivatives with respect to time and space can then be approximated as combinations of the values at a set of nodes. For instance, the partial derivative with respect to time

$$\left.\frac{\partial \phi}{\partial t}\right|_j^n := \frac{\partial \phi}{\partial t}(x = x_j, t = t_n) \qquad [12]$$

can be approximated as:

$$\left.\frac{\partial \phi}{\partial t}\right|_j^n = \frac{\phi_j^{n+1} - \phi_j^n}{\Delta t} + O(\Delta t) \qquad [13]$$

Another alternative is:

$$\left.\frac{\partial \phi}{\partial t}\right|_j^n = \frac{\phi_j^{n+1} - \phi_j^{n-1}}{2\Delta t} + O(\Delta t^2) \qquad [14]$$

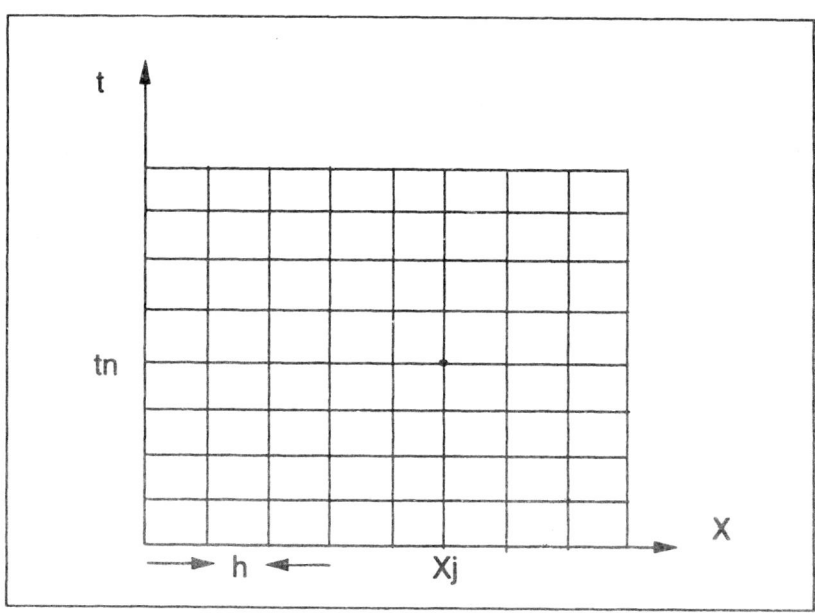

Figure 1. *Finite Difference Grid for the 1D heat conduction problem in a bar*

Second derivatives with respect to space can be obtained in the same way:

$$\frac{\partial^2 \phi}{\partial x^2}\bigg|_j^n = \frac{\phi_{j+1}^n - 2\phi_j^n + \phi_{j-1}^n}{\Delta x^2} + O(\Delta x^2) \tag{15}$$

If we now substitute (13) and (15) into (11), we obtain:

$$\frac{\phi_j^{n+1} - \phi_j^n}{\Delta t} = \frac{D}{\rho c} \frac{\phi_{j+1}^n - 2\phi_j^n + \phi_{j-1}^n}{\Delta x^2} \tag{16}$$

from where we get:

$$\phi_j^{n+1} = \phi_j^n + D^*(\phi_{j+1}^n - 2\phi_j^n + \phi_{j-1}^n) \tag{17}$$

where we have introduced $D^* = D\Delta t / \rho c \Delta x^2$

The finite difference scheme (17) is explicit, as the solution at t^{n+1} can be obtained directly without having to solve any system of equations, and it is conditionally stable as it will be shown later on, as the timestep Δt has to be smaller than a critical value to avoid oscillations growing with time.

Of course, both the problem and the finite difference solution are complemented by suitable boundary and initial conditions.

Stability analysis by Von Neumann method

We will assume that the FD scheme is linear, and can be cast in the form

$$L(\Phi) = L(\Phi_j^{n+1}, \Phi_{j-1}^n, \Phi_j^n, ...) = 0 \qquad [18]$$

The solution of the scheme $\bar{\Phi}$ will also satisfy the condition

$$L(\bar{\Phi}) = L(\bar{\Phi}_j^{n+1}, \bar{\Phi}_{j-1}^n, \bar{\Phi}_j^n, ...) = 0 \qquad [19]$$

If we now introduce the error ε_j^n at $x = x_0 + j\triangle x$, $t = t_0 + n\triangle t$

$$\varepsilon_j^n = \Phi_j^n - \bar{\Phi}_j^n \qquad [20]$$

we can easily see that $L(\Phi - \bar{\Phi}) = 0$, and therefore, $L(\varepsilon) = 0$, i.e., the error satisfies the finite difference equation.

We will assume that the error at time t_n can be written as

$$\varepsilon_j^n = \sum_{l=-N}^{N} E_l^n \exp(I.\kappa_l.j\triangle x) \qquad [21]$$

where $I = \sqrt{-1}$. Because of the linearity of the scheme, we can analyze the behaviour of a particular component $E^n \exp(I.\kappa.j\triangle x)$

It is interesting to note that the wave lengths which can be represented by the grid have a maximum value $\lambda_{\max} = 2L = 2N\triangle x$, and a minimum value $\lambda_{\min} = 2\triangle x$ where $N = L/\triangle x$ is the number of divisions along X. Therefore, a particular wavelength can be written as

$$\lambda_l = 2l\triangle x \quad l = 1..N$$

Wavenumbers are limited by $\kappa_{\min} = \frac{2\pi}{\lambda_{\max}} = \frac{\pi}{N\triangle x}$ and $\kappa_{\max} = \frac{2\pi}{\lambda_{\min}} = \frac{\pi}{\triangle x}$. The $l - th$ component will have

$$\kappa_l = \frac{\pi}{l\triangle x} \quad l = 1..N \qquad [22]$$

It is useful also to introduce a phase lag angle $\varphi = \kappa\triangle x$. The error can be written as

$$E^n \exp(I.\kappa.j\triangle x) = E^n \exp(I.j\varphi)$$

We will write now the errors at points $(j, n+1)$, $(j-1, n)$, (j, n) and $(j+1, n)$:

$$
\begin{array}{rcl}
\varepsilon_j^{n+1} & = & E^{n+1} \exp(I.j\varphi) \\
\varepsilon_{j+1}^n & = & E^n \exp(I.(j+1)\varphi) \\
\varepsilon_j^n & = & E^n \exp(I.j\varphi) \\
\varepsilon_{j-1}^n & & E^n \exp(I.(j-1)\varphi)
\end{array}
\qquad [23]
$$

and we will substitute them into the FD scheme. The result is:

$$E^{n+1} = E^n \left[1 + D^* \left(\exp(I\varphi) - 2 + \exp(-I\varphi)\right)\right]$$

which can be simplified as

$$= E^n \left[1 - 2D^* + 2D^* \cos\varphi\right]$$

$$= E^n \left[1 - 2D^* \left(1 - \cos\varphi\right)\right]$$

$$= E^n \left[1 - 2D^* . 2\sin^2 \frac{\varphi}{2}\right]$$

from where we finally arrive at:

$$E^{n+1} = E^n \left[1 - 4D^* . \sin^2 \frac{\varphi}{2}\right] \qquad [24]$$

If we introduce now ξ^n, which measures the error variation between times t^{n+1} and t^n

$$\xi^n = \frac{E^{n+1}}{E^n} \qquad [25]$$

we obtain

$$\xi^n = 1 - 4D^* . \sin^2 \frac{\varphi}{2}$$

If we want to avoid growing the error, the following condition should be fulfilled:

$$-1 \leq 1 - 4D^* . \sin^2 \frac{\varphi}{2} \leq 1$$

from where we obtain

$$0 \leq D^* \leq \frac{1}{2\sin^2 \frac{\varphi}{2}} \qquad [26]$$

and

$$0 \leq D^* \leq \frac{1}{2} \qquad [27]$$

which is the stability condition. If we choose Δx and Δt such that $D^* = 0.60$, the scheme will be unstable, while $D^* = 0.48$ will result in a stable scheme.

– Example 1

Given a one dimensional bar of constant cross section and length $L = 10$ with boundary conditions $\phi(0, t) = 0$ and $\phi(L, t) = 0$ and the initial distribution of temperature

$$\phi(x, 0) = x/10 \qquad 0 \leq x \leq 5$$
$$\phi(x, 0) = 1 - x/10 \quad 5 \leq x \leq 10$$

obtain the evolution with time of the bar temperature using the finite difference scheme given in this section using a grid with $\Delta x = 1.0$ in the two cases $\Delta t = 0.49$ and $\Delta t = 0.60$. Specific heat, density and thermal conductivity will be taken as unity.

Solution

We will use the FD scheme $\phi_j^{n+1} = \phi_j^n + D^*(\phi_{j+1}^n - 2\phi_j^n + \phi_{j-1}^n)$ with $j = 1..9$

Figures 2 and 3 depict the results obtained in both cases. As expected, the scheme with $\Delta t = 0.60$ is unstable for $D^* = 0.60$.

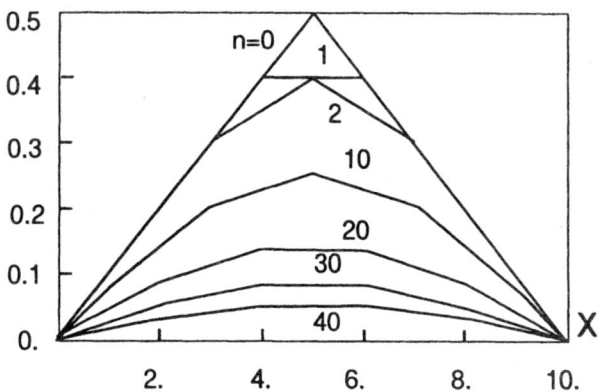

Figure 2. *Explicit scheme with* $D^* = 0.49$

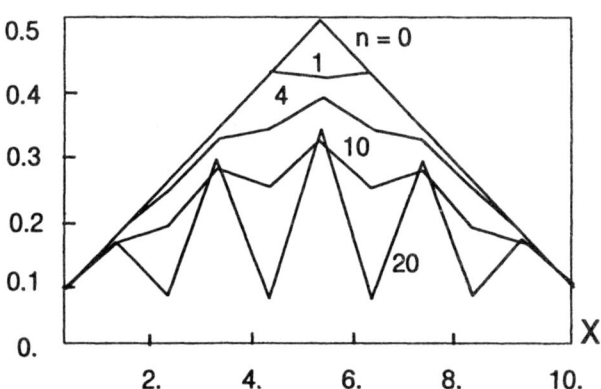

Figure 3. *Explicit scheme with* $D^* = 0.6$

3. Finite Element discretization

3.1. *Introduction*

Discretization of the weak form (8) will be done using Galerkin method. We will assume that the trial functions $u^h(t)$, with $t \in \bar{I}$ belong to the space U^h

$$U^h = \left\{ u^h \in H^1\left(\bar{\Omega}\right) \;\middle|\; u^h = v^h + w^h,\; v^h \in V^h \text{and } w^h - \bar{u} = 0 \text{ on } \partial\Omega_u \right\},$$

where the space of test functions V^h is

$$V^h = \left\{ v^h \in H^1\left(\bar{\Omega}\right) \;\middle|\; v^h = 0 \; x \in \partial\Omega_u \right\}$$

If the dimension of V^h is nh, and we choose a basis $\{N_j\}$, the approximate solution u^h can be expressed as

$$u^h\left(x, t\right) = \sum_{j=1}^{nh} \hat{\phi}_j(t) N_j(x) + w^h \qquad [28]$$

We arrive at:

Find $u^h \in U^h$ such that

$$\left((C_s + 1/Q)\,\dot{u}^h, v^h\right) + a(u^h, v^h) = \int_{\partial\Omega_q} v^h q\, d\Gamma \qquad \forall v^h \in V$$

with $\left(u^h\left(x, 0\right), v^h\right) = \left(u_0, v^h\right)$ for all $v^h \in V^h$

The nh unknowns $\left\{ \hat{\phi}_j \right\}$ can be obtained applying the nh conditions:

$$\left((C_s + 1/Q)\sum_{j=1}^{nh} \dot{\hat{\phi}}_j N_j + \dot{w}^h, N_i\right) + a\left(\sum_{j=1}^{nh} \hat{\phi}_j(t) N_j + w^h, N_i\right) = \int_{\partial\Omega_q} N_i q\, d\Gamma$$

$$[29]$$

from which we obtain the following linear system:

$$\mathbf{K}.\mathbf{\Phi} + \mathbf{C}\frac{d\mathbf{\Phi}}{dt} = \mathbf{f} \qquad [30]$$

with

$$\mathbf{\Phi}(0) = \sum_{j=1}^{nh} \hat{\phi}_j(0) N_j(x) + w^h(0)$$

where

$$\begin{aligned}
K_{ij} &= a\left(N_i, N_j\right) \\
C_{ij} &= \left((C_s + 1/Q) N_j, N_i\right) \qquad\qquad [31] \\
f_i &= \int_{\partial\Omega_q} N_i q\, d\Gamma - a\left(N_i, w^h\right) - \left(N_i, (C_s + 1/Q)\dot{w}^h\right)
\end{aligned}$$

The problem has been transformed into solving the system of ODEs in (30). A most simple method is the Forward Euler, where we approximate the derivatives with respect to time as:

$$\frac{d}{dt}\hat{\mathbf{\Phi}}(t) = \frac{\hat{\mathbf{\Phi}}^{n+1} - \hat{\mathbf{\Phi}}^n}{\Delta t} \qquad [32]$$

from where we arrive at:

$$\hat{\pmb{\Phi}}^{n+1} = \hat{\pmb{\Phi}}^n + \triangle t \mathbf{C}^{-1}\left(\mathbf{f}^n - \mathbf{K}.\hat{\pmb{\Phi}}^n\right) \tag{33}$$

The scheme is of explicit type, and conditionally stable.

– Example 2

A one dimensional bar of constant cross section A and length L has its left end at $\phi = 1^0C$, and the right end is isolated. The initial temperature is given by $\phi(x, 0) = 1 - x/L$. Obtain the finite element equations using three linear elements and the lumped form of matrix $\mathbf{C}$

Solution:

Matrix $\mathbf{C}$ at elements is obtained as:

$$C_{ij}^{(e)} = \int_{(e)} \rho c N_i N_j A dx = \frac{1}{18} A L \rho c \begin{pmatrix} 2 & 1 \\ 1 & 2 \end{pmatrix}$$

from where $\mathbf{C}_L^{(e)}$ is

$$\mathbf{C}_L^{(e)} = \frac{1}{18} A L \rho c \begin{pmatrix} 2+1 & 0 \\ 0 & 2+1 \end{pmatrix} = \frac{1}{6} A L \rho c \begin{pmatrix} 1 & 0 \\ 0 & 1 \end{pmatrix}$$

The global matrix $\mathbf{C}_L$ is obtained by assembling element contributions

$$\mathbf{C}_L = \frac{1}{6} A L \rho c \begin{pmatrix} 1 & 0 & 0 & 0 \\ 0 & 2 & 0 & 0 \\ 0 & 0 & 2 & 0 \\ 0 & 0 & 0 & 1 \end{pmatrix}$$

The stiffness matrix is obtained assembling the element matrices

$$\mathbf{K}^{(e)} = \frac{DA}{L/3} \begin{pmatrix} 1 & -1 \\ -1 & 1 \end{pmatrix}$$

as:

$$\mathbf{K} = \frac{DA}{L/3} \begin{pmatrix} 1 & -1 & 0 & 0 \\ -1 & 2 & -1 & 0 \\ 0 & -1 & 2 & -1 \\ 0 & 0 & -1 & 1 \end{pmatrix}$$

The system is therefore,

$$\begin{pmatrix} \hat{\Phi}_1^{n+1} \\ \hat{\Phi}_2^{n+1} \\ \hat{\Phi}_3^{n+1} \\ \hat{\Phi}_4^{n+1} \end{pmatrix} = \begin{pmatrix} \hat{\Phi}_1^n \\ \hat{\Phi}_2^n \\ \hat{\Phi}_3^n \\ \hat{\Phi}_4^n \end{pmatrix} + \frac{3\triangle t}{AL\rho c} \begin{pmatrix} 2 & 0 & 0 & 0 \\ 0 & 1 & 0 & 0 \\ 0 & 0 & 1 & 0 \\ 0 & 0 & 0 & 2 \end{pmatrix} \cdot$$

$$\left\{ \begin{pmatrix} R_1 \\ 0 \\ 0 \\ 0 \end{pmatrix} - \frac{DA}{L/3} \begin{pmatrix} 1 & -1 & 0 & 0 \\ -1 & 2 & -1 & 0 \\ 0 & -1 & 2 & -1 \\ 0 & 0 & -1 & 1 \end{pmatrix} \cdot \begin{pmatrix} \hat{\Phi}_1^n \\ \hat{\Phi}_2^n \\ \hat{\Phi}_3^n \\ \hat{\Phi}_4^n \end{pmatrix} \right\}$$

Taking into account that $\triangle x = L/3$, the second equation is obtained as

$$\hat{\Phi}_2^{n+1} = \hat{\Phi}_2^n + \frac{D\triangle t}{\rho c \triangle x^2} \left(\hat{\Phi}_1^n - 2\hat{\Phi}_2^n + \hat{\Phi}_3^n \right)$$

which coincides with the finite difference equation obtained above.

3.2. Single step algorithms: A Newmark scheme

The finite element scheme presented in the preceding section is a particular case of a general family of schemes referred to as generalized Newmark [ZIE 00] If we approximate $\hat{\Phi}$ and its derivative with respect to time as:

$$\hat{\Phi}^{n+1} = \hat{\Phi}^n + \triangle t.\overset{\bullet}{\hat{\Phi}}{}^n + \beta \triangle t \triangle \overset{\bullet}{\hat{\Phi}}{}^n \qquad [34]$$

and

$$\overset{\bullet}{\hat{\Phi}}{}^{n+1} = \overset{\bullet}{\hat{\Phi}}{}^n + \triangle \overset{\bullet}{\hat{\Phi}}{}^n$$

where β is a parameter such that:

$$0 \leq \beta \leq 1$$

and substitute them into the equation particularized at time t_{n+1},

$$\mathbf{K}.\hat{\Phi}^{n+1} - \mathbf{f}^{n+1} + \mathbf{C}.\overset{\bullet}{\hat{\Phi}}{}^{n+1} = 0$$

we obtain

$$(\mathbf{C} + \beta \triangle t.\mathbf{K}) \triangle \overset{\bullet}{\hat{\Phi}}{}^n = \mathbf{f}^{n+1} - \left(\mathbf{C}.\overset{\bullet}{\hat{\Phi}}{}^n + \mathbf{K}.\left(\hat{\Phi}^n + \triangle t \overset{\bullet}{\hat{\Phi}}{}^n \right) \right) \qquad [35]$$

which can be expressed in a more compact form as

$$(\mathbf{C} + \beta \triangle t.\mathbf{K}) \triangle \overset{\bullet}{\hat{\Phi}}{}^n = \Psi^{n+1} \qquad [36]$$

where Ψ^{n+1} is

$$\Psi^{n+1} = \mathbf{f}^{n+1} - \mathbf{K}.\hat{\Phi}^{n+1,pred} - \mathbf{C}.\overset{\bullet}{\hat{\Phi}}{}^{n+1,pred}$$

and $\hat{\Phi}^{n+1,pred}, \overset{\bullet}{\hat{\Phi}}{}^{n+1,pred}$

$$\hat{\Phi}^{n+1,pred} = \hat{\Phi}^n + \triangle t \overset{\bullet}{\hat{\Phi}}{}^n$$

$$\overset{\bullet}{\hat{\Phi}}{}^{n+1,pred} = \overset{\bullet}{\hat{\Phi}}{}^n$$

By choosing suitable values of β, we can obtain the following schemes:

β	
0	Forwards Euler
1	Backwards Euler
$\frac{1}{2}$	Crank-Nicolson
$\frac{1}{3}$	Galerkin (time)

For values of the parameter $\beta \geq 0.5$ the scheme is unconditionally stable, being otherwise conditionally stable, i.e., the time step has to be smaller than a critical value to avoid oscillations (and errors) growing with time.

The Forward Euler scheme is said to be explicit as inversion of the matrix C can be easily achieved if a lumped form is used.

The algorithm can be started with the initial values of $\hat{\Phi}^0$ and $\dot{\hat{\Phi}}^0$ The latter can be obtained using

$$\mathbf{K}.\hat{\Phi}^0 - \mathbf{f}^0 + \mathbf{C}.\dot{\hat{\Phi}}^0 = 0$$

from where

$$\dot{\hat{\Phi}}_0 = \mathbf{C}^{-1}.\left(\mathbf{f}_0 - \mathbf{K}.\hat{\Phi}_0\right)$$

3.3. Matrix Stability Analysis

The stability of a finite element scheme can be done analyzing the eigenvalues of the matrix which controls the error evolution. Here we will show an application to the Forward Euler schema. We will consider the problem of heat conduction in a bar with a known initial temperature distribution $\phi_0(x)$ and with both its ends at constant zero temperature:

$$\mathbf{C}\frac{d\hat{\Phi}}{dt} + \mathbf{K}.\hat{\Phi} = 0 \qquad [37]$$

which will be discretized in time with the simple Forward Euler scheme

$$\hat{\Phi}^{n+1} = \hat{\Phi}^n - \Delta t \mathbf{C}^{-1}\mathbf{K}.\hat{\Phi}^n$$

which can be written as

$$\hat{\Phi}^{n+1} = \mathbf{A}.\hat{\Phi}^n \qquad [38]$$

where

$$\mathbf{A} = \left(\mathbf{I} - \Delta t \mathbf{C}^{-1}\mathbf{K}\right)$$

If $\bar{\Phi}^n$ y $\bar{\Phi}^{n+1}$ are the exact solutions of the scheme at times t_n and t_{n+1}, we will have

$$\bar{\Phi}^{n+1} = \mathbf{A}.\bar{\Phi}^n$$

The error evolution is given by

$$\varepsilon^{n+1} = \mathbf{A}.\varepsilon^n \tag{39}$$

from where we obtain

$$\varepsilon^{n+1} = \mathbf{A}^{n+1}.\varepsilon^0$$

To avoid amplification of the error, the absolute values of all the eigenvalues of matrix A should be smaller than 1 :

$$-1 \leq \lambda_A \leq 1$$

and, in order to avoid oscillations,

$$0 \leq \lambda_A \leq 1 \tag{40}$$

Taking into account that

$$\lambda_A = 1 - \Delta t \lambda_{\mathbf{C}^{-1}\mathbf{K}}$$

we arrive to the stability condition

$$0 \leq \Delta t \lambda_{\mathbf{C}^{-1}\mathbf{K}} \leq 1$$

– Example 3

Analyze the stability properties of the finite element problem solved in Example 2

Solution

Matrix $\mathbf{A}$ is

$$\begin{pmatrix} 1 - 2D^* & D^* & 0 & 0 & 0 \\ D^* & 1 - 2D^* & D^* & 0 & 0 \\ 0 & D^* & 1 - 2D^* & D^* & 0 \\ 0 & 0 & D^* & 1 - 2D^* & D^* \\ 0 & 0 & 0 & D^* & 1 - 2D^* \end{pmatrix}$$

and can be written as

$$\mathbf{A} = \mathbf{I} + D^*.\begin{pmatrix} -2 & 1 & 0 & 0 & 0 \\ 1 & -2 & 1 & 0 & 0 \\ 0 & 1 & -2 & 1 & 0 \\ 0 & 0 & 1 & -2 & 1 \\ 0 & 0 & 0 & 1 & -2 \end{pmatrix}$$

The eigenvalues are given by

$$\lambda_A = 1 + D^* \left(-4 \sin^2 \frac{j\pi}{2N} \right)$$

from where

$$-1 \leq 1 - 4D^* \sin^2 \frac{j\pi}{2N} \leq 1$$

or

$$D^* \leq 0.5$$

which coincides with the results of the Von Neumann method.

4. Linear multistep algorithms: Runge Kutta schemes

We will begin by assuming that the time dependent problem has been discretized as:

$$\frac{d\hat{\Phi}}{dt} = \mathbf{H}(\hat{\Phi}, t) \tag{41}$$

where $\mathbf{H}(\hat{\Phi}, t)$ is a matrix operator, which in the cases studied here is given by:

$$\frac{d\hat{\Phi}}{dt} = \mathbf{f} - \mathbf{K}.\hat{\Phi}(t) \tag{42}$$

$$\mathbf{H}(\hat{\Phi}, t) = \mathbf{C}^{-1}. \left(\mathbf{f} - \mathbf{K}.\hat{\Phi}(t) \right) \tag{43}$$

We will assume for the sake of simplicity that $\hat{\Phi}$ and $\mathbf{H}$ are scalar functions $\Phi(t)$ and $H(\Phi, t)$. Writing a series expansion, we have

$$\Phi(t + \Delta t) = \Phi(t) + \Delta t.\frac{d}{dt}\Phi(t) + \frac{1}{2}\Delta t^2 \frac{d^2}{dt}\Phi(t) + \frac{1}{3!}\Delta t^3 \frac{d^3}{dt}\Phi(t) + \dots \tag{44}$$

where derivatives with respect to time can be obtained from the differential equation

$$\frac{d}{dt}\Phi(t) = H(\Phi, t) \tag{45}$$

$$\frac{d^2}{dt}\Phi(t) = \frac{\partial H}{\partial t} + \frac{\partial H}{\partial \Phi}\frac{d}{dt}\Phi(t)$$

$$\frac{d^3}{dt}\Phi(t) = (H_t + H_\Phi H)_t + (H_t + H_\Phi H)_\Phi H$$

$$= H_{tt} + H_{t\Phi} + (H_t + H_\Phi H)H_\Phi + H(H_{\Phi t} + H_{\Phi\Phi}H)$$

where subindexes refer to partial derivatives.

Inserting these values of derivatives into Eqn.(44) we obtain

$$\Phi(t + \Delta t) = \Phi(t) + \Delta t.H + \frac{1}{2}\Delta t^2(H_t + H_\Phi H) + O(\Delta t^3)$$

$$= \Phi(t) + \frac{1}{2}\Delta t.H + \frac{1}{2}\Delta t [H + \Delta t H_t + \Delta t H H_\Phi] + O(\Delta t^3)$$

where it can be easily verified that the third term can be written as

$$H(t + \Delta t, \Phi + \Delta t.H) = H(t, \Phi) + \Delta t H_t + \Delta t H H_\Phi + O(\Delta t^2)$$

We arrive finally at

$$\Phi(t + \Delta t) = \Phi(t) + \frac{1}{2} \Delta t.H + \frac{1}{2} \Delta t.H(t + \Delta t, \Phi + \Delta t.H) \qquad [46]$$

which can be expressed in a more compact form as

$$\Phi(t + \Delta t) = \Phi(t) + \frac{1}{2}(F_1 + F_2) \qquad [47]$$

with

$$F_1 = \Delta t.H(t, \Phi)$$

$$F_2 = \Delta t.H(t + \Delta t, \Phi + \Delta t.H)$$

which is the Runge-Kutta scheme of second order.

Two alternative forms are:

– Henn's scheme

$$\Phi^* = \Phi^n + \Delta t.H^n \qquad [48]$$

$$\Phi^{n+1} = \Phi^n + \frac{1}{2} \Delta t(H^* + H^n)$$

$$\Phi^* = \Phi^n + \frac{1}{2} \Delta t.H^n \qquad [49]$$

$$\Phi^{n+1} = \Phi^n + \Delta t.H^{*n}$$

If we apply the second algorithm to the heat conduction equation, we would obtain:

$$\mathbf{C}.\hat{\Phi}^* = \mathbf{C}.\hat{\Phi}^n + \frac{1}{2} \Delta t.(\mathbf{f}^n - \mathbf{K}.\hat{\Phi}^n) \qquad [50]$$

$$\mathbf{C}.\hat{\Phi}^{n+1} = \mathbf{C}.\hat{\Phi}^n + \Delta t.(\mathbf{f}^n - \mathbf{K}.\hat{\Phi}^*)$$

4.1. *Fourth-order Runge-Kutta scheme*

This scheme is written as

$$\Phi(t + \Delta t) = \Phi(t) + \frac{1}{6}(F_1 + 2F_2 + 2F_3 + F_4) \qquad [51]$$

where

$$F_1 = \Delta t.H$$

$$F_2 = \triangle t.H\left(t + \frac{1}{2}\triangle t, \Phi + \frac{1}{2}F_1\right)$$

$$F_3 = \triangle t.H\left(t + \frac{1}{2}\triangle t, \Phi + \frac{1}{2}F_2\right)$$

$$F_4 = \triangle t.H\left(t + \frac{1}{2}\triangle t, \Phi + \frac{1}{2}F_3\right)$$

The scheme is fourth-order as it is able to reproduce terms in the series development up to $\triangle t^4$, which is included. The corresponding scheme for finite elements is

$$C.\hat{\Phi}^{n+1} = C.\hat{\Phi}^n + \frac{1}{6}(F_1 + 2F_2 + 2F_3 + F_4) \qquad [52]$$

with

$$F_1 = \triangle t.\left(f^n - K.\hat{\Phi}^n\right)$$

$$F_2 = \triangle t.\left(f^{n+\frac{1}{2}} - K.(\hat{\Phi}^n + \frac{1}{2}F_1)\right)$$

$$F_3 = \triangle t.\left(f^{n+\frac{1}{2}} - K.(\hat{\Phi}^n + \frac{1}{2}F_2)\right)$$

$$F_4 = \triangle t.\left(f^{n+1} - K.(\hat{\Phi}^n + F_3)\right)$$

4.2. Error estimation and adaptive time-stepping

We will consider here the fourth-order Runge-Kutta scheme described in the above section, for which the error can be written as

$$E = C_1 \triangle t^5$$

for small values of $\triangle t$.

To obtain an approximation of the error, we will assume that C_1 does not change between time t_n and t_{n+2}. Then we will obtain $\Phi(t + 2.\triangle t)$ in two different ways: (i) using a time increment of $2.\triangle t$ and (ii) using two increments of $\triangle t$. We will write the first solution as $\bar{\Phi}$ and the second as $\bar{\bar{\Phi}}$. If the exact solution is Φ^*, we will have the following relations:

$$\Phi^*(t + 2.\triangle t) = \bar{\Phi} + C_1(2\triangle t)^5$$

and

$$\Phi^*(t + 2.\triangle t) = +2C_1 \triangle t^5$$

Substracting the second from the first, we have

$$0 = \bar{\Phi} - \bar{\bar{\Phi}} + C_1 \triangle t^5(32 - 2)$$

and

$$C_1 \, \triangle \, t^5 = \frac{\bar{\bar{\Phi}} - \bar{\Phi}}{30}$$

Therefore, the error is given by

$$E = \Phi^* - \bar{\bar{\Phi}} = 2C_1 \, \triangle \, t^5$$

$$= \frac{\bar{\bar{\Phi}} - \bar{\Phi}}{15} \tag{53}$$

Once we have estimated the error, and knowing that it is proportional to $\triangle t^5$, it is possible to adjust the time increment in such a way that the error is some target value E_{obj}.

The time step is given by

$$\triangle t_{obj} = \triangle t \left(\frac{E_{obj}}{E} \right) \tag{54}$$

5. References

[HIR 88] C. Hirsch, Numerical Computation of Internal and External Flows, John Wiley and Sons, 1988.

[HUG 87] T.J.R. Hughes, The Finite Element Method. Linear static and Dynamic Finite Element Analysis, Prentice-Hall Int.Ed. London, 1987.

[ZIE 00] O.C. Zienkiewicz and R.L. Taylor, The Finite Element Method, Vol.I, Butterworth and Heinemann, Oxford, 2000.

[PRE 86] W.H. Press, B.P. Flannery, S.A:Teuloski and W.T.Vetterling, Numerical Recipes: The Art of Scientific Computing, Cambridge University Press, 1986.

Chapter 3

Constitutive Modelling for Rate-independent Soils: A Review

Claudio Tamagnini
Dipartimento di Ingegneria Civile e Ambientale, Università di Perugia, Italy

Gioacchino Viggiani
Laboratoire Sols Solides Structures, UJF, INPG, CNRS, Grenoble, France

1. Introduction

In the application of continuum theories to the analysis of any solid mechanics problem, a fundamental role is played by the *constitutive equations*, which are introduced to define the specific properties of the material of which the body under consideration is made. As the mechanical behavior of engineering materials can be described by different and often contradictory constitutive assumptions, constitutive equations do not represent universal laws of nature. Rather, they can be considered definitions of *ideal materials*, i.e., what is usually referred to as *constitutive models*. Although constitutive models may possess only to a limited extent the properties of the actual materials they are intended to model, this do not lessen their worth, which is to produce a mathematical model of the physical system which allows one to predict its behavior under any possible circumstance, from the limited knowledge gathered in a few experimental observations. Of course, the quality of the predictions depends crucially on the ability to define a suitable idealization for the real material which is capable to capture, from a quantitative point of view, these experimentally observed features which are thought to be of relevance for the practical problem at hand. This is particularly true in computational geomechanics, where the materials under consideration – i.e., soil layers or rock masses – are usually characterized by a complex multi–phase structure and by a highly non–linear, irreversible and path–dependent response to the applied loads.

In the application of advanced numerical methods to the analysis of a geotechnical structure – be it, for example, a foundation, an excavation, an earth dam, or a natural slope – the choice of the particular theoretical framework in which the mechanical behavior of the material is described, and the specific mathematical properties of the constitutive equation employed, can have a major impact with respect to the following two aspects:

– first, the characteristics of the constitutive model adopted may affect the response of the system to the applied loads not only from a *quantitative*, but also – and perhaps more importantly – from a *qualitative* point of view;

– second, in selecting an efficient, accurate and robust numerical strategy for the practical implementation of the theory in a general–purpose numerical procedure, such as the finite element method, due account must be taken of the details of the constitutive equations, particularly when material behavior is characterized by strong non–linearity, irreversibility and path–dependence.

As for the numerical aspects related to the implementation of the various classes of constitutive models considered in the following, the reader is referred to the contribution by Tamagnini et al. [TAM 02a] in this volume, as well as to some recent publications on this subject, such as [SIM 97, SIM 98, JIR 02b].

The main objective of this paper is to provide an outline of the different classes of constitutive equations for soils – from the early, pioneering works in standard, perfect plasticity, to more recent developments such as incrementally non–linear models and hypoplasticity – which, in the light of the first aspect just mentioned, might be of use in assessing, from a qualitative point of view, the predictive capabilities of the different theoretical frameworks with respect to the actual performance of real geotechnical structures.

By noting that almost all non–linear constitutive models for geomaterials employed today – both in design applications and for research purposes – are cast in *incremental* form, in the presentation of the various theories the incremental nature of the constitutive equation will be adopted as the main classification criterion, as first proposed by Darve [DAR 90a]. Central to this particular approach is the concept of *incremental non–linearity*, which is a distinct and in most cases more relevant feature of soil behavior than the commonly accepted notion of non–linearity, as derived from the stress–strain response in loading paths of finite size [DAR 78, DAR 90a].

Of course, the present paper is not intended to represent a comprehensive review of the outstanding number of studies which have been carried out on constitutive modeling over the last decades, which would be an almost impossible task, and would require a volume on its own. Rather, the presentation will be limited to those constitutive theories which reflect the authors' own experience and interests. In particular, the discussion will be limited to constitutive equations for *rate–independent, saturated soils* in *isothermal conditions*, obeying the *principle of effective stress* as stated by Terzaghi [TER 48a]. Details on how the constitutive models for saturated soils should be extended to account for partially saturated conditions can be found in the paper by Schrefler & Sanavia [SCH 02, SAN 02] in this volume, or, e.g., in [GEN 96, WHE 96, LEW 98]. Moreover, the constitutive equations for brittle materials – such as rocks or concrete – developed in the framework of damage mechanics are deliberately left out of this exposition, as this subject is dealt with in detail by Pijaudier–Cabot [PIJ 02]. Finally, only constitutive equations for *simple materials*, according to Truesdell & Noll [TRU 65], will be considered in the following. Although non–local or weakly non–local theories for materials with microstructure – such as polar, second gradient or micromorphic materials – have been recently the subject of a considerable amount of research in geomechanics, mainly in relation to the study of strain localization into shear bands, they are outside the scope of the present work. For this interesting subject, the reader is referred to the papers by Pijaudier–Cabot and Jirasek [PIJ 02, JIR 02a] in this volume, or to the book by Vardoulakis & Sulem [VAR 95], and references therein.

2. Notation

In the following, boldface lower– and upper–case letters are used to represent vector and tensor quantities. The symbols $\mathbf{1}$ and $\boldsymbol{I}^s$ are used for the second–order and fourth–order identity tensors, with components:

$$(\mathbf{1})_{ij} = \delta_{ij} \qquad\qquad (\boldsymbol{I}^s)_{ijkl} = \frac{1}{2}(\delta_{ik}\delta_{jl} + \delta_{il}\delta_{jk}) \qquad (1)$$

The symmetric and skew–symmetric parts of a second–order tensor $\boldsymbol{X}$ are denoted as: sym $\boldsymbol{X} := (\boldsymbol{X} + \boldsymbol{X}^T)/2$ and skw $\boldsymbol{X} := (\boldsymbol{X} - \boldsymbol{X}^T)/2$, respectively. The dot product is defined as follows: $v \cdot w := v_i w_i$ for any two vectors v and w; $X \cdot Y := X_{ij} Y_{ij}$ for any two second-order tensors X and Y. The dyadic product is defined as follows: $[v \otimes w]_{ij} := v_i w_j$ for any two vectors v and w; $[X \otimes Y]_{ijkl} := X_{ij} Y_{kl}$ for any two second-order tensors X and Y. The quantity $\|X\| := \sqrt{X \cdot X}$ denotes the Euclidean norm of X. The usual sign convention of soil mechanics (compression positive) is adopted throughout. In line with Terzaghi's principle of effective stress, all stresses are *effective* stresses, unless otherwise stated. In the representation of stress and strain states, use will sometimes be made of the invariant quantities: p (mean stress), q (deviator stress), and θ (Lode angle), defined as:

$$p := \frac{1}{3}(\sigma \cdot 1); \quad q := \sqrt{\frac{3}{2}} \|s\|; \quad \sin(3\theta) := \sqrt{6} \frac{(s^3) \cdot 1}{[(s^2) \cdot 1]^{3/2}} \tag{2}$$

and: ϵ_v (volumetric strain), ϵ_s (deviatoric strain), $\dot{\epsilon}_v$ (volumetric strain rate), and $\dot{\epsilon}_s$ (deviatoric strain rate), defined as:

$$\epsilon_v := \epsilon \cdot 1; \quad \epsilon_s := \sqrt{\frac{2}{3}} \|e\|; \quad \dot{\epsilon}_v := \dot{\epsilon} \cdot 1; \quad \dot{\epsilon}_s := \sqrt{\frac{2}{3}} \|\dot{e}\| \tag{3}$$

In eqs. (2) and (3), $s := \sigma - p\,1$ is the deviatoric part of the stress tensor; $e := \epsilon - (1/3)\epsilon_v\,1$ and $\dot{e} := \dot{\epsilon} - (1/3)\dot{\epsilon}_v\,1$ are the deviatoric parts of the strain and the strain rate tensors, respectively, while s^2 and s^3 are the square and the cube of the deviatoric stress tensor, with components $(s^2)_{ij} := s_{ik} s_{kj}$ and $(s^3)_{ij} := s_{ik} s_{kl} s_{lj}$. It is worth noting that in eq. (3)$_4$, with a slight abuse of notation, the symbol $\dot{\epsilon}_s$ has been employed to denote the second (deviatoric) invariant of the strain rate tensor, which generally does not coincide with the time rate of ϵ_s, as defined in in eq. (3)$_2$.

3. Constitutive equations in rate form: general principles

According to the principles of *determinism* and *local action* [TRU 65], the most general expression for the constitutive equation of a *simple* material is given by:

$$\sigma(x, t) = \mathop{\mathcal{G}}_{\tau=0}^{\infty} \left[F^{(t)}(X, \tau) \right] \tag{4}$$

where $\mathcal{G}$ is a *functional* of the *history* up to time t of the *deformation gradient* associated with the motion $x = \phi(X, t)$ carrying the material point X in the reference configuration to its position x in the current configuration at time t, defined as:

$$F^{(t)}(X, s) := F(X, t-s) \qquad F(X, t) := \frac{\partial \phi}{\partial X}(X, t) \qquad (s \geq 0) \tag{5}$$

Eq. (4) essentially states that the (effective) stress tensor σ is a function of the *entire deformation history*, i.e., that the knowledge of the state of strain at a given time t is in general not sufficient to determine the stress state. This is essential in order to deal with irreversible, anelastic behavior.

A third fundamental principle, the *principle of material frame indifference*, implies the following restriction to the functional $\mathcal{G}$: for every orthogonal tensor function $Q(s)$ and every history $F^{(t)}(X, s)$, the relation:

$$Q_0 \mathop{\mathcal{G}}_{\tau=0}^{\infty} \left[F^{(t)}(X, \tau) \right] Q_0^T = \mathop{\mathcal{G}}_{\tau=0}^{\infty} \left[Q(s) F^{(t)}(X, \tau) \right] \qquad Q_0 := Q(0) \qquad (6)$$

must hold. Conversely, any such functional $\mathcal{G}$ satisfying eq. (6) can be considered as defining the constittutive equation of a particular material.

The fundamental properties of the functional $\mathcal{G}$ should be defined according to our knowledge of the main characteristic of the mechanical behavior of the materials we intend to model. As far as geomaterials – and soils in particular – are concerned, a long standing experimental evidence indicates that the mechanical response of such materials is strongly non–linear and dependent on such factors as current state, previous loading history, load increment size and loading direction. Even the simplest and most common laboratory tests, such as a one–dimensional compression test or a axisymmetric (triaxial) drained compression test, can highlight such features in both fine and coarse–grained soils. A main consequence of this observation is that the constitutive functional $\mathcal{G}$ must be *non–linear* and *non–differentiable*, see [OWE 69]. However, working with non–linear, non–differentiable functionals poses formidable mathematical problems, even in the simplest cases. An alternative strategy, which overcomes this difficulty and is commonly adopted in nonlinear solid mechanics, is to avoid formulating the constitutive equation in *global terms*, as in eq. (4), and rather adopt an incremental (or *rate–type*) formulation, in which the (objective) stress rate is given as a *function* of the rate of deformation $d := \operatorname{sym} \nabla v$ ($v := d\phi/dt$ being the spatial velocity) and of the current state of the material:

$$\overset{\triangledown}{\sigma} = G(\sigma, q, d) \qquad (7)$$

In eq. (7), $\overset{\triangledown}{\sigma}$ is the Jaumann–Zaremba stress rate, defined as:

$$\overset{\triangledown}{\sigma} := \dot{\sigma} + \sigma\omega - \omega\sigma \qquad (8)$$

where $\omega := \operatorname{skw} \nabla v$ is the spin tensor, and q represents a set of *internal* state variables, which are introduced to account for the effects of the previous loading history. An additional set of rate equations is then required to define the evolution of the internal variables in time (*hardening laws* in the framework of classical elastoplasticity).

Restricting our discussion to linear kinematics, the Jaumann stress rate $\overset{\triangledown}{\sigma}$ can be replaced by the standard time rate $\dot{\sigma}$, and the rate of deformation d with the (linearized) strain rate tensor $\dot{\epsilon}$. Thus, eq. (7) can be rewritten as:

$$\dot{\sigma} = G(\sigma, q, \dot{\epsilon}) \qquad (9)$$

Rate–independence means that a change in the time scale does not affect the material response, e.g., doubling the strain rate doubles the stress rate. More generally:

$$G(\sigma, q, \lambda\dot{\epsilon}) = \lambda G(\sigma, q, \dot{\epsilon}) \qquad \forall \lambda > 0 \qquad (10)$$

A direct consequence of the above equation is that the function G is *positively homogeneous* of degree one in $\dot{\epsilon}$. This latter property yields the following alternative expression for the constitutive equation (9):

$$\dot{\sigma} = D\left(\sigma, q, \eta\right)\dot{\epsilon} \tag{11}$$

where D represents the (fourth–order) tangent stiffness tensor at the current state, which depends on the strain rate only through its *direction*, defined by the unit tensor $\eta := \dot{\epsilon}/\|\dot{\epsilon}\|$. Eq. (11) provides a general representation for rate–independent constitutive equations which encompasses as particular cases all the constitutive theories mentioned in sect. 2 and is of paramount importance for the following developments[1].

4. Non–linearity and incremental non–linearity

Let (σ_0, q_0) be the initial state of the material at time $t = 0$. For a given strain path $\mathcal{E}$ from ϵ_0 to $\epsilon(t)$, the state of stress at time t, $\sigma(t)$, is obtained by integrating eq. (11):

$$\sigma(t) = \hat{\sigma}\left(\sigma_0, q_0, \mathcal{E}\right) = \sigma_0 + \int_{\mathcal{E}} D\left(\sigma, q, \eta\right)\frac{d\epsilon}{ds}\,ds \tag{12}$$

From the above equation, it is immediately apparent that the dependence of the tangent stiffness D on the current state (σ, q) renders the function $\hat{\sigma}$ *non–linear*, e.g., doubling the strain increment does not result in doubling the stress increment. This is the notion of non–linearity to be invoked when describing a material response for which the observed stress–strain curve (e.g., in a triaxial compression path) is not a straight line. Another, independent concept of non–linearity can be introduced by considering the functional relation between stress rate and strain rate, as first suggested by Darve [DAR 78]. If the constitutive function G is *linear* in $\dot{\epsilon}$, then the material is said to be *incrementally linear*. In this case, the tangent stiffness tensor D does not depend on the strain rate direction η, and eq. (11) reduces to:

$$\dot{\sigma} = D\left(\sigma, q\right)\dot{\epsilon} \tag{13}$$

It should be stressed that while a linear behavior implies incremental linearity, the opposite is not true. That is, incremental linearity does not imply linearity of the

---

1. As shown in, e.g., [CHA 84] or [CHA 00], a rigorous derivation of eq. (11) can be obtained by considering that, in the vicinity of a given strain rate $\dot{\epsilon}^*$, the linearization of the function G with respect to $\dot{\epsilon}$ yields:

$$\dot{\sigma} = D\left(\sigma, q, \eta^*\right)\dot{\epsilon} + t$$

where $\eta^* := \dot{\epsilon}^*/\|\dot{\epsilon}^*\|$, and t is a symmetric second–order tensor such that:

$$\lim_{\eta \to \eta^*} \frac{t}{\|\eta^* - \eta\|} = 0$$

Eq. (11) is then recovered in the limit $\eta \to \eta^*$.

stress–strain response over a finite load increment. On the other hand, when G is a *non–linear* function of the strain rate, i.e., for any $\dot{\epsilon}_1$ and $\dot{\epsilon}_2$ and $a, b \in \mathbb{R}$:

$$G\left(\sigma, q, a\dot{\epsilon}_1 + b\dot{\epsilon}_2\right) \neq aG\left(\sigma, q, \dot{\epsilon}_1\right) + bG\left(\sigma, q, \dot{\epsilon}_2\right) \tag{14}$$

the material behavior is said to be *incrementally non–linear*. In this case, the tangent stiffness D explicitly depends on the strain rate direction, see eq. (11).

From eq. (14) it follows that:

$$G\left(\sigma, q, \dot{\epsilon}\right) \neq -G\left(\sigma, q, -\dot{\epsilon}\right) \tag{15}$$

which, in turn, implies:

$$D(\sigma, q, \eta) \neq D(\sigma, q, -\eta) \tag{16}$$

Equation (16) expresses a fundamental feature of incrementally non–linear models: for any strain rate direction, the reversal of the loading path is always associated with a change in the tangent stiffness D. Indeed, such a feature is *necessary* in order to correctly describe irreversible behavior. In fact, although eq. (13) is in general non–integrable, the response of an incrementally linear material remains completely *reversible* in any closed loading–unloading program following the same path in two opposite directions.

When discussing the dependence of D on η, it is useful to introduce the concept of *tensorial zone*, as defined by Darve [DAR 78, DAR 90a]. A tensorial zone Z is a portion of the strain rate space in which G is a linear function of $\dot{\epsilon}$. Accordingly, in a particular tensorial zone the tangent stiffness is *independent of* η:

$$\forall \eta \in Z \qquad D(\sigma, q, \eta) = D^Z(\sigma, q) \tag{17}$$

As G is positively homogeneous of degree one in $\dot{\epsilon}$, Z is a cone in the strain rate space with the vertex at the origin. Following [DAR 90a], incrementally non–linear, rate–independent constitutive equations can be classified according to the number of associated tensorial zones. When the number of tensorial zones of G is finite, the constitutive equation is *incrementally multi–linear* (*bi–linear* in the particular case of only two zones). This is the case of classical elastoplasticity, with one or more loading mechanisms, in which each loading/unloading condition defines a boundary between two different tensorial zones. In incrementally multi–linear materials, a crucial issue is represented by the *continuity* of the response at the boundary between any two tensorial zones [GUD 79]. Let ∂Z_{AB} be such a boundary between the tensorial zones Z_A and Z_B. If $\dot{\epsilon}^* \in \partial Z_{AB}$, then, continuity of the response requires that:

$$\dot{\sigma} = D^{Z_A} \dot{\epsilon}^* = D^{Z_B} \dot{\epsilon}^* \qquad \Rightarrow \qquad \left[D^{Z_A} - D^{Z_B}\right] \dot{\epsilon}^* = 0 \tag{18}$$

Equation $(18)_2$ represents a generalization of the continuity condition established by [GRE 56] for hypoelastic materials (see sect. 5). As opposed to multi–linearity, a *strictly* incrementally non–linear behavior is provided by constitutive models for which a *continuous* dependence of D on η is assumed. This can be considered as a generalization of multi–linearity, when the number of tensorial zones goes to infinity.

5. Incrementally linear models

In the early application of continuum mechanics to geotechnical engineering, the enormous analytical difficulties posed by the design of even the simplest geotechnical structures led to the traditional distinction between "deformation" and "failure" problems, for which different, very simple constitutive equations could be used, see e.g., [TER 48b]. The rationale behind this approach being that only some very specific features of soil behavior were of interest for the particular problem at hand, while the others could be neglected without affecting the quality of the prediction in a substantial way. In particular, the only possible constitutive framework for which (analytical) solutions to deformation problems could be obtained at that time – in lack of suitable numerical methods and powerful computer platforms – was provided by the theory of *linear elasticity*. Its successful application then relied on the "proper" selection of the relevant soil constants (in essence, the Young's modulus), which had to be assumed to depend on such primary factors as current stress state, previous stress history, and nature of the applied stress path – in terms of magnitude and, possibly, direction.

In the subsequent attempts to extend this approach to take into account global non–linearity of soil response, two main lines of development can be identified, which led both to incrementally linear formulations. The first approach is directly derived from Truesdell theory of *hypoelasticity* [TRU 56], and is due to the pioneering works of Romano [ROM 74] and Davis & Mullenger [DAV 78]. In this case, the constitutive equation takes the general form (13). A thorough discussion of the relative merits and drawbacks of hypoelasticity as compared to elastoplasticity can be found in [MRO 80]. The second approach can be traced back to the works of Kondner & Zelasko [KON 63] and Duncan & Chang [DUN 70]. In essence, it consists in adopting an isotropic elastic constitutive equation in rate form, in which the elastic stiffness parameters are assumed to depend on the strain level. The stiffness tensor D is typically given by general expressions of the form:

$$D\left(\sigma,\epsilon\right) = K_t\left(p,\epsilon_v\right) \mathbf{1} \otimes \mathbf{1} + 2G_t\left(p,\epsilon_s\right)\left[I^s - \frac{1}{3}\mathbf{1} \otimes \mathbf{1}\right] \qquad (19)$$

In constitutive models of this class, the dependence of the tangent bulk and shear moduli, $K_t\left(p,\epsilon_v\right)$, and $G_t\left(p,\epsilon_s\right)$, on the strain invariants is obtained by curve–fitting the observed stress–strain response in standard loading paths, such as drained (or undrained) triaxial compression, and isotropic compression. For this reason, these constitutive equations are also referred to as *variable–moduli models*. Empirical expressions of the form:

$$K_s\left(p,\epsilon_v\right) = K_{s0}(p)F_K\left(\epsilon_v\right) \qquad\qquad G_s\left(p,\epsilon_s\right) = G_{s0}(p)F_G\left(\epsilon_s\right) \qquad (20)$$

are typically suggested in terms of *secant* moduli, K_s and G_s, which are easier to determine directly. In eq. (20), F_K and F_G are non–increasing functions of the corresponding strain invariants. The simplest expression for the function F_G was introduced by [KON 63], by assuming that the stress–strain curve in drained triaxial compression might be interpolated by a hyperbola ("hyperbolic" model). A dependence of G_{s0} on mean stress was subsequently introduced in the hyperbolic model by [DUN 70].

The variable–moduli approach has been recently revived by various authors, see e.g. [JAR 86, JAR 88, TAT 97, GOT 99], and has largely inspired much of the experimental research carried out on pre–failure deformation of geomaterials over the last decade. Major improvements of laboratory measurement techniques [JAR 84, GOT 91] have allowed one to obtain very high quality stress–strain data, which in turn have motivated new, more refined analytical expressions for the decay function F_G.

Since their introduction, variable–moduli models have become relatively popular in the geotechnical community, and many of them have been actually used in practical design [JAR 88, ST. 93]. The reason for their appeal stems from their relative simplicity and ease of incorporation into standard finite element codes. However, these models suffer from a number of important drawbacks, as discussed, e.g., in [PYK 86, BUR 91]. Herein, the following additional remarks can be made about the consequences of the assumption (19).

1) In standard implementations of variable–moduli models, one of the arguments of D in eq. (19) is actually redundant. In fact, the stress invariant p is typically used to define the material response at the *initial state only* (K_{s0}, G_{s0}), whereas volumetric and deviatoric strains are introduced to describe *indirectly*, through the decay functions F_K and F_G, the effects of the evolution of the state of the material along the (prescribed) loading path. As the stress–strain curves from which the decay functions are obtained are invertible, the state of stress is actually the only state variable of the material and no further state variables exist which can account for the effects of the previous loading history.

2) Typically, the functions F_K and F_G are derived from a single, specific loading path (axisymmetric or isotropic compression), which, although easily accessible in the laboratory, might be quite far from those commonly experienced by the soil in practical applications. In this case, the use of eq. (19) for *all* possible loading paths represents a potentially dangerous extrapolation.

3) Regardless of the particular choice for the functions K_t and G_t, the assumed structure of tangent stiffness rules out any volumetric–deviatoric coupling, that is, a change in volume associated with purely deviatoric loading (or, conversely, distortion associated with purely isotropic loading). This appears to be questionable, on the basis of currently available experimental data. For example, an indirect evidence of such a coupling can be found in the resonant column test reported by Georgiannou et al. [GEO 91], where residual excess pore pressures were measured at strain levels as small as $10^{-2}\%$.

4) Eq. (19) implies that, as in isotropic linear elasticity, stress and strain rates are always coaxial, regardless of the current state of stress. Again, available experimental evidence suggests that this property is not a general feature of soil behavior for all possible loading conditions [HON 89, FRY 95]. In any case, it is important to observe

that variable–moduli models are systematically developed based on results from laboratory tests in which the principal directions of stress, strain, stress rate and strain rate are all coincident and remain fixed throughout. Thus, they do not allow any conclusion to be drawn concerning this particular point.

Another very important issue related to the practical use of hypoelastic or variable–moduli models concerns their extension to include a dependence of stiffness on loading direction. This particular aspect will be discussed in detail in the following sect. 6.1.

6. Incrementally bi–linear models

6.1. Variable moduli models with loading–unloading conditions

A major problem associated with the application of variable–moduli models to practical engineering problems arises when the stress–paths in the soil present one or more stress–reversals (cyclic loading), or when different zones of soil undergo significantly different stress paths, in terms of size and direction. A typical example might be provided by the analysis of excavations in stiff cohesive soils. In these cases, some regions of the soil mass can experience stress–path reversal, while others follow a continuation of the loading path associated with the previous history of the deposit. Available experimental evidence, see e.g. [ATK 86, SMI 92b, ROY 98, CAL 98], shows that in this case, widely different responses in terms of tangent stiffness might be expected. This particular aspect of soil behavior is essentially related to the development of irreversible strains, which, as already discussed in sect. 4, cannot be modelled by incrementally linear formulations. In order to extend hypoelastic and variable–moduli formulations to overcome this problem, two alternative strategies have been typically followed in the literature, namely:

– coupling the variable–moduli approach with classical plasticity, which, overall, leads to a non–linear elastic–plastic formulation, where irreversibility is associated with the development of plastic strains. This approach has been used, e.g., in [JAR 86, JAR 88, JAR 91];

– defining one or more fictitious loading/unloading criteria, which is equivalent to the introduction of multiple tensorial zones. In each zone, a constitutive equation like eq. (13) holds, but different material constants apply to different tensorial zones.

As for the first approach, it must be noted that coupling variable–moduli (or hypoelastic) models with plasticity is effective only when the stress state is on the yield surface – failure surface in the case of perfect plasticity. For stress states far from limit conditions, these formulations do not differ from incrementally linear models. The second approach has been used, for example, by Romano [ROM 74], who introduced in his hypoelastic constitutive equation a single loading–unloading criterion based on the sign of the stress power $\dot{W} := \sigma \cdot \dot{\epsilon}$. A more refined hypoelastic model with four tensorial zones has been proposed by Davis & Mullenger [DAV 78], who adopted two such criteria, based on the sign of the volumetric stress power, $\dot{W}_v := p\dot{\epsilon}_v$, and of the deviatoric stress power, $\dot{W}_s := s \cdot \dot{e}$. Examples of applications of variable–moduli models with two tensorial zones are provided, e.g., by the various implementation of the Duncan & Chang model discussed by Pike [PYK 86]. In these approaches, the switch condition is based either on the sign of the maximum principal stress rate, $\dot{\sigma}_1$, or on the sign of the rate of the normalized deviator stress, $d(q/q_{lim})$, q_{lim} representing the limiting value of the deviatoric stress for the current mean stress and Lode angle. As first observed by Gudehus [GUD 79], a major drawback of this approach is that it usually violates the continuity condition (18) at the boundary between the different tensorial zones. This can give rise to serious instabilities in numerical applications, when the computed direction of the stress increment approaches the boundary between the loading and unloading zones.

6.2. Single–mechanism plasticity

6.2.1. Basic principles and constitutive equations

The most important case of constitutive equations with two tensorial zones is provided by the classical *theory of plasticity* with a *single plastic mechanism*, and its various generalizations to describe, for example, induced anisotropy and cyclic behavior. The general framework of the theory of plasticity is now well established and a thorough treatment of this subject can be found in many excellent textbooks, such as, for example, [LUB 90, SIM 97, JIR 02b]. As for plasticity in soil mechanics, good references are provided, e.g., by [DES 84, LOR 90b, LOR 90a, ZIE 99]. A detailed presentation of this subject, for which the interested reader is referred to the aforementioned works, is clearly out of the scope of the present work. Herein, only a few remarks are made on the structure of elastoplastic constitutive equations in the light of the notions introduced in sect. 4.

Under the hypothesis of linear kinematics, the strain rate is decomposed additively in an elastic, reversible part, $\dot{\epsilon}^e$, and a plastic, irreversible part, $\dot{\epsilon}^p$:

$$\dot{\epsilon} = \dot{\epsilon}^e + \dot{\epsilon}^p \tag{21}$$

The elastic strain rate is usually linked to the stress rate, $\dot{\sigma}$ by assuming a suitable elastic constitutive equation in rate form:

$$\dot{\sigma} = D^e(\sigma)\dot{\epsilon}^e = D^e(\sigma)[\dot{\epsilon} - \dot{\epsilon}^p] \tag{22}$$

where D^e is the fourth–order elastic stiffness tensor, generally dependent on the current stress. In those cases in which the existence of a strain energy function $\psi(\epsilon^e)$ can be postulated, the elastic constitutive equation and the elastic stiffness tensor are given by:

$$\sigma(\epsilon^e) = \frac{\partial \psi}{\partial \epsilon^e}(\epsilon^e) \qquad D^e = \frac{\partial^2 \psi}{\partial \epsilon^e \otimes \partial \epsilon^e} \tag{23}$$

Irreversibility is introduced by requiring that the state of the material (σ, q) belongs to the convex set:

$$\mathbb{E}_\sigma := \left\{ (\sigma, q) \mid f(\sigma, q) \leq 0 \right\} \tag{24}$$

where $f(\sigma, q)$ is the so-called *yield function*. The evolution of plastic strains is prescribed by the following *flow rule*:

$$\dot{\epsilon}^p = \dot{\gamma} \frac{\partial g}{\partial \sigma}(\sigma, q) \tag{25}$$

where $g(\sigma, q)$ is the *plastic potential*, and $\dot{\gamma} \geq 0$ is the *plastic multiplier*. Finally, the evolution of the internal variables is provided by assigning a suitable *hardening law*:

$$\dot{q} = \dot{\gamma} h(\sigma, q) \tag{26}$$

where $h(\sigma, q)$ is a prescribed hardening function. The plastic multiplier appearing in eqs. (25) and (26) is subject to the so-called *Kuhn–Tucker complementarity conditions*:

$$\dot{\gamma} \geq 0 , \quad f(\sigma, q) \leq 0 , \quad \dot{\gamma} f(\sigma, q) = 0 \tag{27}$$

stating that plastic deformations may occur only for states on the yield surface. From the *consistency condition* $\dot{\gamma} f(\sigma, q) = 0$, the elastic constitutive equation (22) and the flow rule (25), the following expression for the plastic multiplier is obtained:

$$\dot{\gamma} = \frac{1}{K_p} \left\langle \frac{\partial f}{\partial \sigma} \cdot D^e \dot{\epsilon} \right\rangle \tag{28}$$

where $\langle x \rangle := (x + |x|)/2$ denotes the ramp function, and:

$$K_p := \frac{\partial f}{\partial \sigma} \cdot D^e \frac{\partial g}{\partial \sigma} + H_p > 0 \qquad H_p := -\frac{\partial f}{\partial q} \cdot h \tag{29}$$

The scalar function $H_p(\sigma, q)$ is called *hardening modulus* of the material. A positive value of H_p denotes *hardening*, a negative value indicates *softening*, while $H_p = 0$ characterize the special case of *perfect plasticity*. As thoroughly discussed in, e.g., [SIM 97, JIR 02b], the assumption that $K_p > 0$ is crucial in the establishment of the correct formulation of the loading/unloading conditions in presence of softening. Its effect is essentially to place a restriction on the amount of allowable softening. Substituting the above expression (28) for the plastic multiplier in eqs. (22) and (25), we obtain:

$$\dot{\sigma} = D^{ep} \dot{\epsilon} \tag{30}$$

$$D^{ep} := D^e - \frac{\hbar(\dot{\gamma})}{K_p} \left(D^e \frac{\partial g}{\partial \sigma} \right) \otimes \left(\frac{\partial f}{\partial \sigma} D^e \right) \tag{31}$$

where $\hbar(\dot{\gamma})$ denotes the Heaviside step function, equal to one if $\dot{\gamma} > 0$ and zero otherwise.

Introducing the two unit tensors:

$$n_f := \left\| \frac{\partial f}{\partial \sigma} \right\|^{-1} \frac{\partial f}{\partial \sigma} \qquad\qquad n_g := \left\| \frac{\partial g}{\partial \sigma} \right\|^{-1} \frac{\partial g}{\partial \sigma} \qquad (32)$$

representing the unit normals to the yield surface and the plastic flow direction, eqs. (25), (28) and (31) can be cast in the following alternative form:

$$\dot{\epsilon}^p = \dot{\lambda}\, n_g \qquad (33)$$

$$\dot{\lambda} = \frac{1}{\widehat{K}_p} \langle n_f \cdot D^e \dot{\epsilon} \rangle \qquad (34)$$

$$D^{ep} = D^e - \frac{\hbar(\dot{\lambda})}{\widehat{K}_p} (D^e\, n_g) \otimes (n_f\, D^e) \qquad (35)$$

where:

$$\widehat{K}_p := n_f \cdot D^e n_g + \widehat{H}_p \qquad \widehat{H}_p := \left(\left\| \frac{\partial f}{\partial \sigma} \right\| \left\| \frac{\partial g}{\partial \sigma} \right\| \right)^{-1} H_p \qquad (36)$$

According to the expression (28) – or (34) – for the plastic multiplier, the switch between elastic and plastic response is ruled by the sign of the scalar product $n_f \cdot D^e \dot{\epsilon}$. Material response is thus incrementally *bi–linear*, and the two tensorial zones resulting from the switch conditions are the regions of the the strain rate space separated by the plane Π passing through the origin and normal to the unit vector:

$$\nu := \frac{D^e n_f}{\|D^e n_f\|} \qquad (37)$$

It is worth noting that the loading/unloading condition based on the sign of the plastic multiplier $\dot{\gamma}$ guarantees the continuity at the boundaries of the tensorial zones. In fact:

$$\forall\, \dot{\epsilon} \in \Pi \qquad \dot{\sigma} = \left\{ D^e - \frac{1}{\widehat{K}_p} (D^e n_g) \otimes (n_f D^e) \right\} \dot{\epsilon} = D^e \dot{\epsilon} \qquad (38)$$

since $\nu \cdot \dot{\epsilon} = 0$ on Π.

6.2.2. Perfect plasticity

According to the definitions laid out in the previous sections, *elastic–perfectly plastic* models are those for which the yield function and the plastic potential are given functions of the stress tensor σ only. Thus, in this case, no internal variables are present and $H_p = 0$. The first applications of perfect plasticity to soil mechanics can be traced back to the various solutions of failure problems for foundations, retaining walls, or slopes obtained through the method of characteristics (slip line theory) [SOK 65], or the application of the upper and lower bound theorems of limit analysis

[CHE 76]. In both these approaches, the soil is modelled as a rigid–perfectly plastic medium, with a failure condition provided, e.g., by the so–called *Mohr–Coulomb criterion*:

$$f(\sigma) = (\sigma_1 - \sigma_3) - 2c\cos\phi - (\sigma_1 + \sigma_3)\sin\phi = 0 \qquad (39)$$

where σ_1 and σ_3 are the maximum and minimum principal stresses, and c and ϕ are two material constants defining the *cohesion* and the *friction angle* of the soil, respectively. Rather than in attempting to model the complete stress–strain law of the material, the emphasis was placed on the determination of the stress field in limit conditions, in order to evaluate the stability of the system with respect to a particular collapse mechanism. The success of this simple and elegant approach to failure problems is witnessed by the fact that most of the design methods currently in use for geotechnical structures are still based on such limit solutions, and specific numerical techniques have been developed to extend limit analysis to those cases for which no sufficiently accurate analytic solution can be found, see, e.g., [SLO 95, PON 97, LYA 02]. Extension of the above concepts to the analysis of more complex deformation problems, such as soil–structure interaction or the modelling of the transition from the small–strain regime up to failure conditions, had to wait until the pioneering application of the finite element method to soil mechanics. Examples of the use of elastic–perfectly plastic models with pressure–dependent failure criteria such as Mohr–Coulomb or Drucker–Prager, are given, e.g., in [ZIE 77, SIR 83] for shallow foundations, [ZIE 75] for slopes, [SMI 92a, SHA 95] for flexible retaining structures, and [ROW 83, WON 91] for tunnels.

6.2.3. *Isotropic hardening plasticity*

The experience gathered in using classical perfect plasticity in the analysis of deformation problems has shown how these formulations provide a too crude description of the actual behavior of natural soils in pre–failure conditions. A radical change of perspective in soil plasticity occurred after the pioneering work of Roscoe and coworkers in Cambridge, which lead, in the sixties, to the basic principles of the so–called "Critical State Soil Mechanics" (CSSM) [ROS 68, SCH 68]. The practical use of CSSM in geotechnical applications started in the early seventies, when CSSM was interpreted as a particular application of isotropic–hardening elastoplasticity, see e.g., [ZIE 71], and generalized to full six–dimensional stress and strain states. The road was then open to a new approach to geotechnical engineering practice, in which no such distinction between failure and deformation problems, or elastic response and plastic collapse was needed any longer.

The prototype of isotropic hardening elastoplastic models for cohesive soils is the so–called "Modified Cam–Clay" (MCC) [ROS 68], which assumes an associative flow rule ($f \equiv g$). The yield surface adopted in the original MCC formulation has the following equation:

$$f(p, q, p_s) = p(p - p_s) + \frac{q^2}{M^2} = 0 \qquad (40)$$

which, for any possible value of p_s, describes a family of ellipses with principal axes parallel to the coordinate axes and passing through the origin, see Figure 1a. In eq. (40)

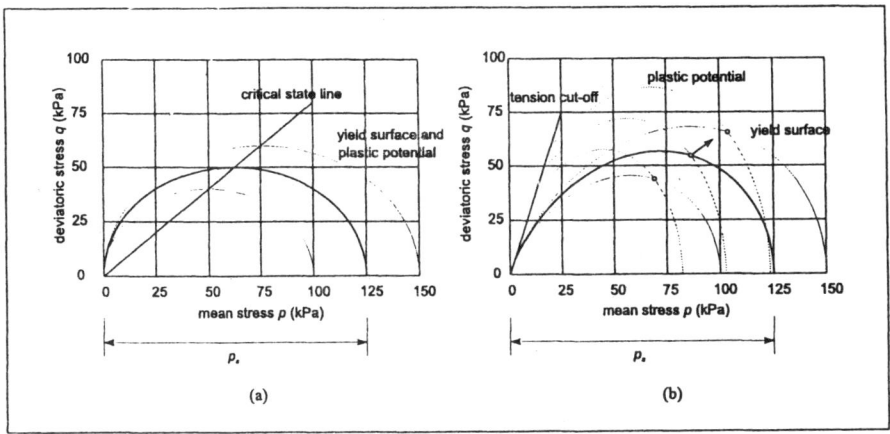

Figure 1. *Yield surfaces and plastic potentials for isotropic hardening elastoplastic models: a) Modified Cam–Clay [ROS 68]; b) Sinfonietta Classica [NOV 88].*

M is a constant defining the ratio between the principal axes of the ellipse. The scalar quantity p_s controls the size of the yield surface, and represents the only internal state variable of the material. Its evolution is provided by an empirically derived logarithmic law of the type[2]:

$$\dot{p}_s = \rho_s\, p_s\, \dot{\epsilon}_v^p \qquad (41)$$

with $\rho_s = $ const. In the notation of eqs. (26),(29):

$$\dot{p}_s = \dot{\gamma}\, h_s(p,q,p_s) \qquad h_s := \rho_s\, p_s\, \frac{\partial f}{\partial p} \qquad H_p = -\rho_s\, p_s\, \frac{\partial f}{\partial p}\, \frac{\partial f}{\partial p_s} \qquad (42)$$

From eq. (26) it is clear that the "failure" conditions for the material ($p_s = $ const. and $H_p = 0$) are characterized by purely distortional plastic strain rates, i.e., the material can be deformed indefinitely at constant stress and constant volume. Such particular states, the existence of which is experimentally observed in clayey soils, are defined *critical states*, and form the basis of all subsequent modern treatments of hardening plasticity for cohesive soils. Modifications of MCC to improve its predictive capabilities have been discussed by numerous authors. Among them, we recall the extension of the yield function (40) to include the third stress invariant θ [ZIE 73], the adoption of a composite yield surface to include the so called *Hvorslev surface* for yield points on the "supercritical side" of the critical state line [ZIE 73, HOU 84], and the adoption of a hyperelastic constitutive equation [HOU 85]. The class of isotropic hardening models known in the literature as *cap models* [DIM 71, SAN 76] can be considered essentially as CS models with a modified supercritical yield function, the position of which, however, does not change with plastic strains.

---

2. Note that, to avoid using too many different symbols, the notations employed in this work can sometimes be different from the one adopted in the original works cited.

In the application of the concepts of isotropic hardening plasticity to non–cohesive soils, two major limitations of classical critical state models have been pointed out. First, the assumption of an associated flow rule is generally not supported by available experimental data on sand dilatancy, see e.g., [POO 66, POO 67], and do not allow the modelling of the experimentally observed phenomenon of *static liquefaction* in the *hardening regime* observed in loose sands under undrained conditions [NOV 96]. Second, the hypothesis of purely volumetric hardening does not allow one to describe the so–called *phase transition* effect – i.e., the transition from contractant to dilatant behavior – typically observed in dense sand under undrained compression. Non-associative isotropic hardening models for sands have been proposed since the pioneering work of Pooroshasb et al. [POO 66, POO 67] who coupled a Cam–Clay type plastic potential with a classical Mohr-Coulomb yield locus. Subsequent improvements were proposed, e.g., by Nova & Wood [NOV 79] and Kim & Lade [KIM 88, LAD 88]. As for the hardening function, Nova [NOV 77] and [WIL 77] independently proposed an extension of the volumetric hardening rule (41) with the following form:

$$\dot{p}_s = \rho_s \, p_s \left\{ \dot{\epsilon}_v^p + \xi_s \dot{\epsilon}_s^p \right\} \qquad H_p := -\rho_s \, p_s \left\{ \frac{\partial g}{\partial p} + \xi_s \frac{\partial g}{\partial q} \right\} \frac{\partial f}{\partial p_s} \qquad (43)$$

The parameter ξ_s appearing in eq. (43) can be considered either a constant [NOV 77] or a monotonically decreasing function of the accumulated plastic deviatoric strains [WIL 77]. In this last case, a critical state is recovered in the ultimate conditions at very large plastic strains.

An example of isotropic hardening models for sands which combines good predictive capabilities for monotonic loading with a relatively simple mathematical structure, with a limited number of material constants easily linked to observed material behavior in standard tests is provided by the model proposed by Nova under the name *Sinfonietta Classica* [NOV 88]. For this model, the adopted yield function and plastic potential are given by the following equations:

$$f(p, r, p_s) = 3\beta \, (\gamma - 3) \, \ln \left(\frac{p}{p_s} \right) - \gamma \, \mathrm{tr} \left(r^3 \right) + \frac{9}{4} \, (\gamma - 1) \, \mathrm{tr} \left(r^2 \right) = 0 \qquad (44)$$

$$g(p, r, p_s^*) = 9 \, (\gamma - 3) \, \ln \left(\frac{p}{p_s^*} \right) - \gamma \, \mathrm{tr} \left(r^3 \right) + \frac{9}{4} \, (\gamma - 1) \, \mathrm{tr} \left(r^2 \right) = 0 \qquad (45)$$

where $r := s/p$ is the stress–ratio tensor, s being the deviatoric part of the stress tensor, while β and γ are material parameters ($\beta \neq 3$ denotes non–associative behavior). The corresponding curves are shown in the $q : p$ plane in Figure 1b. A hardening law with volumetric and deviatoric hardening similar to eq. (43) is assumed for the internal variable p_s.

Subsequent developments gave rise to a number of constitutive models which progressively diverged from the basis assumptions of CSSM in the attempt to cover further aspects of experimentally observed soil behavior, as well as to tackle other, more challenging classes of engineering problems. The breath and depth of such scientific production is well portrayed, for example, by the proceedings of the workshops held in Grenoble in 1982 [GUD 84], Cleveland in 1988 [SAA 89], and Horton in 1992 [KOL 93a]. Another useful source of references is provided by the special volume published on the occasion of the XI ICSMFE [MUR 85].

6.2.4. Anisotropic hardening plasticity

Almost all geotechnical materials such as rocks, coarse–grained soils and fine–grained soils are characterized – to a certain extent – by the existence of some preferential orientations at the microstructural level. In granular soils, for example, such preferential orientations can be associated with the spatial distributions of the contact normals, to grain shape and to void shape, see [ODA 85]. Moreover, the directional properties of the microstructure might remain more or less stable during the deformation of the solid skeleton (as, e.g., the distribution of grain orientations in the tests performed by Oda et al. [ODA 85]), or they might evolve as a consequence of grain rearrangements upon applied loading (as, e.g., the distribution of contact normals [ODA 85]). From these observations, it follows naturally that the phenomenological response of the material – reflecting the properties of the microstructure – can be characterized by a more or less marked *anisotropy*, both in terms of stress–strain response in pre–failure conditions, and in terms of shear strength. According to the possibility that superimposed loading histories may change the directional properties of the microstructure, two different kind of anisotropy can be distinguished at the macroscopic level, see [CAS 44]:

– *inherent anisotropy*, "[. . .] a physical characteristics inherent in the material and entirely independent of the applied strains";

– *induced anisotropy*, "[. . .] a physical characteristic due exclusively to the strain associated with the applied stress".

Inherent anisotropy is usually relevant, e.g., in stratified rocks, where strong intergranular bonds prevent the occurrence of strong rearrangements of the microstructure, or in coarse–grained soils with strongly non–circular particles, the orientation of which cannot be modified easily unless a substantial amount of grain crushing occurs. On the contrary, induced anisotropy plays a major role in non–cemented granular soils with rounded particles or in clays, where the applied loading can modify and, in some cases, even erase the effects of the previous loading history. Direct and indirect experimental evidence of inherent and induced anisotropy is reported, e.g., in [ALL 77] for soft rocks, in [ODA 78, WON 85, YAS 91, YOS 98] for sands, and in [DUN 66, MIT 70, TAV 77, GRA 83b, SMI 92b] for clays.

In the framework of the classical theory of plasticity, inherent anisotropy can be dealt with in the formulation of the elastic constitutive equation – as in, e.g., [BOE 75, GRA 83a] – and/or in the definition of yield and plastic potential functions. Constitutive equations for inherent anisotropy have been proposed, e.g., by Nova [NOV 86] or Pastor [PAS 91], based on a approach first suggested by Hill [HIL 50]. Essentially, these models are derived from existing isotropic hardening formulations by replacing the standard invariants of the stress tensor with corresponding anisotropic invariants defined by means of suitably chosen (constant) *structure tensors*, which are employed as metric tensors in the construction of the scalar invariants entering in the constitutive functions.

The description of induced anisotropy – i.e., the evolution of the directional properties of the material with the loading history – requires that the set of internal state variables q includes at least one tensor–valued quantity. In most of the existing anisotropic hardening plasticity models, this is usually assumed to be a symmetric second–order tensor, with the character of a *microstructure tensor*. Although this limits the degree of symmetry of the material to *orthotropy*, see [BOE 87], it is considered sufficient for most geomaterials of relevant practical interest. In the presence of a symmetric second–order microstructure tensor in the set of internal variables, the general restrictions imposed on the yield and plastic potential functions by the principle of material frame indifference, as well as the consequences of induced anisotropy on the relative orientation between the principal directions of the stress and the plastic strain tensors are discussed in detail in [BAK 84]. Plasticity models with anisotropic hardening can be broadly grouped into two different classes, according to the experimental evidence which they were intended to reproduce, namely:

– constitutive models with *kinematic hardening*, capable of modelling soil behavior under cyclic loading paths, see, e.g., [WOO 82] and references therein;

– constitutive models with *rotational hardening*, which are capable of describing the changes in the orientation of the yield surface with the evolution of plastic strains, as observed, e.g., in [YAS 91, TAV 77, GRA 83b, SMI 92b].

Kinematic hardening models for soils originate from the pioneering work of Mroz [MRO 67], Iwan [IWA 67], and Dafalias & Popov [DAF 75]. In such models, a yield function of the form:

$$f(\sigma, \alpha, q_k) = \hat{f}(\hat{\sigma}, q_k) = 0 \qquad\qquad \hat{\sigma} := \sigma - \alpha \qquad (46)$$

is assumed, in which the so–called *back–stress* α is the microstructure tensor, responsible for the induced anisotropy, and q_k ($k = 1, \ldots, n$) indicates the other *scalar* internal variables. As α changes during the loading process, the yield surface is dragged by the stress–path as indicated qualitatively in Figure 2. However, the motion of the yield surface is restricted by larger, outer surface, usually referred to as *Bounding Surface* (BS), of equation:

$$F(\sigma, \bar{q}_k) = 0 \qquad\qquad \{\bar{q}_k\} \subset \{q_k\} \qquad (47)$$

The BS separates admissible states from impossible ones, and is generally similar in shape to the yield locus. Models of this kind have been proposed by various authors; among them we recall the works of Prevost [PRE 77, PRE 86], Mroz et al. [MRO 78, MRO 81], Hashiguchi [HAS 85, HAS 88], Wood and coworkers [ALT 89, GAJ 99, ROU 00]. Most of these works represent a straightforward extension of classical Mod-

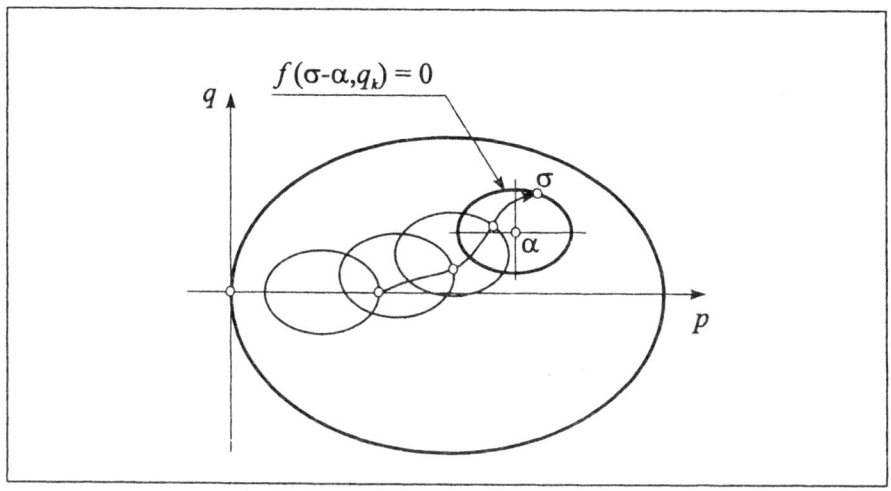

Figure 2. *Kinematic hardening inside the Bounding Surface.*

ified Cam–Clay, see sect. 6.2.3. As an example, in the model of Al–Tabbaa & Wood [ALT 89] the yield and Bounding Surface functions are given by:

$$F(\sigma, p_c) = \frac{3}{2M_\theta} s \cdot s + (p - p_c)^2 - p_c^2 = 0 \tag{48}$$

$$f(\sigma, p_c) = \frac{3}{2M_\theta} (s - \text{dev}\, \alpha) \cdot (s - \text{dev}\, \alpha) + (p - p_\alpha)^2 - R^2 p_c^2 = 0 \tag{49}$$

where $p_c = p_s/2$, $p_\alpha - \text{tr}\, \alpha/3$ and $R \ll 1$ is a material constant representing the ratio between the sizes of the two surfaces.

In this class of models, the hardening function adopted for p_c (or p_s) is similar to the one adopted in critical state models, see eq. (41). As for α, rather than prescribing explicitly the hardening function, the hardening modulus H_p is assigned as a monotonically decreasing function of the distance δ between the current state and a *image state* $\overline{\sigma}$ on the BS, defined as the point at which the unit normals to $f = 0$ and $F = 0$ have the same direction (see Figure 3a):

$$H_p = \widehat{H}\left(\overline{H}_p, \delta\right) \qquad \frac{\partial \widehat{H}}{\partial \delta} > 0 \qquad \widehat{H}\left(\overline{H}_p, 0\right) = \overline{H}_p \tag{50}$$

In eq. (50), $\delta := \|\overline{\sigma} - \sigma\|$ and $\overline{H}_p$ is the plastic modulus at $\overline{\sigma}$:

$$\overline{H}_p := -\frac{\partial F}{\partial p_c} h_c \tag{51}$$

obtained from the consistency condition on the BS: $\dot{F}(\overline{\sigma}, \overline{q}_k) = 0$. When the stress–path touches the BS, the two surfaces must share the same tangent, otherwise some

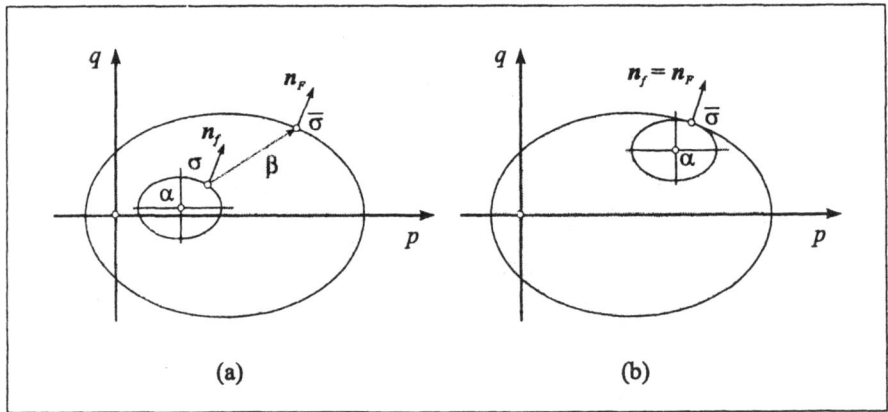

Figure 3. *Kinematic hardening models: definition of the image point.*

admissible states would fall *outside* the BS, see Figure 3b. As shown by Hashiguchi [HAS 85], this is obtained through an appropriate definition of the evolution equation for α. For the Al–Tabbaa & Wood model [ALT 89], the non–intersection condition requires that:

$$\dot{\alpha} = \dot{\overline{\alpha}} + (\alpha - \overline{\alpha})\, \frac{\dot{p}_c}{p_c} + \frac{n \cdot \left[\dot{\overline{\sigma}} - (\dot{p}_c/p_c)\,\overline{\sigma}\right]}{n \cdot (\overline{\sigma} - \sigma)} (\overline{\sigma} - \sigma) \tag{52}$$

In eq. (52), the first term is related to the translation of the center of the BS, the second represents the effect of the change in size of the BS (and of the yield surface), and the third a net translation in the direction of the tensor β: $= \overline{\sigma} - \sigma$, see Figure 3a.

Anisotropic plasticity models with rotational hardening are more suitable for describing the anisotropy induced by loading histories associated with depositional processes in natural deposits, such as one–dimensional compression and, possibly, swelling. These models can be traced back to the pioneering works of Sekiguchi & Ohta [SEK 77] for clays, or Ghaboussi & Momen [GHA 82] for sands. Constitutive equations of this kind proposed for sands are usually intended to model irreversible processes associated with deviatoric loading paths, and therefore adopt conical–shaped yield surfaces, *open* towards the range of high mean pressures (Figure 4a). Among them we recall the models proposed in [GHA 82, POO 85, GAJ 99]. Rotational hardening models for fine–grained soils, on the contrary, adopt *closed* yield surfaces (Figures 4b, c), in order to reproduce the irreversible deformations usually observed in these materials along isotropic or proportional loading paths (q / p = const.). Examples of rotational hardening models for clays are given in [HAS 79, BAN 86, ANA 86]. Exceptions to this general trend are provided by the models of di Prisco et al. [PRI 93] – actually a generalization of the Sinfonietta Classica model discussed in the previous section – and Pestana & Whittle [PES 99], which can be employed for coarse as well as fine–grained materials. It is worth noting that in most of the afore-mentioned models, the rotational anistropy is employed in connection to some form

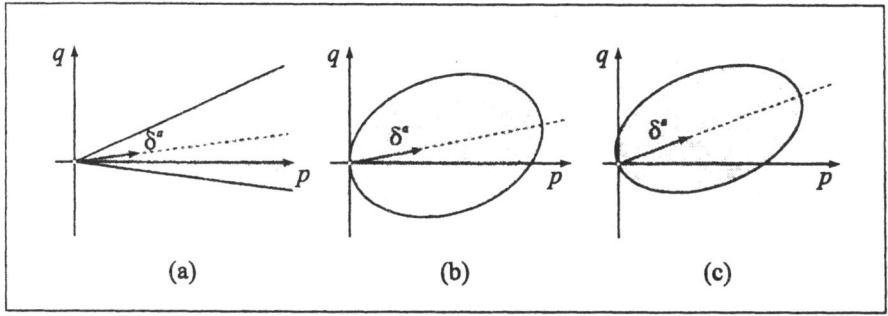

Figure 4. *Typical yield surfaces adopted in rotational hardening models.*

of generalized plasticity allowing plastic flow inside the main state boundary surface, which will be discussed in sect. 6.3.

Rotational hardening models can be easily derived as generalizations of classical isotropic hardening formulations (e.g., Modified Cam–Clay) by simply replacing the stress invariants entering in the yield and plastic potential functions with appropriate *mixed invariants* which take into due account the microstructure tensor. A possible way of defining such mixed invariants is as follows (after [ANA 86], slightly modified):

$$p^a := \frac{1}{3}\,\sigma \cdot \delta^a \qquad q^a := \sqrt{\frac{3}{2}}\,\|s^a\| \qquad \sin(3\theta^a) := \sqrt{6}\frac{(d^a)^3 \cdot 1}{[(d^a)^2 \cdot 1]^{3/2}} \qquad (53)$$

where δ^a is the microstructure tensor, and:

$$s^a := \sigma - p^a\,\delta^a \qquad d^a := \mathrm{dev}\,(s^a) \qquad \delta^a \cdot \delta^a = 3 \qquad (54)$$

In eqs. (54), the projections of σ on the isotropic axis, commonly used to construct the isotropic and deviatoric invariants of the stress tensor, are replaced by the corresponding projections on the microstructure tensor δ^a, now playing the role of the unit tensor – compare eqs. (54) with eqs. (2) – and, in fact, defining the rotation of the surface with respect to the isotropic axis, see Figure 4. Several alternative strategies have been proposed to link the evolution of the microstructure tensor (i.e., the rotation of the yield surface) with the plastic strain rate. All of them must, however, satisfy the orthogonality condition $\delta^a \cdot \dot{\delta}^a = 0$, required by the assumption (54)$_3$. Anandarajah & Dafalias [ANA 86] proposed the following hardening law:

$$\dot{\delta}^a = \lambda^*\,(\dot{e}^p)^a \qquad (\dot{e}^p)^a := \dot{e}^p - \frac{1}{3}\,(\dot{e}^p \cdot \delta^a)\,\delta^a \qquad (55)$$

where λ^* is a scalar function of the invariants of σ and δ^a, asymptotically going to zero as the norm of $\mathrm{dev}(\delta^a)$ increases, to model the phenomenon of *hardening saturation*. A different approach is adopted by di Prisco et al. [PRI 93], who assume:

$$\dot{\delta}^a = C_p\,\{\hat{\chi} - (\hat{\chi} \cdot \delta^a)\,\delta^a\}\,\|\dot{\epsilon}^p\| \qquad (56)$$

with C_p a material parameter, and $\widehat{\chi}$ a given, tensor–valued function of the current stress state, with the property $\widehat{\chi} \cdot \widehat{\chi} = 3$. As $\dot{\delta}^a = 0$ for $\delta^a = \widehat{\chi}$, this last quantity represents the asymptotic value for δ^a at very large accumulated plastic strains.

6.3. Bounding Surface models and generalized plasticity

An important limitation of classical elastoplasticity as applied to geomaterials is represented by the assumption of a large elastic domain, inside which the response of the material is purely reversible. In light of the concepts introduced in sect. 4, classical – perfect or hardening – elastoplasticity is characterized by an incrementally bi–linear constitutive equation *only* for states *on the yield surface*. All elastic states are, by definition, endowed with an incrementally linear response. However, a large body of experimental evidence suggests that soil behavior can be irreversible and path–dependent *even for strongly preloaded states*, and that plastic yielding is a rather gradual process. Although such effects can be considered of secondary importance in the simulation of monotonic loading paths, it must be noted that a strong dependence of the small–strain stiffness on the loading path direction has been observed, e.g., by [ATK 86, STA 90] in heavily overconsolidated soils, and that such a feature of soil behavior – which cannot be reproduced by any incrementally linear model – can be of great importance in all practical applications in which strong variations of the stress–path direction are expected in different zones of the soil mass, e.g., in the analysis of excavations. Moreover, irreversible (plastic) strains occurring well inside the locus of admissible stress states are obviously of great importance in cyclic loading processes, and the accurate description of such phenomena as cyclic mobility or liquefaction under repeated loading (see, e.g., [WOO 82]) requires to take them into proper account.

The kinematic hardening models discussed in the previous section – mostly developed during the early 1980s, in response to the problems posed by the design of cyclic loaded structures such as off–shore platforms, or by the quantitative prediction of soil response during earthquakes – are certainly capable to deal successfully with this particular issue. However, a number of alternative strategies have also been proposed for the same purpose, which represent genuine generalizations of the classical framework. Among them, definitely worthy of mention are the so–called *bounding surface models*, originally developed by Dafalias and his coworkers, and the models developed in the framework of *Generalized Plasticity*, as defined by Pastor et al. [PAS 90].

The key concept in the formulation of a Bounding Surface model is the fact that, as in kinematic hardening elastoplastic models mentioned before, there exists a surface in stress space – the Bounding Surface (BS), defined by an equation similar to eq. (47) – which separates admissible from impossible states. Such a surface is subject to hardening processes which may change its size, shape and orientation due to the development of plastic strains, exactly as a standard yield surface in classical plasticity. However, such a surface is *not* a yield surface, as plastic strains can occur for stress states located in its interior. In particular, at each admissible state (inside or on

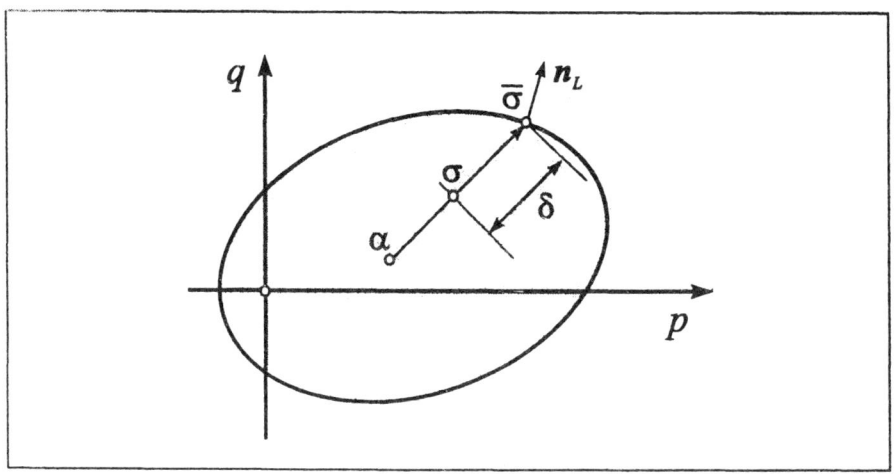

Figure 5. *Radial mapping rule in Bounding Surface models.*

the BS), a flow rule identical to eq. (33) is assumed, in which the plastic multiplier $\dot{\lambda}$ is replaced by:

$$\dot{\lambda} = \frac{1}{\widetilde{K}_p} \langle n_L \cdot D^e \dot{\epsilon} \rangle \tag{57}$$

where:

$$\widetilde{K}_p := n_L \cdot D^e n_g + \widetilde{H}_p \tag{58}$$

in which n_L is a unit tensor defining the loading direction, and $\widetilde{H}_p$, by analogy with the standard formulation, plays the role of the plastic modulus. The definition of these last two quantites relies crucially on the possibility of associating to each stress state σ inside the BS a corresponding *image state* $\overline{\sigma}$ on the BS, through a non–invertible *mapping rule*. In the so–called *radial mapping* BS models, see [DAF 86a], this is accomplished by simply projecting the current stress onto the BS from a given *projection center* α, see Figure 5. Once the image state is found, the loading direction is taken as the gradient of the BS at $\overline{\sigma}$, while the plastic modulus H_p is assumed to be a monotonically decreasing function of the distance $\delta := \|\overline{\sigma} - \sigma\|$ between the current state and the image state, and of the plastic modulus $\overline{H}_p$ at $\overline{\sigma}$:

$$\widetilde{H}_p = \widetilde{H}\left(\overline{H}_p, \delta\right) \qquad \frac{\partial \widetilde{H}}{\partial \delta} > 0 \qquad \widetilde{H}\left(\overline{H}_p, 0\right) = \overline{H}_p \tag{59}$$

The constitutive equation in rate form is then given by an equation similar to eq. (35), with $\dot{\lambda}$ provided by eq. (57) and $\widehat{K}_p$ replaced by $\widetilde{K}_p$ of eq. (58). The analogies existing between this procedure for defining the loading direction and the plastic modulus and the one outlined for kinematic hardening models in sect. 6.2.4 are apparent. As a matter of fact, Dafalias [DAF 86a] considered kinematic hardening models as a special class of BS models, characterized by a special form of mapping rule. However, differently than in kinematic hardening plasticity, in radial mapping BS models, no elastic region exists anymore, and an incrementally bi–linear response is obtained at *any* state. A comprehensive review of Bounding Surface models is provided in [DAF 86a]. Applications of the Bounding Surface Concept to the modelling of clays are reported, e.g., in [ZIE 85, DAF 86b, ANA 86, WHI 94, TAM 99] while applications to coarse–grained soils are given, e.g., by [PAS 85, BAR 86, CRO 94].

Starting from the works of Zienkiewicz & Mroz [ZIE 84], Pastor et al. [PAS 90] developed the framework of *Generalized Plasticity* as a further generalization of the Bounding Surface concept, where the concepts of plastic potential, yield function and consistency condition are completely abandoned. In the incrementally bi–linear version of the theory, the plastic strain rate is provided by the following equations:

$$\dot{\epsilon}^p = \dot{\lambda}_L\, n_{gL} \qquad \text{if}: \quad n_L \cdot D^e \dot{\epsilon} > 0 \quad \text{(loading)} \tag{60}$$

$$\dot{\epsilon}^p = \dot{\lambda}_U\, n_{gU} \qquad \text{if}: \quad n_L \cdot D^e \dot{\epsilon} < 0 \quad \text{(unloading)} \tag{61}$$

$$\dot{\epsilon}^p = 0 \qquad \text{if}: \quad n_L \cdot D^e \dot{\epsilon} = 0 \quad \text{(neutral loading)} \tag{62}$$

in which:

$$\dot{\lambda}_L = \frac{1}{\widehat{K}_{p,L}}\, n_L \cdot D^e \dot{\epsilon} \qquad \widehat{K}_{pL} := n_L \cdot D^e n_{gL} + \widehat{H}_{p,L} \tag{63}$$

$$\dot{\lambda}_U = \frac{1}{\widehat{K}_{p,U}}\, n_L \cdot D^e \dot{\epsilon} \qquad \widehat{K}_{pU} := n_L \cdot D^e n_{gU} + \widehat{H}_{p,U} \tag{64}$$

In eqs. (60)–(64), n_L, n_{gL} and n_{gU} are three unit second–order tensors representing the loading direction, the plastic flow direction for plastic loading and the plastic flow direction for plastic unloading (*reverse loading*), respectively, while the scalars $\widehat{H}_{p,L}$ and $\widehat{H}_{p,U}$ are the corresponding plastic moduli for (plastic) loading and unloading. All these quantities are considered as prescribed functions of the state variables (σ, q), and, in general, their definition do not require any yield function, plastic potential or consistency condition to be assumed. The corresponding expressions for the elastoplastic tangent stiffness tensor are given by:

$$D^{ep} = \begin{cases} D^e - (1/\widehat{K}_{p,L})\,(D^e\, n_{gL}) \otimes (n_L\, D^e) & \text{(plastic loading)} \\ D^e - (1/\widehat{K}_{p,U})\,(D^e\, n_{gU}) \otimes (n_L\, D^e) & \text{(plastic unloading)} \end{cases} \tag{65}$$

It is worth noting that both classical plasticity and Bounding Surface plasticity are recovered from generalized plasticity as special cases, with suitable choices for the constitutive functions n_L, n_{gL}, n_{gU}, $\widehat{H}_{p,L}$ and $\widehat{H}_{p,U}$, see [PAS 90] for further details.

6.4. Plasticity with extended hardening rules

A last, notable case of incrementally bilinear formulations is provided by elastoplastic models with extended hardening rules – as defined by [TAM 02b] – proposed in the geomechanics context to describe a number of practically relevant aspects of the mechanical behav, or of geomaterials. A common, distinctive feature of those constitutive theories is that the size and shape of the yield locus, as well as its evolution with the loading process are assumed to depend, in addition to accumulated plastic strains, on some other non–mechanical state variable, usually of scalar nature. Among them, we recall:

– the simple thermoplastic model proposed by Nova [NOV 86] to describe the influence of temperature on the brittle–ductile transition of rocks in geophysical applications, in which the preconsolidation pressure depend on the temperature T;

– the elastoplastic models for unsaturated soil (formulated in terms of Bishop effective stresses) in which an explicit dependence of the size of the yield surface on the degree of saturation is assumed to simulate the phenomenon of collapse upon wetting for partially saturated soil, see, e.g., [JOM 00];

– the extension of classical elastoplasticity advocated by [NOV 00] to describe the effects of weathering on cemented soils or weak rocks, in which some bonding–related internal variables are subject to both mechanical and chemical degradation, described through a normalized, scalar weathering function X_d.

These approaches share also some similarities with a number of viscoplastic models based on the concept of a non–stationary yield locus, see e.g., [FLA 90],[BOR 92], and to chemoplastic models proposed for early–age concrete [ULM 96] or clays subject to environmental loading [HUE 92, HUE 97].

Let θ denote the additional (scalar) variable affecting the mechanical response of the material, i.e., temperature, suction or chemical degradation. A first modification of the classical theory to account for the changes in θ is introduced in the elastic law, which now reads, in rate form:

$$\dot{\sigma} = D^e\,(\sigma,\theta)\,.[\dot{\epsilon} - \dot{\epsilon}^p] + m\,(\sigma,\theta)\,\dot{\theta} \tag{66}$$

where $m(\sigma,\theta)$ is a coupling coefficient (thermal stress coefficient for $\theta \equiv T$). While the definitions of elastic domain, flow rule and loading/unloading conditions are identical to those of the classical theory – eqs. (24), (25) and (27) – the evolution equation for the internal variables now assumes the following generalized form:

$$\dot{q} = \dot{\gamma}h(\sigma,q,\theta) + \dot{\theta}\eta(\sigma,q,\theta) \tag{67}$$

where: $h(\sigma,q,\theta)$ and $\eta(\sigma,q,\theta)$ are suitable hardening functions. The first term on the RHS of eq. (67) quantifies the changes in the internal variables associated with plastic strains, the second term accounts for all non–mechanical hardening/softening processes induced by a change of θ. From the consistency condition $\dot{\gamma}\dot{f}(\sigma,q) = 0$, the elastic constitutive equation (66) and the flow rule (25), the following generalized expression for the plastic multiplier is obtained:

$$\dot{\gamma} = \frac{1}{K_p}\left\langle \frac{\partial f}{\partial \sigma} \cdot D^e \dot{\epsilon} + \left(\frac{\partial f}{\partial q} \cdot \eta + \frac{\partial f}{\partial \sigma} \cdot m \right) \dot{\theta} \right\rangle \tag{68}$$

with K_p given by eq. (29). This in turns provides the following constitutive equations in rate form:

$$\dot{\sigma} = D^{ep}\,\dot{\epsilon} + m^{ep}\,\dot{\theta} \qquad (69)$$

$$\dot{q} = G\,\dot{\epsilon} + G_\theta\,\dot{\theta} \qquad (70)$$

where:

$$D^{ep} := D^e - \frac{\hbar(\dot{\gamma})}{K_p}\left(D^e\frac{\partial g}{\partial\sigma}\right)\otimes\left(\frac{\partial f}{\partial\sigma}D^e\right) \qquad (71)$$

$$m^{ep} := m - \frac{\hbar(\dot{\gamma})}{K_p}\left(\frac{\partial f}{\partial q}\cdot\eta + \frac{\partial f}{\partial\sigma}\cdot m\right)D^e\frac{\partial g}{\partial\sigma} \qquad (72)$$

$$G := \frac{\hbar(\dot{\gamma})}{K_p}\,h\otimes\left(\frac{\partial f}{\partial\sigma}D^e\right) \qquad (73)$$

$$G_\theta := \frac{\hbar(\dot{\gamma})}{K_p}\left(\frac{\partial f}{\partial q}\cdot\eta + \frac{\partial f}{\partial\sigma}\cdot m\right)h + \eta \qquad (74)$$

According to eq. (68), the plastic multiplier $\dot{\gamma}$ can be considered as the sum of the following two terms:

$$\dot{\gamma}_m := \frac{1}{K_p}\frac{\partial f}{\partial\sigma}\cdot D^e\dot{\epsilon} \qquad \dot{\gamma}_\theta := \frac{1}{K_p}\left(\frac{\partial f}{\partial q}\cdot\eta + \frac{\partial f}{\partial\sigma}\cdot m\right)\dot{\theta} \qquad (75)$$

The first, $\dot{\gamma}_m$, coincides with the plastic multiplier of classical elastoplasticity – see eq. (28) – while the second, $\dot{\gamma}_\theta$, accounts for the effect of non–mechanical hardening/softening processes. Note that, for plastic loading to occur, only the sum of $\dot{\gamma}_m$ and $\dot{\gamma}_\theta$ needs to be positive. In particular, plastic strains may occur even for a trial stress rate $\dot{\sigma}^{tr} := D^e\dot{\epsilon}$ pointing *inwards* the current yield locus ($\dot{\gamma}_m < 0$), provided that the change in θ gives rise to a reduction in size of the elastic domain sufficiently large to keep the plastic multiplier positive, as, for example, in the case of chemical degradation. Examples of application of this general framework to the modelling of mechanical and chemical degradation processes in weak rocks or bonded soils are provided in, e.g., [NOV 01, CAS 02a], and by Castellanza et al. [CAS 02b] in this volume.

7. Incrementally multi–linear models

A natural generalization of single–mechanism plasticity is provided by multi–mechanism plasticity, in which two or more yield surfaces are assumed to exist. Let $f^A(\sigma, q)$, with $A = 1, \ldots, m$ denote the yield surfaces associated with the m plastic mechanisms, and $g^A(\sigma, q)$ the corresponding plastic potentials. At a prescribed state (σ, q), define the set $\mathbb{A}$ as:

$$\mathbb{A} := \left\{A \in \{1, \ldots, m\} \quad \text{such that:} \quad f^A(\sigma, q) = 0\right\} \qquad (76)$$

i.e., the set of indices corresponding to potentially *active* plastic mechanisms. The above definition allows one to determine the plastic multiplier $\dot{\gamma}_A$ for all $A \in \mathbb{A}$ as:

$$\dot{\gamma}_A = \sum_{A \in \mathbb{A}} (\hat{K}_p^{-1})_{AB}\, n_f^B \cdot \boldsymbol{D}^e \dot{\epsilon} \qquad (77)$$

in which n_f^A and n_g^A denote as before the unit normals to the yield function f^A and plastic potential g^A, the (positive–definite) matrix $\hat{K}_p$, whose components are given by:

$$(\hat{K}_p)_{AB} := n_f^A \cdot \boldsymbol{D}^e n_g^B + (\hat{H}_p)_{AB} \qquad (78)$$

and $\hat{H}_p$ is a suitable matrix of hardening moduli, see [LOR 90a] for details. The tangent stiffness tensor is then given by:

$$\boldsymbol{D}(\sigma, q) := \boldsymbol{D}^e - \sum_{A \in \mathbb{P}} \sum_{B \in \mathbb{P}} (K_p^{-1})_{AB}\, \left(\boldsymbol{D}^e n_g^A\right) \otimes \left(n_f^B \boldsymbol{D}^e\right) \qquad (79)$$

in which:

$$\mathbb{P} := \{ A \in \mathbb{A} \quad \text{such that:} \quad \dot{\gamma}_A > 0 \} \qquad (80)$$

denotes the subset of those active mechanisms which undergo plastic loading. For each plastic mechanism, the loading/unloading condition provided by the definitions (77) and (80) is formally identical to that of single–mechanism plasticity. Note that $\boldsymbol{D} \equiv \boldsymbol{D}^e$ when no plastic loading occurs in any of the active mechanisms (i.e., $\mathbb{P} \equiv \varnothing$).

In multi–mechanism plasticity, incremental non–linearity takes the form of multi-linearity. Note that each plastic mechanism can be active *only* if the stress state is on the corresponding yield locus. Therefore, the material response is characterized by two tensorial zones if the current stress state belongs to just one yield surface, four tensorial zones if the state is at the intersection of two yield surfaces, and so on, see Figure 6. The maximum number of available tensorial zones is 2^m, where m is the total number of plastic mechanisms at the common intersections (if any) of all the yield surfaces. Well–known examples of two–mechanisms plasticity models for geomaterials are those proposed by Lade [LAD 77] and Vermeer [VER 78]. Extension of the framework of generalized plasticity (sect. 6.3) to the multi–linear case is also possible along the same lines, as discussed in [PAS 90]. The interested reader is referred to this work for further details. Finally, an example of incrementally–multilinear model which does not belong to the framework of elastoplasticity is provided by the incrementally octo-linear model by Darve & Labanieh [DAR 82]. In this model, no decomposition of the strain rate into a reversible and an irreversible part is postulated, and the constitutive equation in rate form is constructed by interpolating the experimentally observed response along special stress–paths, called *generalized triaxial compression paths*, see [DAR 82].

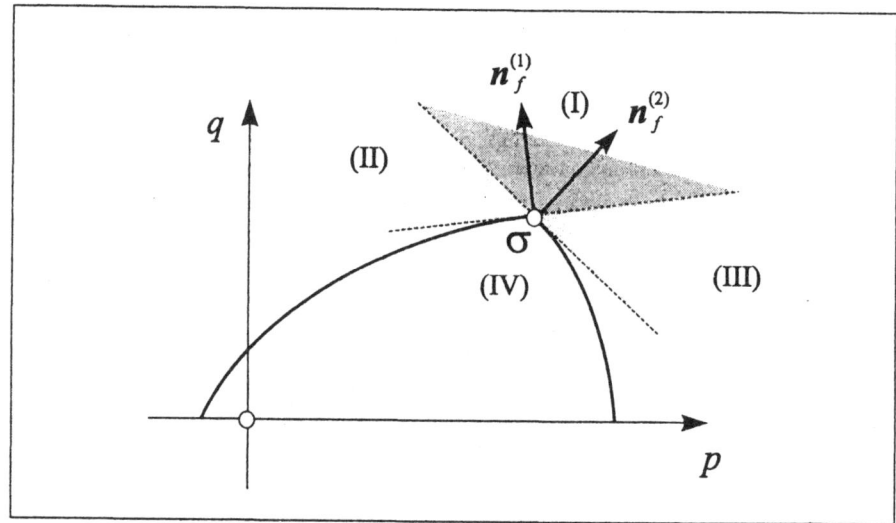

Figure 6. *Tensorial zones for a material with two active plastic mechanisms.*

8. Incrementally non–linear models

8.1. General concepts

In recent years, an alternative approach for modeling soil behavior has been pro-posed, which departs from the framework of plasticity theories and rather can be con-sidered as a generalization of Truesdell theory of hypoelasticity [TRU 56], obtained by assuming that the constitutive function G in eq. (7) is non–differentiable. The distinctive features of this approach are:

1) the absence of any kinematic decomposition of strain rates into reversible and irreversible parts;

2) the continuously non–linear dependence of the tangent stiffness tensor on the strain rate direction[3].

For this class of theories, the most general expression of the rate constitutive equation is provided by the so–called *incrementally non–linear models of the second order*, for which:

$$\dot{\sigma} = L\left(\sigma, q\right)\dot{\epsilon} + \frac{1}{\|\dot{\epsilon}\|}Q\left(\sigma, q\right)\left[\dot{\epsilon} \otimes \dot{\epsilon}\right] \tag{81}$$

see [DAR 82, DAR 99] and references therein. In eq. (81), the first term on the right hand side is a linear function of the strain rate $\dot{\epsilon}$, while the second is quadratic in $\dot{\epsilon}$,

---

3. Note that feature (1), but not feature (2), is also present in some bounding surface models [DAF 86a, WAN 90], and in the endochronic models developed by Bažant and Valanis and coworkers [BAZ 78, VAL 84].

Q being a sixth–order tensor. Incremental non–linearity is accounted for by the term $(1/\|\dot{\epsilon}\|)Q[\dot{\epsilon} \otimes \dot{\epsilon}]$. In fact, the tangent stiffness D corresponding to eq. (81) is given, in components, by:

$$D_{ijkl} = L_{ijkl} + (Q_{ijabkl} + Q_{ijklab})\eta_{ab} - Q_{ijabcd}\eta_{ab}\eta_{cd}\eta_{kl} \qquad (82)$$

and therefore is a *continuous* function of the strain rate direction. In this case, the number of possible tensorial zones goes to *infinity*, and a *strictly*, or thoroughly incrementally non–linear behavior is obtained.

8.2. The theory of hypoplasticity

A particular class of incrementally non–linear models which has received special attention in recent times is provided by the theory of *hypoplasticity*, as defined by [KOL 91][4]. In hypoplasticity, the general constitutive equation (9) takes the following particular form:

$$\dot{\sigma} = L(\sigma, q)\dot{\epsilon} + N(\sigma, q)\|\dot{\epsilon}\| \qquad (83)$$

where L is a fourth–order tensor and N is a symmetric second–order tensor, both depending on the current state of the material. Note that eq. (83) is, in fact, a particular case of eq. (81), obtained by setting $Q := N \otimes I^s$. An essential feature of equation (83), as compared to the more general eq. (81), is that the non–linearity in $\dot{\epsilon}$ is accounted for through the *scalar* quantity $\|\dot{\epsilon}\|$. A general outline of the theory of hypoplasticity was laid down by [KOL 91], and several review papers followed thereafter, the most recent of which are those by [WU 00] and [TAM 00]. In fact, two different formulations of hypoplasticity have been given over the last decade, with the specific objective of modeling the behavior of granular, coarse–grained materials. The first was developed in Karlsruhe after the pioneering work of Kolymbas [KOL 93b] and will be referred to in the following as K–hypoplasticity. The second originated in Grenoble from the work of Chambon and coworkers [CHA 89, CHA 94] under the general name of *CLoE* hypoplasticity. Although these two approaches share a number of similarities, the motivations for their independent development were different in various respects. As a result, some important differences are apparent in their original formulation, as well as in their respective subsequent developments, see [TAM 00].

The particular structure of the basic constitutive equation (83) of hypoplastic models allows a simple and effective graphical illustration of incremental non–linearity. This can be done by employing the so–called *stress response envelopes* (SRE), first proposed by [GUD 79] as a tool for visualizing the properties of a given constitutive equation in rate–form. A stress response envelope is defined as the image in the stress rate space of the unit sphere in the strain rate space, under the map defined by the constitutive equation[5]. In the general case, a SRE is a "surface" in a six–dimensional

---

4. Note that the term hypoplasticity was also used by [DAF 86a] to refer to a different class of incrementally non–linear models, developed in the framework of hardening elastoplasticity.
5. For rate–independent materials, the choice of the strain rate norm is arbitrary.

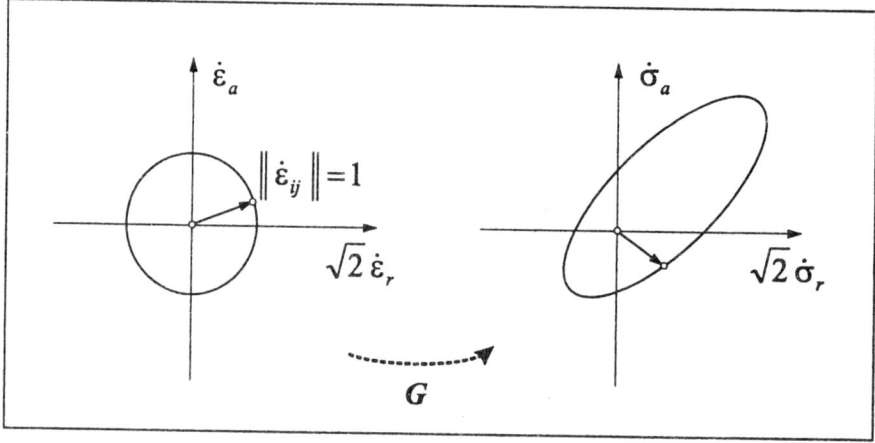

Figure 7. *Stress Response Envelope for axisymmetric loading.*

space. However, in the particular – yet quite interesting – case of axisymmetric (tri-axial) loading, the number of independent stress and strain variables reduces to two, and a convenient graphical representation of the SRE can be given in the so–called Rendulic plane of stress rates: $\dot{\sigma}_a : \sqrt{2}\dot{\sigma}_r$, where σ_a and σ_r are the axial and radial (principal) stresses, respectively, see Figure 7. Upon the map defined by eq. (83), the SRE of a hypoplastic material is obtained as sketched in Figure 8. The effect of the linear operator L is to transform the unit circle in Figure 8a into an ellipse *centered at the origin* of the stress rate plane (Figure 8b). The subsequent application of the nonlinear term $N \|\dot{\epsilon}\|$ – in which only the strain rate norm is involved – results in a *translation* of the ellipse along the direction defined by the tensor N (Figure 8c). In its final configuration, the SRE is *non–symmetric* with respect to the origin of the stress rate space. This lack of symmetry expresses in graphical terms the concept of incremental non-linearity as given in sect. 4.

In the first generation of hypoplastic models, the set of state variables is limited to the stress tensor, i.e., the constitutive tensors L and N are function of σ only:

$$\dot{\sigma} = L\left(\sigma\right)\dot{\epsilon} + N\left(\sigma\right)\|\dot{\epsilon}\| \tag{84}$$

Despite its relative simplicity, eq. (84) has proven to provide a good qualitative and quantitative description of the experimentally observed behavior of loose and dense sands under *monotonic* loading [BAU 96, CHA 94]. However, a major limitation of this class of hypoplastic constitutive equations is represented by their inherent inabil-ity to correctly describe soil behavior upon non–symmetric cyclic paths, since the response predicted for any cycle is *identical* to that of the first cycle (thus leading to the so–called *ratcheting effect*). Subsequent improvements of K–hypoplasticity were achieved by extending the space of state variables to include: i) the void ratio, in order to obtain a unified formulation for a wide range of initial densities and to embody the concept of critical state [GUD 96, BAU 96]; ii) a *structure* tensor to introduce inher-

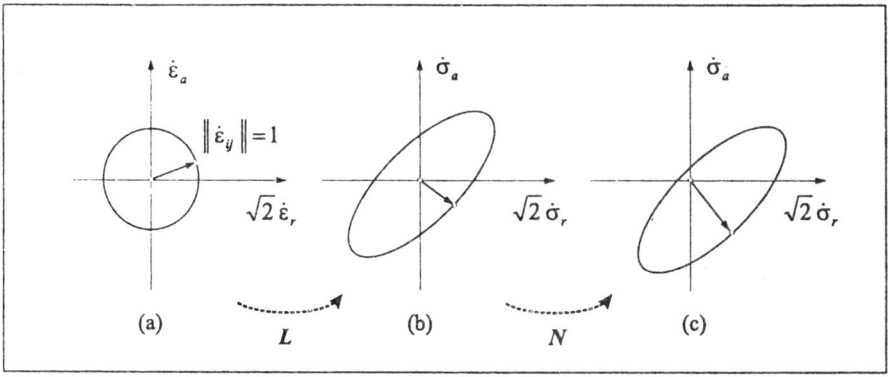

Figure 8. *Graphical interpretation of hypoplastic constitutive equation: a) unit circle in the strain rate plane; b) effect of the linear term; c) effect of the non–linear term.*

ent anisotropy [WU 98]; iii) a strain–like tensorial internal variable, to model cyclic loading [NIE 97]. The aspects covered in these developments are beyond the scope of the present work, and will not be discussed further. Only the development referred to in point (iii) deserves some more detailed comment, as it is important in the context of the present discussion. In the approach recently proposed by [NIE 97], the basic eq. (83) is modified into:

$$\dot{\sigma} = D\left(\sigma, e, \eta_\delta, \eta\right) \dot{\epsilon} \tag{85}$$

where e is the void ratio, η is the strain rate direction, and:

$$\eta_\delta := \begin{cases} \delta/\|\delta\| & \text{if: } \delta \neq 0 \\ 0 & \text{if: } \delta = 0 \end{cases} \tag{86}$$

is the direction of the internal variable δ. The symmetric second–order tensor δ, referred to as *intergranular strain*, is equipped with the following evolution equation:

$$\dot{\delta} := \begin{cases} \left(I - \rho^{\beta_r} \eta_\delta \otimes \eta_\delta\right) \cdot \dot{\epsilon} & \text{if } \eta_\delta \cdot \dot{\epsilon} > 0 \\ \dot{\epsilon} & \text{if } \eta_\delta \cdot \dot{\epsilon} \leq 0 \end{cases} \tag{87}$$

At any given state, the current value of δ records the effects of the recent deformation history of the material. By denoting with $\rho := \|\delta\|/R \in [0, 1]$ the normalized magnitude of the intergranular strain (where $R := \max\|\delta\|$ is a material constant), the tangent stiffness tensor takes the following expressions:

$$D = [\rho^\chi m_T + (1 - \rho^\chi)m_R]L(\sigma, e) +$$
$$\rho^\chi(1 - m_T)\left[L(\sigma, e)\eta_\delta\right] \otimes \eta_\delta + \rho^\chi N(\sigma, e) \otimes \eta_\delta \tag{88}$$

if $\eta_\delta \cdot \dot{\epsilon} > 0$, or:

$$D = [\rho^\chi m_T + (1 - \rho^\chi)m_R]L(\sigma, e) + \rho^\chi(m_R - m_T)\left[L(\sigma, e)\eta_\delta\right)] \otimes \eta_\delta \tag{89}$$

if $\eta_\delta \cdot \dot{\epsilon} \leq 0$. The scalar quantities β_r, χ, m_T and m_R appearing in eqs. (87)–(89) are material constants. As already noted by Niemunis & Herle [NIE 97], the strain rate direction η does not appear explicitly neither in eq. (88) nor (89). This is to be contrasted with the corresponding expression for the tangent stiffness tensor of the general hypoplastic model (83), which reads:

$$D(\sigma, q, \eta) = L(\sigma, q) + N(\sigma, q) \otimes \eta \qquad (90)$$

As a matter of fact, in hypoplasticity with intergranular strains, incremental non-linearity is only associated with the switch condition ruled by the sign of the scalar product $\eta_\delta \cdot \dot{\epsilon}$. Therefore, this formulation can be considered as incrementally bi-linear, and in this respect it is closer to generalized or bounding surface plasticity theories, rather than to true incrementally non–linear models.

That soils are thoroughly non–linear, rather than bi– or multi–linear, is first of all suggested by qualitative considerations on the mechanisms responsible for inelastic deformations at the microstructural level. For example, Tobita [TOB 97] remarks that the overall response of a granular frictional material is incrementally non–linear even if only one micro–mechanism (e.g., intergranular slip, or grain crushing) is recognized as the source of inelastic deformation. Indeed, any consideration on the micromechanical origin of inelastic deformation of granular soils leads naturally to the need for a fully non–linear incremental behavior at the macroscopic level. Even more important, from a practical point of view, is the observation that bi– or multi–linear incremental models, while performing well in the simulation of standard laboratory test results, fail to capture soil response under more complex loading paths which include rotations of principal stress and strain directions, and/or changes of the Lode angle at constant mean and deviatoric stresses (collectively denoted as *rotational shear*). In fact, according to Wang et al. [WAN 90], the successful simulation of soil behavior under rotational shear can be achieved only by adopting a fully incrementally non–linear constitutive formulation. As a matter of fact, a remarkably good prediction of the circular loading test proposed to the competitors at the 1988 Cleveland Workshop [SAA 89] was obtained by Darve & Dendani [DAR 88b], using an incrementally non–linear model of the second order. It is worth noting that the accurate description of soil response under rotational shear is of more than an academic interest, because the loading paths associated with many field loading conditions result in widespread changes of principal stress values and directions. Besides the obvious case of cyclic loading – of paramount interest in earthquake and offshore geotechnical engineering – rotational shear is a major component of the loading path experienced by the soil surrounding more common geotechnical structures such as shallow and deep excavations, as well as deep foundations. A third important motivation behind the development of incre-mentally non–linear theories can be found in the inadequacy of classical flow theories of plasticity with a single loading mechanism to provide a satisfactory description of experimentally observed shear banding processes, interpreted as a *bifurcation* prob-lem, see [VAR 95]. Bifurcation analysis shows that the predicted onset of bifurcation and shear band orientation are strongly affected by the way in which the constitutive model describes the response of the material for deformation paths which, upon bifurcation, change abruptly their direction (sometimes referred to as "loading to the side" paths). As a matter of fact, this was one of the major motivations behind the development of the theory, since a general, closed form solution for the (non–linear) shear band bifurcation problem can be found for hypoplastic models without the need of additional restrictive assumptions, while this is not possible, in general, for constitutive equations described by eq. (81), see [TAM 01] for details. As for the recent generalizations of the theory of plasticity discussed in sect. 6.3 and 6.4, the theory of hypoplasticity remains mainly confined to research applications. However, its applicability to the solution of practical geotechnical problems has been demonstrated in a number of cases, including excavations [ELH 97, VIG 99, KUD 99, VIG 00], shallow foundations [TEJ 99], embankments and piled foundations [CUD 00].

9. References

[ALL 77] ALLIROT D., BOEHLER J. P., SAWCZUK A., "Irreversible deformations of anisotropic rock under hydrostatic pressure", *Int. J. Rock Mech. Min. Sci. & Geomech. Abstr.*, vol. 14, 1977, p. 77–83.

[ALT 89] AL-TABBAA A., WOOD D. M., "An experimentally based bubble model for clay", PIETRUSCZCZAK S., PANDE G. N., Eds., *NUMOG III*, Elsevier Applied Science, 1989, p. 91–99.

[ANA 86] ANANDARAJAH A., DAFALIAS Y. F., "Bounding surface plasticity. III: Application to anisotropic cohesive soils", *J. Engng. Mech., ASCE*, vol. 112, num. 12, 1986, p. 1292–1318.

[ATK 86] ATKINSON J. H., RICHARDSON D., STALLEBRASS S. E., "Effect of recent stress history on the stiffness of overconsolidated clay", *Géotechnique*, vol. 40, num. 4, 1986, p. 531–540.

[BAK 84] BAKER R., DESAI C. S., "Induced anisotropy during plastic straining", *Int. J. Num. Anal. Meth. Geomech.*, vol. 8, 1984, p. 167–185.

[BAN 86] BANERJEE P. K., YOUSIF N. B., "A plasticity model for the mechanical behavior of anisotropically consolidated clay", *Int. J. Num. Anal. Meth. Geomech.*, vol. 10, 1986, p. 521–541.

[BAR 86] BARDET J. P., "Modelling of sand behaviour with bounding surface plasticity", PANDE G. N., VAN IMPE W. F., Eds., *Numerical Models in Geomechanics*, Jackson and Sons., 1986, p. 131–150.

[BAU 96] BAUER E., "Calibration of a comprehensive hypoplastic equation for granular materials", *Soils and Foundations*, vol. 36, num. 1, 1996, p. 13–26.

[BAZ 78] BAZANT Z. P., "Endochronic inelasticity and incremental plasticity", *Int. Journal of Solids and Structures*, vol. 14, 1978, p. 691–714.

[BOE 75] BOEHLER J. P., "Sur les formes invariantes dans les sous–groupe orthotrope de revolution", *Zeits. Angew. Math. Mech.*, vol. 55, 1975, p. 609–611.

[BOE 87] BOEHLER J. P., *Application of tensor functions in solid mechanics*, CISM Courses and Lectures n. 292, Springer Verlag, New York, 1987.

[BOR 92] BORJA R. I., "Generalized creep and stress–relaxation model for clays", *J. Geotech. Engng., ASCE*, vol. 118, num. 11, 1992, p. 1765–1786.

[BUR 91] BURGHIGNOLI A., PANE V., CAVALERA L., "Modelling stress–strain–time behaviour of natural soils. Monotonic loading", AGI, Ed., *X ECSMFE*, Firenze, Balkema, Rotterdam, 1991, p. 961–979.

[CAL 98] CALLISTO L., CALABRESI G., "Mechanical behaviour of a natural soft clay", *Géotechnique*, vol. 48, num. 4, 1998, p. 495–513.

[CAS 44] CASAGRANDE A., CARILLO N., "Shear failure of anisotropic materials", *Proc. Boston Soc. Civil Engrs.*, vol. 31, 1944, p. 74–87.

[CAS 02a] CASTELLANZA R., "Weathering Effects on the Mechanical Behaviour of Bonded Geomaterials: an Experimental, Theoretical and Numerical Study", PhD thesis, Politecnico di Milano, 2002.

[CAS 02b] CASTELLANZA R., NOVA R., TAMAGNINI C., "Mechanical effects of chemical degradation of bonded geomaterials in boundary value problems", *Revue Francaise de Génie Civil*, , 2002.

[CHA 84] CHAMBON R., "Une loi rhéologique incrémentale non linéaire pour les sols non visqueux", *Journal de Mécanique Théorique et Appliquée*, vol. 3, num. 4, 1984, p. 521–544.

[CHA 89] CHAMBON R., "Une classe de lois de comportement incrémentalement nonlinéaires pour les sols non visqueux, résolution de quelques problèmes de cohérence", *C. R. Acad. Sci.*, vol. 308, num. II, 1989, p. 1571–1576.

[CHA 94] CHAMBON R., DESRUES J., HAMMAD W., CHARLIER R., "CLoE, a new rate–type constitutive model for geomaterials: theoretical basis and implementation", *Int. J. Num. Anal. Meth. Geomech.*, vol. 18, 1994, p. 253-278.

[CHA 00] CHAMBON R., CROCHEPEYRE S., DESRUES J., "Localization criteria for non-linear constitutive equations of geomaterials", *Mech. Cohesive–Frictional Materials*, vol. 5, 2000, p. 61–82.

[CHE 76] CHEN W. F., *Limit Analysis and Soil Plasticity*, Elsevier, 1976.

[CRO 94] CROUCH R. S., WOLF J. P., "Unified 3d critical state bounding–surface plasticity model for soils incorporating continuous plastic loading under cyclic paths. Part I: constitutive relations", *Int. J. Num. Anal. Meth. Geomech.*, vol. 18, 1994, p. 735–758.

[CUD 00] CUDMANI R., NÜBEL K., "Examples of finite element calculations with the hypoplastic law", KOLYMBAS D., Ed., *Constitutive Modelling of Granular Materials*, Springer, Berlin, 2000, p. 523–538.

[DAF 75] DAFALIAS Y. F., POPOV E. V., "A model of non–linearly hardening materials for complex loading", *Arch. Mech.*, vol. 21, 1975, p. 173–192.

[DAF 86a] DAFALIAS Y. F., "Bounding surface plasticity. I: Mathematical foundation and hypoplasticity", *J. Engng. Mech., ASCE*, vol. 112, num. 9, 1986, p. 966–987.

[DAF 86b] DAFALIAS Y. F., HERRMANN L. R., "Bounding surface plasticity. II: Application to isotropic cohesive soils", *J. Engng. Mech., ASCE*, vol. 112, num. 12, 1986, p. 1263–1291.

[DAR 78] DARVE F., "Une formulation incrémentale non–lineaire des lois rhéologiques; application aux sols", Thèse d'Etat, Grenoble, 1978.

[DAR 82] DARVE F., LABANIEH S., "Incremental constitutive law for sands and clays. Simulations of monotonic and cyclic tests", *Int. J. Num. Anal. Meth. Geomech.*, vol. 6, 1982, p. 243–275.

[DAR 88a] DARVE F., CHAU B., DENDANI H., "Incrementally multi–linear and non–linear constitutive relations: a comparative study for practical use", SWOBODA G., Ed., *Numerical Methods in Geomechanics*, Balkema, Rotterdam, 1988, p. 37–43.

[DAR 88b] DARVE F., DENDANI H., "An incrementally non–linear constitutive relation and its predictions", SAADA A. S., BIANCHINI G. F., Eds., *Constitutive Equations for Granular Non–Cohesive Soils*, Balkema, Rotterdam, 1988, p. 237–254.

[DAR 90a] DARVE F., "The expression of rheological laws in incremental form and the main classes of constitutive equations", DARVE F., Ed., *Geomaterials: Constitutive Equations and Modelling*, Elsevier, 1990, p. 123–148.

[DAR 90b] DARVE F., "Incrementally non–linear constitutive relationships", DARVE F., Ed., *Geomaterials: Constitutive Equations and Modelling*, Elsevier, 1990, p. 213-237.

[DAR 99] DARVE F., ROGUIEZ X., "Constitutive relations for soils: new challenges", *Rivista Italiana di Geotecnica*, vol. 4, 1999, p. 9–35.

[DAV 78] DAVIS R. O., MULLENGER G., "A rate–type constitutive model for soil with a critical state", *Int. J. Num. Anal. Meth. Geomech.*, vol. 2, 1978, p. 255–282.

[DES 84] DESAI C. S., SIRIWARDANE H. J., *Constitutive Laws for Engineering Materials, with Emphasis on Geologic Materials*, Prentice–Hall, 1984.

[DIM 71] DI MAGGIO F. L., SANDLER I. S., "Material model for granular soil", *J. Engng. Mech. Div., ASCE*, vol. 97, 1971, p. 935–950.

[DUN 66] DUNCAN J. M., SEED H. B., "Anisotropy and stress reorientation of clays", *J. Soil Mech. Found. Div., ASCE*, vol. 92, num. SM5, 1966, p. 21–50.

[DUN 70] DUNCAN J. M., CHANG C. Y., "Nonlinear analysis of stress and strain in soils", *J. Soil Mech. Found. Div., ASCE*, vol. 96, num. SM5, 1970, p. 1629–1653.

[ELH 97] EL HASSAN N., DESRUES J., CHAMBON R., "Numerical modelling of borehole instability using a non linear incremental model with bifurcation analysis", OKA F. et al., Eds., *Int. Symp. on Deformation and Progressive Failure in Geomechanics*, Nagoya, Japan, 1997, Pergamon Press, p. 677-682.

[FLA 90] FLAVIGNY E., NOVA R., "Viscous properties of geomaterials", DARVE F., Ed., *Geomaterials: Constitutive Equations and Modelling*, Elsevier, 1990, p. 27–54.

[FRY 95] FRYDMAN S., TALESNICK M., PUZRIN A., "Colinearity of stresses, strains and strain increments during shearing of soft clay", *J. Geotech. Engng., ASCE*, vol. 121, num. 2, 1995, p. 174–184.

[GAJ 99] GAJO A., WOOD D. M., "A kinematic hardening constitutive model for sands: the multiaxial formulation", *Int. J. Num. Anal. Meth. Geomech.*, vol. 23, 1999, p. 925–965.

[GEN 96] GENS A., "Constitutive modelling: application to compacted soils", ALONSO E., DELAGE P., Eds., *Unsaturated Soils/Sols Non Saturés*, Balkema, Rotterdam, 1996, p. 1179–1200.

[GEO 91] GEORGIANNOU V. N., RAMPELLO S., SILVESTRI F., "Static and dynamic measurements of undrained stiffness on natural overconsolidated clays", AGI, Ed., *X ECSMFE*, Firenze, vol. 1, Balkema, Rotterdam, 1991, p. 91–96.

[GHA 82] GHABOUSSI J., MOMEN H., "Modelling and analysis of cyclic behavior of sands", PANDE G. N., ZIENKIEWICZ O. C., Eds., *Soil Mechanics: Transient and Cyclic Loads*, Wiley, New York, 1982, p. 313–342.

[GOT 91] GOTO S., TATSUOKA F., SHIBUYA S., KIM Y. S., SATO T., "A simple gauge for local small strain measurements in the laboratory", *Soils and Foundations*, vol. 31, num. 1, 1991, p. 169–180.

[GOT 99] GOTO S., BURLAND J. B., TATSUOKA F., "Non-linear soil model with various strain levels and its application to axisymmetric excavation problems", *Soils and Foundations*, vol. 39, num. 4, 1999, p. 111–119.

[GRA 83a] GRAHAM J., HOULSBY G. T., "Anisotropic elasticity of a natural clay", *Géotechnique*, vol. 33, num. 2, 1983, p. 165–180.

[GRA 83b] GRAHAM J., NOONAN M. L., LEW K. V., "Yield states and stress–strain relationship in a natural plastic clay", *Can. Geotech. J.*, vol. 20, 1983, p. 502–516.

[GRE 56] GREEN A. E., "Hypoelasticity and plasticity", *Archive for Rational Mechanics and Analysis*, vol. 5, 1956, p. 725–734.

[GUD 79] GUDEHUS G., "A comparison of some constitutive laws for soils under radially symmetric loading and unloading", WITTKE, Ed., 3rd *Int. Conf. Num. Meth. Geomech.*, Aachen, Balkema, Rotterdam, 1979, p. 1309–1324.

[GUD 84] GUDEHUS G., DARVE F., VARDOULAKIS I., *Constitutive Relations for Soils*, Balkema, Rotterdam, 1984.

[GUD 96] GUDEHUS G., "A comprehensive constitutive equation for granular materials", *Soils and Foundations*, vol. 36, num. 1, 1996, p. 1–12.

[HAS 79] HASHIGUCHI K., "Constitutive equations of granular media with an anisotropic hardening", *III Int. Conf. Num. Meth. in Geomechanics*, Aachen, Germany, 1979, Balkema, Rotterdam, p. 435–439.

[HAS 85] HASHIGUCHI K., "Two- and three–surface models of plasticity", *V Int. Conf. Num. Meth. in Geomechanics*, Nagoya, Japan, 1985, Balkema, Rotterdam, p. 285–292.

[HAS 88] HASHIGUCHI K., "Mathematically consistent formulation of elastoplastic constitutive equations", *VI Int. Conf. Num. Meth. in Geomechanics*, Innsbruck, Austria, 1988, Balkema, Rotterdam, p. 467–472.

[HIL 50] HILL R., *The Mathematical Theory of Plasticity*, Oxford University Press, Oxford, 1950.

[HON 89] HONG W. P., LADE P. V., "Strain increment and stress directions in torsion shear tests", *J. Geotech. Engng., ASCE*, vol. 115, num. 10, 1989, p. 1388–1401.

[HOU 84] HOULSBY G. T., WROTH C. P., WOOD D. M., "Prediction of the results of laboratory tests on a clay using a critical state model", GUDEHUS G., DARVE F., VARDOULAKIS I., Eds., *Constitutive Relations for Soils*, Balkema, Rotterdam, 1984.

[HOU 85] HOULSBY G. T., "The use of a variable shear modulus in elastic–plastic models for clays", *Comp. & Geotechnics*, vol. 1, 1985, p. 3–13.

[HUE 92] HUECKEL T., "Water mineral interactions in hygromechanics of clay exposed to environmental load: a mixture theory approach", *Can. Geotech. J.*, vol. 29, 1992, p. 1071–1086.

[HUE 97] HUECKEL T., "Chemo–plasticity of clays subjected to stress and flow of a single contaminant", *Int. J. Num. Anal. Meth. Geomech.*, vol. 21, 1997, p. 43–72.

[IWA 67] IWAN W. D., "On a class of models for the yield behaviour of continuous and composite systems", *J. Appl. Mechanics, ASME*, vol. 34, 1967, p. 612–617.

[JAR 84] JARDINE R. J., SYMES M. J., BURLAND J. B., "The measurement of soil stiffness in the triaxial apparatus", *Géotechnique*, vol. 34, 1984, p. 323–340.

[JAR 86] JARDINE R. J., POTTS D. M., FOURIE A. B., BURLAND J. B., "Studies of the influence of non-linear stress-strain characteristics in soil-structure interaction", *Géotechnique*, vol. 36, num. 3, 1986, p. 377–396.

[JAR 88] JARDINE R. J., POTTS D. M., "Hutton tension leg platform foundations: an approach to the prediction of driven pile behaviour", *Géotechnique*, vol. 38, num. 2, 1988, p. 231–252.

[JAR 91] JARDINE R. J., POTTS D. M., ST. JOHN H. D., HIGHT D. W., "Some practical applications of a non–linear ground model", AGI, Ed., *X ECSMFE*, Firenze, vol. 1, Balkema, Rotterdam, 1991, p. 223–228.

[JIR 02a] JIRASEK M., "Titolo", *Revue Francaise de Génie Civil*, , 2002.

[JIR 02b] JIRASEK M., BAZANT Z. P., *Inelastic Analysis of Structures*, Wiley, Chichester, 2002.

[JOM 00] JOMMI C., "Remarks on the constitutive modelling of unsaturated soils", *Experimental evidence and theoretical approaches in unsaturated soils*, Balkema, Rotterdam, 2000, p. 139–153.

[KIM 88] KIM M. K., LADE P. V., "Single hardening constitutive model for frictional materials. I: Plastic potential function", *Comp. & Geotechnics*, vol. 5, 1988, p. 307–324.

[KOL 91] KOLYMBAS D., "An outline of hypoplasticity", *Archive of Applied Mechanics*, vol. 61, 1991, p. 143–151.

[KOL 93a] KOLYMBAS D., *Modern Approaches to Plasticity*, Elsevier, 1993.

[KOL 93b] KOLYMBAS D., WU W., "Introduction to hypoplasticity", KOLYMBAS D., Ed., *Modern Approaches to Plasticity*, Elsevier, 1993, p. 213–224.

[KON 63] KONDNER R. L., ZELASKO J. S., "A hyperbolic stress–strain formulation for sands", *II Pan.–Am. Conf. SMFE*, vol. 1, 1963, p. 289–324.

[KUD 99] KUDELLA P., MAYER P. M., "Calculation of soil displacements due to retaining wall construction", PANDE G., PIETRUSCZCZAK S., SCHWEIGER H., Eds., *NUMOG VII*, Balkema, Rotterdam, 1999, p. 573–580.

[LAD 77] LADE P. V., "Elastoplastic stress–strain theory for cohesionless soil with curved yield surfaces", *Int. Journal of Solids and Structures*, vol. 13, num. 11, 1977, p. 1019–1035.

[LAD 88] LADE P. V., KIM M. K., "Single hardening constitutive model for frictional materials. II: Yield criterion and plastic work contours", *Comp. & Geotechnics*, vol. 6, 1988, p. 13–29.

[LAN 88] LANIER J., "Generation of a data bank by the use of Grenoble true triaxial apparatus", SAADA A. S., BIANCHINI G., Eds., *Constitutive Equations for Granular Non-Cohesive Soils*, Balkema, Rotterdam, 1988, p. 47–58.

[LEW 98] LEWIS R. W., SCHREFLER B. A., *The Finite Element Method in the Static and Dynamic Deformation and Consolidation of Porous Media*, Wiley, Chichester, 1998.

[LOR 90a] LORET B., "Geomechanical applications of the theory of multimechanisms", DARVE F., Ed., *Geomaterials: Constitutive Equations and Modelling*, Elsevier, 1990, p. 187–211.

[LOR 90b] LORET B., "An introduction to the classical theory of elastoplasticity", DARVE F., Ed., *Geomaterials: Constitutive Equations and Modelling*, Elsevier, 1990, p. 149–186.

[LUB 90] LUBLINER J., *Plasticity Theory*, Mac Millan, London, 1990.

[LYA 02] LYAMIN A. V., SLOAN S. W., "Upper bound limit analysis using linear finite elements and nonlinear programming", *Int. J. Num. Anal. Meth. Geomech.*, vol. 26, 2002, p. 181–216.

[MIT 70] MITCHELL R. J., "On the yielding and mechanical strength of Leda clays", *Can. Geotech. J.*, vol. 7, 1970, p. 297–312.

[MRO 67] MROZ Z., "On the description of anisotropic work–hardening", *Journal of the Mechanics and Physics of Solids*, vol. 15, 1967, p. 163–175.

[MRO 78] MROZ Z., NORRIS V. A., ZIENKIEWICZ O. C., "An anisotropic hardening model for soils and its application to cyclic loading", *Int. J. Num. Anal. Meth. Geomech.*, vol. 2, 1978, p. 203–221.

[MRO 80] MROZ Z., "On hypoelastic and plasticity approaches to constitutive modelling of inelastic behaviour of soils", *Int. J. Num. Anal. Meth. Geomech.*, vol. 4, 1980, p. 45–56.

[MRO 81] MROZ Z., NORRIS V. A., ZIENKIEWICZ O. C., "An anisotropical critical state model for soils subject to cyclic loading", *Géotechnique*, vol. 31, num. 4, 1981, p. 451–469.

[MUR 85] MURUYAMA S., "Constitutive Laws of Soils", Jap. Soc. SMFE, 1985.

[NEL 77] NELSON I., "Constitutive models for use in numerical computations", *Plastic and Long Term Effects, DMSR '77, Karlsruhe*, vol. 2, 1977, p. 45–97.

[NIE 97] NIEMUNIS A., HERLE I., "Hypoplastic model for cohesionless soils with elastic strain range", *Mech. Cohesive–Frictional Materials*, vol. 2, 1997, p. 279–299.

[NOV 77] NOVA R., "On the hardening of soils", *Archiwum Mechaniki Stosowanej*, vol. 29, 1977, p. 445–458.

[NOV 79] NOVA R., WOOD D. M., "A constitutive model for sand in triaxial compression", *Int. J. Num. Anal. Meth. Geomech.*, vol. 3, 1979, p. 255–278.

[NOV 86] NOVA R., "Soil models as a basis for modelling the behaviour of geophysical materials", *Arch. Mech.*, vol. 64, 1986, p. 31–44.

[NOV 88] NOVA R., "Sinfonietta Classica: an exercise on classical soil modelling", SAADA, BIANCHINI, Eds., *Constitutive Equations for Granular Non–Cohesive Soils*, Cleveland, 1988, Balkema, Rotterdam.

[NOV 96] NOVA R., "Modelling: classical elastoplastic models", CHAMBON R., Ed., *8th ALERT School on Bifurcation and Localization in Geomaterials*, ALERT Geomaterials, 1996.

[NOV 00] NOVA R., "Modelling the weathering effects on the mechanical behaviour of granite", KOLYMBAS D., Ed., *Constitutive Modelling of Granular Materials*, Horton, Greece, 2000, Springer, Berlin.

[NOV 01] NOVA R., CASTELLANZA R., "Modelling weathering effects on the mechanical on the mechanical behaviour of soft rocks", *Int. Conf. on Civil Engineering*, Bangalore, India, 2001, Interline Publishing, p. 157–167.

[ODA 78] ODA M., KOISHIKAWA I., HIGUCHI T., "Experimental study of anisotropic shear strength of sand by plane strain test", *Soils and Foundations*, vol. 18, num. 1, 1978, p. 25–38.

[ODA 85] ODA M., NEMAT-NASSER S., KONISHI J., "Stress–induced anisotropy in granular masses", *Soils and Foundations*, vol. 25, num. 3, 1985, p. 85–97.

[OWE 69] OWEN D. R., WILLIAMS W. O., "On the time derivatives of equilibrated response functions", *Archive for Rational Mechanics and Analysis*, vol. 33, num. 4, 1969, p. 288–306.

[PAS 85] PASTOR M., ZIENKIEWICZ O. C., LEUNG K. H., "Simple model for transient soil loading in earthquake analysis. II: non–associative model for sands", *Int. J. Num. Anal. Meth. Geomech.*, vol. 9, 1985, p. 477–498.

[PAS 90] PASTOR M., ZIENKIEWICZ O. C., CHAN A. H. C., "Generalized plasticity and the modelling of soil behaviour", *Int. J. Num. Anal. Meth. Geomech.*, vol. 14, 1990, p. 151–190.

[PAS 91] PASTOR M., "Modelling of anisotropic sand behaviour", *Comp. & Geotechnics*, vol. 11, 1991, p. 173–208.

[PES 99] PESTANA A., WHITTLE A. J., "Formulation of a unified constitutive model for clays and sands", *Int. J. Num. Anal. Meth. Geomech.*, vol. 23, 1999, p. 1215–1243.

[PIJ 02] PIJAUDIER-CABOT J., "Titolo", *Revue Francaise de Génie Civil*, , 2002.

[PON 97] PONTES I. D. S., BORGES L. A., ZOUAIN N., LOPES F. R., "An approach to limit analysis with cone–shaped yield surfaces", *Int. J. Num. Meth. Engng.*, vol. 40, 1997, p. 4011–4032.

[POO 66] POOROSHASB H. B., HOLUBEC I., SHERBOURNE A. N., "yielding and flow of sand in triaxial compression (part I)", *Can. Geotech. J.*, vol. 3, 1966, p. 179–190.

[POO 67] POOROSHASB H. B., HOLUBEC I., SHERBOURNE A. N., "yielding and flow of sand in triaxial compression (part II)", *Can. Geotech. J.*, vol. 4, 1967, p. 376–397.

[POO 85] POOROSHASB H. B., PIETRUSZCZAK S., "On the yielding and flow of sand: a generalized two–surface model", *Comp. & Geotechnics*, vol. 1, 1985, p. 33–58.

[PRE 77] PREVOST J. H., "Mathematical modelling of monotonic and cyclic undrained clay behaviour", *Int. J. Num. Anal. Meth. Geomech.*, vol. 1, 1977, p. 195–216.

[PRE 86] PREVOST J. H., "Constitutive equations for pressure–sensitive soils: theory, numerical implementation, and examples", DUNGAR R., STUDER J. A., Eds., *Geomechanical Modelling in Engineering Practice*, Balkema, Rotterdam, 1986, p. 331-350.

[PRI 93] DI PRISCO C., NOVA R., LANIER J., "A mixed isotropic–kinematic hardening constitutive law for sand", KOLYMBAS D., Ed., *Modern Approaches to Plasticity*, Elsevier, 1993, p. 83–124.

[PYK 86] PYKE R., "The use of linear elastic and piecewise linear models in finite element analyses", DUNGAR, STUDER, Eds., *Geomechanical Modelling in Engineering Practice*, Balkema, Rotterdam, 1986, p. 167–188.

[ROM 74] ROMANO M., "A continuum theory for granular media with a critical state", *Arch. Mech.*, vol. 26, 1974, p. 1011–1028.

[ROS 68] ROSCOE K. H., BURLAND J. B., "On the generalised stress–strain behaviour of 'wet' clay", HEYMAN J., LECKIE F. A., Eds., *Engineering Plasticity*, Cambridge Univ. Press, Cambridge, 1968, p. 535–609.

[ROU 00] ROUAINIA M., WOOD D. M., "A kinematic hardening constitutive model for natural clays with loss of structure", *Géotechnique*, vol. 50, num. 2, 2000, p. 153–164.

[ROW 83] ROWE R. K., KACK G. J., "A theoretical examination of the settlements induced by tunnelling: four case histories", *Can. Geotech. J.*, vol. 20, 1983, p. 299–314.

[ROY 98] ROYIS P., DOANH T., "Theoretical analysis of strain response envelopes using incrementally non-linear constitutive equations", *Int. J. Num. Anal. Meth. Geomech.*, vol. 22, 1998, p. 97–132.

[SAA 89] SAADA A. S., BIANCHINI G. F., *Constitutive equations for granular non-cohesive soils*, Balkema, Rotterdam, 1989.

[SAN 76] SANDLER I. S., DI MAGGIO F. L., BALADI G. Y., "Generalised cap model for geologic materials", *J. Soil Mech. Found. Div., ASCE*, vol. 102, 1976, p. 683–697.

[SAN 02] SANAVIA L., SCHREFLER B. A., "A finite element model for water saturated and partially saturated geomaterials", *Revue Francaise de Génie Civil*, , 2002.

[SCH 68] SCHOFIELD A. N., WROTH C. P., *Critical State Soil Mechanics*, McGraw–Hill, London, 1968.

[SCH 02] SCHREFLER B. A., SANAVIA L., "Coupling equations for water saturated and partially saturated geomaterials", *Revue Francaise de Génie Civil*, , 2002.

[SEK 77] SEKIGUCHI H., OHTA H., "Induced anisotropy and time dependency in clays", *IX ICSMFE, Specialty Session 9*, Balkema, Rotterdam, 1977, p. 229–238.

[SHA 95] SHAROUR I., GHORBANBEIGI S., VON WOLFFERSDORDFF P. A., "Three-dimensional finite element analysis of diaphragm wall construction", *Revue Francaise de Géotechnique*, vol. 71, 1995, p. 39–47.

[SIM 97] SIMO J. C., HUGHES T. J. R., *Computational Inelasticity*, Springer, 1997.

[SIM 98] SIMO J. C., *"Numerical Analysis and Simulation of Plasticity"*, vol. VI of *Handbook of Numerical Analysis*, p. 183–499, Elsevier Science, 1998.

[SIR 83] SIRIWARDANE H. J., DESAI C. S., "Computational procedures for nonlinear three-dimensional analysis with some advanced constitutive laws", *Int. J. Num. Anal. Meth. Geomech.*, vol. 7, 1983, p. 143–171.

[SLO 95] SLOAN S. W., KLEEMAN P. W., "Upper bound limit analysis with discontinuous velocity fields", *Comp. Meth. Appl. Mech. Engng.*, vol. 127, 1995, p. 293–314.

[SMI 92a] SMITH I. M., HO D. K. H., "Influence of construction technique on the performance of a braced excavation in marine clay", *Int. J. Num. Anal. Meth. Geomech.*, vol. 16, 1992, p. 845–867.

[SMI 92b] SMITH P. R., JARDINE R. J., HIGHT D. W., "The yielding of Bothkennar clay", *Géotechnique*, vol. 42, num. 2, 1992, p. 257–274.

[SOK 65] SOKOLOWSKI V. V., *Statics of Granular Media*, Pergamon, Oxford, 1965.

[ST. 93] ST. JOHN H. D., POTTS D. M., JARDINE R. J., HIGGINS K. G., "Prediction and performance of ground response due to construction of a deep basement at 60 Victoria Embankment", HOULSBY G. T., SCHOFIELD A. N., Eds., *Predictive Soil Mechanics (Wroth Mem. Symp.)*, Thomas Telford, London, 1993.

[STA 90] STALLEBRASS S. E., "Modelling the effect of recent stress history on the behaviour of overconsolidated soils", PhD thesis, The City University, London, 1990.

[TAM 99] TAMAGNINI C., D'ELIA M., "A simple bounding surface model for bonded clays", JAMIOLKOWSKI M., LANCELLOTTA R., LO PRESTI D., Eds., *Pre-failure Deformation Characteristics of Geomaterials*, Balkema, Rotterdam, 1999, p. 565–572.

[TAM 00] TAMAGNINI C., VIGGIANI G., CHAMBON R., "A Review of Two Different Approaches to Hypoplasticity", KOLYMBAS D., Ed., *Constitutive Modelling of Granular Materials*, Springer, Berlin, 2000, p. 107–145.

[TAM 01] TAMAGNINI C., VIGGIANI G., CHAMBON R., "Some remarks on shear band analysis in hypoplasticity", MUHLHAUS H.-B., DYSKIN A., PASTERNAK E., Eds., *Localization and Bifurcation Theory for Soils and Rocks*, Balkema, 2001.

[TAM 02a] TAMAGNINI C., CASTELLANZA R., NOVA R., "Implicit integration of constitutive equations in computational plasticity", *Revue Francaise de Génie Civil*, , 2002.

[TAM 02b] TAMAGNINI C., CASTELLANZA R., NOVA R., "Numerical integration of elastoplastic constitutive equations for geomaterials with extended hardening rules", PANDE G. N., PETRUSZCZAK S., Eds., *NUMOG VIII*, Balkema, Rotterdam, 2002, p. 213–218.

[TAT 97] TATSUOKA F., JARDINE R. J., LO PRESTI D., DI BENEDETTO H., KODAKA T., "Theme Lecture: Characterizing the pre–failure deformation properties of geomaterials", *XIV ICSMFE*, vol. 4, Balkema, Rotterdam, 1997, p. 2129–2164.

[TAV 77] TAVENAS F., LEROUEIL S., "Effects of stresses and time on on yielding of clays", *IX ICSMFE*, vol. 1, Balkema, Rotterdam, 1977, p. 319–326.

[TEJ 99] TEJCHMAN J., HERLE I., "A class "A" prediction of the bearing capacity of plane strain footings on sand", *Soils and Foundations*, vol. 39, num. 5, 1999, p. 47–60.

[TER 48a] TERZAGHI K., *Theoretical Soil Mechanics*, John Wiley, New York, 1948.

[TER 48b] TERZAGHI K., PECK R. B., *Soil Mechanics in Engineering Practice*, John Wiley, New York, 1948.

[TOB 97] TOBITA Y., "Importance of incremental nonlinearity in the deformation of granular materials", FLECK N. A., COCKS A. C. F., Eds., *IUTAM Symp. on Mech. of Granular and Porous Materials*, Kluwer Acad. Publisher, 1997, p. 139–150.

[TRU 56] TRUESDELL C. A., "Hypo–elastic shear", *J. Appl. Physics*, vol. 27, 1956, p. 441–447.

[TRU 65] TRUESDELL C. A., NOLL W., "The non–linear field theories of mechanics", FLÜGGE S., Ed., *Encyclopedia of Physics*, vol. III/3, Springer, Berlin, 1965.

[ULM 96] ULM F. J., COUSSY O., "Strength growth as chemo–plastic hardening in early age concrete", *J. Engng. Mech., ASCE*, vol. 122, num. 12, 1996, p. 1123–1132.

[VAL 84] VALANIS K. C., LEE C. F., "Endochronic theory of cyclic plasticity with applications", *J. Appl. Mechanics, ASME*, vol. 51, 1984, p. 789–794.

[VAR 95] VARDOULAKIS I., SULEM J., *Bifurcation Analysis in Geomechanics*, Blackie Acad. & Professional, New York, 1995.

[VER 78] VERMEER P. A., "A double hardening model for sand", *Géotechnique*, vol. 28, 1978, p. 413–433.

[VIG 99] VIGGIANI G., TAMAGNINI C., "Hypoplasticity for modeling soil non–linearity in excavation problems", JAMIOLKOWSKI M., LANCELLOTTA R., LO PRESTI D., Eds., *Prefailure Deformation Characteristics of Geomaterials*, Balkema, Rotterdam, 1999, p. 581–588.

[VIG 00] VIGGIANI G., TAMAGNINI C., "Ground movements around excavations in granular soils: a few remarks on the influence of the constitutive assumptions on FE predictions", *Mech. Cohesive–Frictional Materials*, vol. 5, num. 5, 2000, p. 399–423.

[WAN 90] WANG Z. L., DAFALIAS Y. F., SHEN C. K., "Bounding surface hypoplasticity model for sand", *J. Engng. Mech., ASCE*, vol. 116, num. 5, 1990, p. 983–1001.

[WHE 96] WHEELER S. J., KARUBE D., "Constitutive modelling", ALONSO E., DELAGE P., Eds., *Unsaturated Soils/Sols Non Saturés*, Balkema, Rotterdam, 1996, p. 1323–1356.

[WHI 94] WHITTLE A. J., KAVVADAS M. J., "Formulation of MIT–E3 constitutive model for overconsolidated clays", *J. Geotech. Engng., ASCE*, vol. 120, num. 1, 1994, p. 173–198.

[WIL 77] WILDE P., "Two invariants depending models of granular media", *Arch. Mech. Stos.*, vol. 29, 1977, p. 799–809.

[WON 85] WONG R. K. S., ARTHUR J. R. F., "Induced and inherent anisotropy in sand", *Géotechnique*, vol. 35, num. 4, 1985, p. 471–481.

[WON 91] WONG R. C. K., KAISER P. K., "Performance assessment of tunnels in cohesionless soils", *J. Geotech. Engng., ASCE*, vol. 117, num. 12, 1991, p. 1880–1901.

[WOO 82] WOOD D. M., "Laboratory investigations of the behaviour of soils under cyclic loading: a review", PANDE G. N., ZIENKIEWICZ O. C., Eds., *Soil Mechanics – Cyclic and Transient Loads*, Wiley, Chichester, 1982, p. 513–582.

[WU 98] WU W., "Rational approach to anisotropy of sand", *Int. J. Num. Anal. Meth. Geomech.*, vol. 22, 1998, p. 921–940.

[WU 00] WU W., KOLYMBAS D., "Hypoplasticity then and now", KOLYMBAS D., Ed., *Constitutive Modelling of Granular Materials*, Springer, Berlin, 2000.

[YAS 91] YASUFUKU N., MURATA H., HYODO M., "Yield characteristics of anisotropically consolidated sand under low and high stresses", *Soils and Foundations*, vol. 31, num. 1, 1991, p. 95–109.

[YOS 98] YOSHIMINE M., ISHIHARA K., VARGAS W., "Effects of principal stress direction and intermediate principal stress on undrained shear behavior of sand", *Soils and Foundations*, vol. 38, num. 3, 1998, p. 179–188.

[ZIE 71] ZIENKIEWICZ O. C., NAYLOR D. J., "An adaptation of Critical State Soil Mechanics theory for use in finite elements", PARRY R. H. G., Ed., *Stress–Strain Behaviour of Soils (Roscoe Mem. Symp.)*, Foulis, Henley–on–Thames, 1971.

[ZIE 73] ZIENKIEWICZ O. C., NAYLOR D. J., "Finite element studies of soils and porous media", ODEN J. T., DE ARANTES E. R., Eds., *Lect. Finite Elements in Continuum Mechanics*, UAH Press, 1973.

[ZIE 75] ZIENKIEWICZ O. C., HUMPHESON C., LEWIS R. W., "Associated and non-associated visco–plasticity and plasticity in soil mechanics", *Géotechnique*, vol. 25, num. 4, 1975, p. 671–689.

[ZIE 77] ZIENKIEWICZ O. C., HUMPHESON C., "Viscoplasticity: a generalized model for soil behaviour", DESAI C. A., CHRISTIAN J. T., Eds., *Numerical Models in Geotechnical Engineering*, McGraw-Hill, New York, 1977.

[ZIE 84] ZIENKIEWICZ O. C., MROZ Z., "Generalized plasticity formulation and applications to geomechanics", DESAI C. S., GALLAGHER R. H., Eds., *Mechanics of Engineering Materials*, Wiley, 1984.

[ZIE 85] ZIENKIEWICZ O. C., LEUNG K. H., PASTOR M., "Simple model for transient soil loading in earthquake analysis. I: basic model and its application", *Int. J. Num. Anal. Meth. Geomech.*, vol. 9, 1985, p. 453–476.

[ZIE 99] ZIENKIEWICZ O. C., CHAN A. H. C., PASTOR M., SCHREFLER B. A., SHIOMI T., *Computational Geomechanics with Special Reference to Earthquake Engineering*, Wiley, Chichester, 1999.

Chapter 4

Coupling Equations for Water Saturated and Partially Saturated Geomaterials

A Mathematical Model for Multiphase Porous Materials: Fundamentals and Formulation

Bernhard A. Schrefler and Lorenzo Sanavia
Department of Structural and Transportation Engineering, University of Padua, Italy

1. Introduction

This paper presents a mathematical model for a saturated and partially saturated porous material capable of sustaining large elastic or elasto-plastic strains. Mechanics of porous materials has a wide spectrum of engineering applications and hence, in recent years, several porous media models and their numerical solutions have appeared in the literature (see [LEW 98], [OCZ 99], [BOE 00] for a comprehensive state of the art). Most of these models are restricted to fluid saturated materials and have been developed using small strain assumptions. Conditions of partial saturation are of importance in engineering practice because many porous materials are in this natural state or can reach this state during deformations. Some simple examples can be found in soils or in concrete and in biological tissues, which can contain air or other gases in the pores together with liquids. For soils, this is the case of the zones above the free surface, or of deep reservoirs of hydrocarbon gas. The partially saturated state can also be reached during the deformation due, for instance, to earthquake in an earth dam or during the particular case of strain localization of dense sands under globally undrained conditions, where negative water pressures are measured and cavitation of the pore water is observed [VAR 95], [MOK 98]. Large strains also can be important. They result when ultimate or serviceability limit state is reached, as for example during slope instability or during the consolidation process in compressible clays. In the laboratory, this can be the case of drained or undrained biaxial tests of sands, where axial logarithmic strains of the order of $0.12 - 0.15$ are reached [VAR 95], [MOK 98], or the case of triaxial tests of peats, where axial strains of the order of 0.15 are measured.

In the model developed in this manuscript, the porous medium is treated as an non isothermal multiphase continuum with the pores of the solid skeleton filled by water and air. The governing equations at macroscopic level are derived in Section 2 in a spatial setting and are based on averaging procedures (hybrid mixture theory). This model follows from [LEW 98], where the interested reader can find further details and remarks. Solid displacements, temperature, water and gas pressures are the primary variables. Water and gas are assumed to obey Darcy's law. In the partially saturated state, the degree of saturation and the relative permeability of water and gas are dependent on the capillary pressure by experimental functions. Within the formulation developed, the elasto-plastic behaviour of the solid skeleton can be described by the multiplicative decomposition of the deformation gradient into an elastic and a plastic part, as shown in [SAN 02]. In this case, the modified effective stress in partially saturated conditions (Bishop like stress) in the form of Kirchhoff measure of the stress tensor and the logarithmic principal strains are used in conjunction with an hyperelastic free energy function. The effective stress state can be limited by a suitable yield surface.

As notation and symbols are concerned, bold-face letters denote tensors; capital or lower case letters are used for tensors in the reference or in actual configuration. The symbol '·' denotes the scalar product between two vectors (e. g. $\mathbf{a} \cdot \mathbf{b} = a_i b_i$), while the symbol ':' denotes a double contraction of (adjacent) indices of two tensors of rank two or/and higher (e. g. c : d $= c_{ij}\, d_{ij}$, e : f $= e_{ijkl}\, f_{kl}$). Cartesian co-ordinates are used throughout the paper.

2. Mathematical model of thermo-hydro-mechanical behaviour of geomaterials

The full mathematical model necessary to simulate thermo-hydro-mechanical transient behaviour of fully and partially saturated porous media is developed in [LEW 98] using averaging theories following Hassanizadeh and Gray [HAS 79a], [HAS 79b], [HAS 80]. The underlying physical model, thermodynamic relations and constitutive equations for the constituents, as well as governing equations are summarized in the following.

The partially saturated porous medium is treated as multiphase system composed of $\pi = 1, \ldots, k$ ($k = 3$) constituents with the voids of the solid skeleton (s) filled with water (w) and gas (g) (see Figure 1). The latter is assumed to behave as an ideal mixture of two species: dry air (non-condensable gas, ga) and water vapour (condensable one, gw).

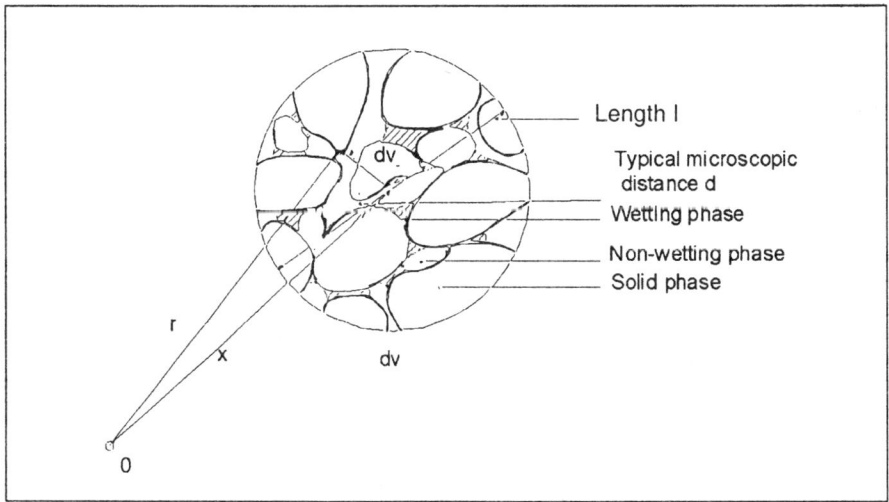

Figure 1. *Averaging volume* $dv(\mathrm{x},\ t)$ *of a three phase porous medium*

2.1. Microscopic balance equations

At the microscopic level, the balance equation of any π-phase may be described by the classical continuum mechanics. At the interfaces with other constituents, the material properties and thermodynamic quantities may present step discontinuities (see e. g. [GRA 01] or [SCH 02] for the jump conditions to be fulfilled). For a thermodynamic property ψ, the balance equation within the π-phase may be written as [HAS 79a] and [HAS 79b]

$$\frac{\partial \rho \psi}{\partial t} + \operatorname{div}(\rho \psi \dot{r}) - \operatorname{div} i - \rho b = \rho G \qquad [1]$$

where $\dot{r}$ is the local value of the velocity field of the π-phase in a fixed point in space, i is the flux vector associated with ψ, b the external supply of ψ and G is the net production of ψ. Fluxes are positive as outflows. The thermodynamic property ψ for the different balance equations and the values assumed by the quantities of (1) are listed in Table 1 following [HAS 79b],

Quantity	ψ	i	b	G
Mass	1	0	0	0
Momentum	$\dot{r}$	t_m	g	0
Energy	$E + 0.5\dot{r} \cdot \dot{r}$	$t_m \dot{r} - q$	$g \cdot \dot{r} + h$	0
Entropy	λ	Φ	S	φ

Table 1. *Thermodynamic properties for the microscopic mass balance equations*

where E is the specific intrinsic energy, λ the specific entropy, t_m the microscopic stress tensor, q a heat flux vector, Φ the entropy flux, g the external momentum supply related to gravitational effects, h the intrinsic heat source, S an intrinsic entropy source and φ denotes an increase of entropy. The angular momentum balance equation has been omitted because the constituents are assumed to be microscopically non polar (the interested reader is referred to [EHL 98] regarding a saturated or empty porous media model with a polar solid skeleton).

Using spatial averaging operators [HAS 79a] defined over a representative elementary volume R. E. V. (of volume $dv(x, t)$ in the deformed configuration $B_t \subset \mathbb{R}^3$, see Figure 1, where x is the vector of the spatial co-ordinates and t is the current time), the microscopic equations are integrated over the R. E. V. giving the macroscopic balance equations.

As a consequence, at the macroscopic level the multiphase porous material is modelled by a substitute continuum of volume B_t with boundary ∂B_t that fills the entire domain simultaneously, instead of the real fluids and the solid which fill only a part of it. In this substitute continuum each constituent π has a reduced density which is obtained through the volume fraction $\eta^\pi(x, t) = dv^\pi(x, t)/dv(x, t)$ with the constraint

$$\sum_{\pi=1}^{k} \eta^\pi = 1 \qquad [2]$$

where $dv^\pi(\mathbf{x}, t)$ is the π-phase volume inside the R. E. V. in the actual placement $\mathbf{x}$.

In the present formulation heat conduction, vapour diffusion, heat convection, water flow due to pressure gradients or capillary effects and latent heat transfer due to water phase change (evaporation and condensation) inside the pores are taken into account. The solid is deformable and non-polar, and the fluid, the solid, and the thermal fields are coupled. All fluids are in contact with the solid phase. The constituents are assumed to be isotropic, homogeneous, immiscible except for dry air and vapour, and chemically non-reacting. Local thermal equilibrium between solid matrix, gas, and liquid phases is assumed so that the temperature is the same for all the constituents. The state of the medium is described by water pressure p^w, gas pressure p^g, temperature θ, and the displacement vector of the solid matrix u.

Before summarizing the macroscopic balance equations, we specify the kinematics introducing the notion of initial and current configuration (Figure 2). In the following, the stress is defined as tension positive for the solid phase, while pore pressure is defined as compressive positive for the fluids.

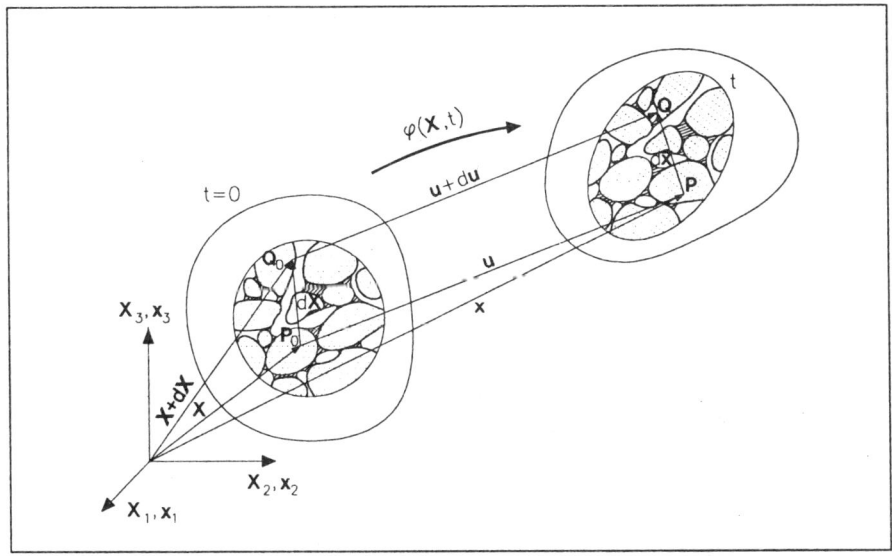

Figure 2. *Initial and current configuration of a multiphase medium*

2.2. Kinematic equations

At the macroscopic level the multiphase medium is described as the superposition of all π-phases, whose material points X^π with co-ordinates $\mathbf{X}^\pi$ in the reference configuration $B_0^\pi \subset \mathbb{R}^3$ at time $t = t_0$ can occupy simultaneously each spatial point $\mathbf{x}$ in the deformed configuration $B_t \subset \mathbb{R}^3$ at time t. In the Lagrangian description of the motion in terms of material co-ordinates the position of each material point in the actual configuration x is a function of its placement $\mathbf{X}^\pi$ in a chosen reference configuration B_0^π and of the current time t:

$$x = \chi^\pi(\mathbf{X}^\pi, t) \qquad [3]$$

with $x = \mathbf{x}^\pi$, or it is given by the sum of the reference position $\mathbf{X}^\pi$ and the displacement $\mathbf{u}^\pi = (\mathbf{X}^\pi, t)$ at time t

$$x = \mathbf{X}^\pi + \mathbf{u}^\pi(\mathbf{X}^\pi, t) \qquad [4]$$

In (3), $\chi^\pi(\mathbf{X}^\pi, t)$ is a continuous and bijective motion function (deformation map) of each phase because the Jacobian J^π of each motion function

$$J^\pi = \det \frac{\partial \chi^\pi(\mathbf{X}^\pi, t)}{\partial \mathbf{X}^\pi} > 0 \qquad [5]$$

is restricted to be a positive value. The deformation gradient $\mathbf{F}^\pi(\mathbf{X}^\pi, t)$ is defined as

$$\mathbf{F}^\pi = \operatorname{Grad}^\pi \chi^\pi(\mathbf{X}^\pi, t) \qquad [6]$$

where the differential operator ' Grad^π ' denotes partial differentiation with respect to the reference position $\mathbf{X}^\pi$. Hence, from (5), $J^\pi = \det \mathbf{F}^\pi$.

The velocity and the acceleration of each constituent are given as

$$\mathbf{V}^\pi = \frac{\partial \chi^\pi(\mathbf{X}^\pi, t)}{\partial t}, \psi \qquad \mathbf{A}^\pi = \frac{\partial^2 \chi^\pi(\mathbf{X}^\pi, t)}{\partial t^2} \qquad [7]$$

Due to the non-singularity of the Lagrangian relationship (3), the existence of its inverse function leads to the description of the motion in terms of spatial co-ordinates:

$$\mathbf{X}^\pi = (\chi^\pi)^{-1}(x, t) \qquad [8]$$

The inverse $(\mathbf{F}^\pi)^{-1}(x, t)$ of the deformation gradient is given by

$$(\mathbf{F}^\pi)^{-1} = \operatorname{grad} \mathbf{X}^\pi(\mathbf{x},\, t) \qquad\qquad [9]$$

where the differential operator 'grad' is now referred to spatial co-ordinates x. The spatial parametrization of the velocity is given by

$$\mathbf{v}^\pi = \mathbf{v}^\pi(\mathbf{x},\, t) = \mathbf{V}^\pi \circ (\chi^\pi)^{-1} \qquad\qquad [10]$$

where 'o' denotes the composition of functions. The parametrization of the spatial acceleration is related to the spatial velocity by the application of the chain rule to (10):

$$\mathbf{a}^\pi = \mathbf{a}^\pi(\mathbf{x},\, t) = \frac{\partial \mathbf{v}^\pi}{\partial t} + (\operatorname{grad} \mathbf{v}^\pi)\, \mathbf{v}^\pi = \mathbf{A}^\pi \circ (\chi^\pi)^{-1} \qquad\qquad [11]$$

Since the individual constituents follow in general different motions, different material time derivatives must be formulated. For an arbitrary scalar-valued function $f^\pi(\mathbf{x},\, t)$, its material time derivative following the velocity of the constituents π is defined by [LEW 98]

$$\frac{\mathrm{D}^\pi f^\pi}{\mathrm{D}t} = \frac{\partial f^\pi}{\partial t} + \operatorname{grad} f^\pi \cdot \mathbf{v}^\pi \qquad\qquad [12]$$

where $f^\pi(\mathbf{x},\, t)$ must be substituted by $\mathbf{f}^\pi(\mathbf{x},\, t)$ in case of vector or tensor valued function $\mathbf{f}^\pi(\mathbf{x},\, t)$. Thus, $\mathbf{a}^\pi = \mathrm{D}^\pi \mathbf{v}^\pi / \mathrm{D}t$.

In the theory of multiphase materials it is common to assume the motion of the solid as a reference and to describe the fluids in terms of motion relative to the solid. This means that a fluid relative velocity and the material time derivative with respect to the solid are introduced. The solid motion can be described both in terms of material or spatial co-ordinates. The second approach is now presented because the most natural numerical formulation of the elasto-plastic initial-boundary-value problem is based on the weak form of the balance equations in the spatial setting.

The fluid relative velocity $\mathbf{v}^{\pi s}(\mathbf{x},\, t)$ in spatial parametrization or diffusion velocity is given by

$$\mathbf{v}^{\pi s}(\mathbf{x},\, t) = \mathbf{v}^\pi(\mathbf{x},\, t) - \mathbf{v}^s(\mathbf{x},\, t) \qquad\qquad [13]$$

and the material time derivative of $f^\pi(\mathbf{x},\, t)$ with respect to the moving solid phase (s) is given by

$$\frac{D^s f^\pi}{Dt} = \frac{D^\pi f^\pi}{Dt} + \text{grad } f^\pi \cdot v^{s\pi} \quad \text{with} \quad v^{s\pi} = -v^{\pi s} \tag{14}$$

For the section closure, the spatial velocity gradient $l^s(x, t)$ of the solid will be recalled, which is defined as the gradient of the velocity (10) with respect to spatial co-ordinates, i. e.

$$l^s = \text{grad } v^s = \frac{\partial F^s}{\partial t} (F^s)^{-1} = d^s + w^s \tag{15}$$

where $d^s(x, t)$ and $w^s(x, t)$ are the symmetric and the skew-symmetric part of $l^s(x, t)$, also called spatial rate of deformation tensor and spin tensor, respectively.

All strain measures and strain rates for each constituent follow similarly to classical non-linear continuum mechanics, but are not reported here because they are not useful for the approach developed in the sequel (see e. g. [MAR 83]).

2.3. Mass balance equations

The averaged macroscopic balance equation for the solid phase is

$$\frac{D^s \rho_s}{Dt} + \rho_s \text{ div } v^s = \frac{\partial \rho_s}{\partial t} + \text{div} (\rho_s v^s) = 0 \tag{16}$$

where $v^s(x, t)$ is the mass averaged solid velocity, $\rho_s(x, t)$ is the averaged density of the solid related to the intrinsic averaged density $\rho^s(x, t)$ by the volume fraction $\eta^s(x, t)$.

For the generic π-phase the relationship between the averaged density and the intrinsic averaged density is

$$\rho_\pi(x, t) = \eta^\pi(x, t) \rho^\pi(x, t) \tag{17}$$

where the intrinsic density $\rho^\pi(x, t)$ is also named real or true density in the so-called Theory of Porous Media, e. g. [BOE 00].

The mass balance equation for water is

$$\frac{D^w \rho_w}{Dt} + \rho_w \text{ div } v^w = \frac{\partial}{\partial t}(n S_w \rho^w) + \text{div} (n S_w \rho^w v^w) = \rho_w e^w \tag{18}$$

where $\rho_w \, e^w(x, \, t)$ is the quantity of water per unit time and volume lost through evaporation. The corresponding equations for dry air and vapour are, respectively,

$$\frac{D^{ga} \rho_{ga}}{Dt} + \rho_{ga} \, \text{div} \, \mathbf{v}^{ga} = \frac{\partial}{\partial t} \left(n \, S_g \, \rho^{ga} \right) + \text{div} \left(n \, S_g \, \rho^{ga} \, \mathbf{v}^{ga} \right) = 0 \qquad [19]$$

$$\frac{D^{gw} \rho_{gw}}{Dt} + \rho_{gw} \, \text{div} \, \mathbf{v}^{gw} = \frac{\partial}{\partial t} \left(n \, S_g \, \rho^{gw} \right) + \text{div} \left(n \, S_g \, \rho^{gw} \, \mathbf{v}^{gw} \right) = \rho_{gw} \, e^{gw} \qquad [20]$$

where $n(x, \, t)$ is the porosity of the medium, defined as

$$n = \frac{dv^w + dv^g}{dv} = \frac{dv^{\text{voids}}}{dv} = 1 - \eta^s \qquad [21]$$

and S_w and S_g are the water and gas degrees of saturation. The following relationships hold:

$$\eta^w = n \, S_w \qquad \text{with} \qquad S_w = \frac{dv^w}{dv^w + dv^g} \, ,$$

$$\eta^g = n \, S_g \qquad \text{with} \qquad S_g = \frac{dv^g}{dv^w + dv^g} \qquad [22]$$

with the saturation constraint $S_w + S_g = 1$. The right-hand side of (18) and (20) sum to zero.

2.4. Linear momentum balance equations

The linear momentum balance equation for the solid and the π-fluid are

$$\text{div} \, \mathbf{t}^s + \rho_s \, (\mathbf{g} - \mathbf{a}^s) + \rho_s \, \hat{\mathbf{t}}^s = 0 \qquad [23]$$

and

$$\text{div} \, \mathbf{t}^\pi + \rho_\pi \, (\mathbf{g} - \mathbf{a}^\pi) + \rho_\pi \, (\mathbf{e}^\pi + \hat{\mathbf{t}}^\pi) = 0 \qquad [24]$$

respectively, where $\mathbf{t}^\pi(x, \, t)$ is the partial *Cauchy* stress tensor defined via the constitutive equation presented in Section 2.7. $\hat{\mathbf{t}}^\pi(x, \, t)$ accounts for the exchange of momentum due to mechanical interaction with other phases, $\rho_\pi \, \mathbf{a}^\pi$ for the volume density of the inertial force, $\rho_\pi \, \mathbf{g}$ for the volume density of gravitational force, and $\mathbf{e}^\pi(\mathbf{x}, t)$ takes into account the momentum exchange due to averaged mass supply or mass exchange between the fluid and the gas phases and the change of density. The linear momentum balance equation of the multiphase medium is subjected to the constraint [LEW 98]:

$$\sum_{\pi=1}^{k} \rho_\pi \left(\mathbf{e}^\pi + \hat{\mathbf{t}}^\pi \right) = 0 \qquad [25]$$

2.5. Angular momentum balance equation

All the phases are considered microscopically non-polar and hence at macroscopic level the angular momentum balance equation states that the partial stress tensor is symmetric [LEW 98]:

$$\mathbf{t}^\pi = (\mathbf{t}^\pi)^T \qquad [26]$$

2.6. Energy balance equation and entropy inequality

The energy balance equation for the π-phase may be written as [LEW 98]

$$\rho_\pi \frac{D^\pi E^\pi}{Dt} = \mathbf{t}^\pi : \mathbf{d}^\pi + \rho_\pi \, h^\pi - \operatorname{div} \mathbf{q}^\pi + \rho_\pi \, R^\pi \qquad [27]$$

where $\rho_\pi \, R^\pi$ represents the exchange of energy between the π-phase and other phases of the medium due to phase change and mechanical interaction, $\mathbf{q}^\pi$ is the internal heat flux, h^π results from the heat sources, and $\mathbf{d}^\pi$ is the spatial rate of the deformation tensor. E^π accounts for the specific internal energy of the volume element.

The entropy inequality for the mixture, useful for the development of the constitutive equations [SCH 02] is

$$\sum_\pi \left(\rho_\pi \frac{D^\pi \lambda^\pi}{Dt} + \rho_\pi \, e^\pi \lambda^\pi + \operatorname{div} \frac{\mathbf{q}^\pi}{\theta^\pi} - \frac{\rho_\pi \, h^\pi}{\theta^\pi} \right) \geq 0 \qquad [28]$$

where θ^π is the absolute temperature, λ^π is the specific entropy of the constituent π, and $e^\pi \lambda^\pi$ the entropy supply due to mass exchange.

2.7. Constitutive equations

The momentum exchange term $\rho_\pi \, \hat{t}^\pi$ of the linear momentum balance equation of the fluid can be expressed as [LEW 98]

$$\rho_\pi \, \hat{t}^\pi = -\mu^\pi \, (\eta^\pi)^2 \, k^{-1} \, v^{\pi s} + p^\pi \, \text{grad} \, \eta^\pi \quad \text{with} \quad \pi = g, \, w\psi \qquad [29]$$

Here, $k = k^{r\pi} \, k^\pi$, where $k^\pi(x, t) = k^\pi(\rho^\pi, \eta^\pi, T)$ is the intrinsic permeability tensor of dimension $[L^2]$ depending in the isotropic case on the porosity of the medium, $k^{r\pi}(S_\pi)$ is the relative permeability parameter, and μ^π is the dynamic viscosity. The relative permeability is a function of the π-phase degree of saturation S_π and is determined in laboratory tests (see e. g. [LEW 98] and [OCZ 99]).

The partial stress tensor in the fluid phase of the linear momentum balance equation (24) is related to the macroscopic pressure $p^\pi(x, t)$ of the π-phase

$$t^\pi = -\eta^\pi \, p^\pi \, 1 \qquad [30]$$

where 1 is the second order unit tensor.

From the entropy inequality it can also be shown that the spatial solid stress tensor $t^s(x, t)$ of the linear momentum balance equation (23) is decomposed as follows:

$$t^s = \eta^s \, (t_e^s - p^s \, 1) \qquad [31]$$

and that the effective *Cauchy* stress tensor $\sigma'(x, t)$, which is responsible for all major deformation in the solid skeleton, is

$$\sigma' = \eta^s \, t_e^s \qquad [32]$$

In (31), $t_e^s(x, t)$ is the dissipative part [GRA 91], [GRA 01] or effective stress tensor of the solid phase, while $p^s(x, t)$ is the equilibrium part, also called solid pressure, with $p^s = S_w \, p^w + S_g \, p^g$.

From the previous equations, it follows that the total *Cauchy* stress tensor $\sigma = t^s + t^w + t^g$ can be written in the usual form used in soil mechanics

$$\sigma = \sigma' - (S_w \, p^w + S_g \, p^g) \, 1 \qquad [33]$$

The elasto-plastic behaviour of the solid skeleton at finite strain can be based on the multiplicative decomposition of the deformation gradient $\mathbf{F}^s(\mathbf{X}^s,\ t)$ into an elastic and plastic part originally proposed by Lee [LEE 69] for crystals (see e. g. [SAN 02] for details concerning also the numerical algorithm):

$$\mathbf{F}^s = \mathbf{F}^{se}\,\mathbf{F}^{sp} \qquad [34]$$

This decomposition states the existence of an intermediate stress free configuration and its validity has been suggested for cohesive-frictional soils by Nemat-Nasser [NEM 83], where the plastic part of the deformation gradient is viewed as an internal variable related to the amount of slipping, crushing, yielding, and, for plate like particles, plastic bending of the granules comprizing the soil.

The pressure $p^g(\mathbf{x},\ t)$ is given in the sequel. For a gaseous mixture of dry air and water vapour, the ideal gas law is introduced because the moist air is assumed to be a perfect mixture of two ideal gases. The equation of state of perfect gas (the *Clapeyron* equation) and *Dalton*'s law applied to dry air (ga), water vapour (gw) and moist air (g), yield:

$$p^{ga} = \rho^{ga}\,R\,\theta/M_a \quad , \qquad p^{gw} = \rho^{gw}\,R\,\theta/M_w \qquad [35]$$

$$p^g = p^{ga} + p^{gw} \quad , \qquad \rho^g = \rho^{ga} + \rho^{gw} \qquad [36]$$

In the partially saturated zones, water is separated from its vapour by a concave meniscus (capillary water). Due to the curvature of this meniscus, the sorption equilibrium equation gives the relationship between the capillary pressure $p^c(\mathbf{x},\ t)$ and the gas $p^g(\mathbf{x},\ t)$ and water pressure $p^w(\mathbf{x},\ t)$ [GRA 91]:

$$p^c = p^g - p^w \qquad [37]$$

The equilibrium water vapour pressure $p^{gw}(\mathbf{x},\ t)$ can be obtained from the *Kelvin-Laplace* equation:

$$p^{gw} = p^{gws}(\theta)\,\exp\left(\frac{p^c\,M_w}{\rho^w\,R\,\theta}\right) \qquad [38]$$

where the water vapour saturation pressure p^{gws}, depending only upon the temperature $\theta(\mathbf{x},\ t)$, can be calculated from the *Clausius-Clapeyron* equation or from an empirical correlation.

The saturation $S_\pi(\mathbf{x}, t)$ is an experimentally determined function of the capillary pressure p^c and the temperature θ:

$$S_\pi = S_\pi(p^c, \theta) \qquad [39]$$

For the binary gas mixture of dry air and water vapour, *Fick*'s law gives the following relative velocities $\mathbf{v}_g^\pi = \mathbf{v}^\pi - \mathbf{v}^g$ ($\pi = ga, gw$) of the diffusing species:

$$\mathbf{v}_g^{ga} = -\frac{M_a M_w}{M_g^2} \mathbf{D}_g \text{ grad} \left(\frac{p^{ga}}{p^g}\right) = -\mathbf{v}_g^{gw} \qquad [40]$$

where $\mathbf{D}_g$ is the effective diffusivity tensor and M_g is the molar mass of the gas mixture:

$$\frac{1}{M_g} = \frac{\rho^{gw}}{\rho^g}\frac{1}{M_w} + \frac{\rho^{ga}}{\rho^g}\frac{1}{M_a} \qquad [41]$$

2.8. Initial and boundary conditions

For model closure it is necessary to define the initial and boundary conditions. The initial conditions specify the full fields of gas pressure, water pressure, temperature, displacements, and velocity:

$$p^g = p_0^g, \quad p^w = p_0^w, \quad \theta = \theta_0, \quad \mathbf{u} = \mathbf{u}_0, \quad \dot{\mathbf{u}} = \dot{\mathbf{u}}_0 \quad \text{at} \quad t = t_0 \qquad [42]$$

The boundary conditions can be imposed values on ∂B_π or fluxes on ∂B_π^q, where the boundary is $\partial B = \partial B_\pi \cup \partial B_\pi^q$. The imposed values on the boundary for gas pressure, water pressure, temperature, and displacements are as follows:

$$p^g = \hat{p}^g \quad \text{on} \quad \partial B_g, \qquad p^w = \hat{p}^w \quad \text{on} \quad \partial B_w,$$

$$\theta = \hat{\theta}\psi \quad \text{on} \quad \partial B_\theta, \psi \qquad \mathbf{u} = \hat{\mathbf{u}} \quad \text{on} \quad \partial B_u \quad \text{for} \quad t \geq t_0 \qquad [43]$$

The volume average flux boundary conditions for dry air and water species conservation equations and the energy equation to be imposed at the interface between the porous media and the surrounding fluid (the natural boundary conditions) are the following:

$$\begin{aligned}
\left(\rho^{ga}\mathbf{v}^g - \rho^g\mathbf{v}_g^{gw}\right)\cdot\mathbf{n} &= q^{ga} & \text{on } \partial B_g^q \\
\left(\rho^{gw}\mathbf{v}^g + \rho^w\mathbf{v}^w + \rho^g\mathbf{v}_g^{gw}\right)\cdot\mathbf{n} &= \beta_c\left(\rho^{gw} - \rho_\infty^{gw}\right) + q^{gw} + q^w & \text{on } \partial B_c^q \\
-\left(\rho^w\mathbf{v}^w\Delta h_{\text{vap}} - \lambda_{\text{eff}}\nabla\theta\right)\cdot\mathbf{n} &= \alpha_c\left(\theta - \theta_\infty\right) + q^\theta & \text{on } \partial B_\theta^q
\end{aligned} \qquad [44]$$

for $t \geq t_0$, where $n(x, t)$ is the vector perpendicular to the surface of the porous medium, pointing towards the surrounding gas, $\rho_\infty^{gw}(x, t)$ and $\theta_\infty(x, t)$ are, respectively, the mass concentration of water vapour and temperature in the undisturbed gas phase distant from the interface, $\alpha_c(x, t)$ and $\beta_c(x, t)$ are convective heat and mass transfer coefficients, while $q^{ga}(x, t)$, $q^{gw}(x, t)$, $q^w(x, t)$, and $q^\theta(x, t)$ are the imposed dry air flux, imposed vapour flux, imposed liquid flux, and imposed heat flux, respectively.

The traction boundary conditions for the displacement field related to the total *Cauchy* stress tensor $\sigma(x, t)$ are

$$\sigma \, n = \bar{t} \quad \text{on} \quad \partial B_u^q \qquad [45]$$

where $\bar{t}(x, t)$ is the imposed *Cauchy* traction vector.

Conclusions

A mathematical formulation for the thermo-hydro-mechanical behaviour of a partially saturated porous material has been presented. This model is suitable for the numerical discretisation via the finite element method, as will be shown in the contribution "*A finite element model for water saturated and partially saturated geomaterials*" by the authors in this publication. For the interested reader, further details of the use of the model in environmental geomechanics and soil dynamics can be found in the textbooks [LEW 98] and [OCZ 99] and in the references contained in. Moreover, a thermodynamic framework of the model is developed in [SCH 02], where interfacial phenomena between the constituents are taken into account. The extension of the model presented here for the simulation of concrete structure at high temperature is developed in [GAW 02].

Acknowledgements

This work has been carried out within the research project *Cofin MM08323597* sponsored by the Italian Ministry of Scientific and Technological Research *MIUR*.

3. References

[BOE 00] DE BOER R., *Theory of Porous Media: Highlight in Historical Development and Current State*, Springer-Verlag, Berlin, 2000.

[EHL 98] EHLERS W., VOLK W. "On theoretical and numerical methods in the theory of porous media based on polar and non-polar elasto-plastic solid materials", *Int. J. Solids and Structures* vol. 35, 1998, p. 4597-4617.

[GAW 02] GAWIN D., PESAVENTO F., SCHREFLER B.A., "Modelling of thermo-chemical and mechanical damage of concrete at high temperature", submitted.

[GRA 91] GRAY W. G., HASSANIZADEH M., "Unsaturated Flow Theory including Interfacial Phenomena", Water Resources Res., vol. 27, num. 8, 1991, p. 1855-1863.

[GRA 01] GRAY W. G., SCHREFLER B.A., "Thermodynamic approach to effective stress in partially saturated porous media", Eur. J. Mech. A/Solids, vol. 20, 2001, p. 521-538.

[HAS 79a] HASSANIZADEH M., GRAY W. G., "General Conservation Equations for Multiphase System: 1. Averaging technique", Adv. Water Res., vol. 2, 1979, p. 131-144.

[HAS 79b] Hassanizadeh M., GRAY W. G., "General Conservation Equations for Multi-Phase System: 2. Mass, Momenta, Energy and Entropy Equations", Adv. Water Res., vol. 2, 1979, p. 191-201.

[HAS 80] HASSANIZADEH M., GRAY W. G., "General Conservation Equations for Multi-Phase System: 3. Constitutive Theory for Porous Media Flow", Adv. Water Res., vol. 3, 1980, p. 25-40.

[LEE 69] LEE E. H., "Elastic-Plastic Deformation at Finite Strains", J. Appl. Mech., vol. 1, num. 6, 1969.

[LEW 98] LEWIS R. W., SCHREFLER B. A., The Finite Element Method in the Static and Dynamic Deformation and Consolidation of Porous Media, John Wiley & Sons, Chichester, 1998.

[MAR 83] MARSDEN J. E., HUGHES T. J. R., Mathematical Foundations of Elasticity, Prentice-Hall, Englewood Cliffs, 1983.

[MOK 98] MOKNI M., DESRUES J., "Strain Localization Measurements in Undrained Plane-strain Biaxial Tests on Hostun RF Sand", Mech. Cohes-Frict. Mater., vol. 4, 1998, p. 419-441.

[NEM 83] NEMAT-NASSER S., "On Finite Plastic Flow of Crystalline Solids and Geomaterials", Transactions of ASME vol. 50, 1983, p. 1114-1126.

[SAN 02] SANAVIA L., SCHREFLER B. A., STEINMANN P., "A formulation for an unsaturated porous medium undergoing large inelastic strains", Computational Mechanics, vol. 28, 2002, p. 25-40.

[SCH 02] SCHREFLER B.A., "Mechanics and thermodynamics of saturated-unsaturated porous materials and quantitative solutions", Archives of Rational Mechanics in print.

[VAR 95] VARDOULAKIS J., SULEM J., Bifurcation Analysis in Geomechanics, Blakie Academic and Professional, London, 1995.

[OCZ 99] ZIENKIEWICZ O. C., CHAN A., PASTOR M., SCHREFLER B. A., SHIOMI T., Computational Geomechanics with special Reference to Earthquake Engineering, John Wiley & Sons, Chichester, 1999.   •

Chapter 5

Continuum Damage Modelling and Some Computational Issues

Gilles Pijaudier-Cabot and Ludovic Jason
R&DO, Laboratoire de Génie Civil de Nantes, Ecole Centrale de Nantes, France

1. Introduction

This paper is concerned with the presentation of continuum damage models and their implementation in non linear finite element analyses. Continuum damage means that the mechanical effects of progressive micro cracking, void nucleation and growth are represented by a set of state variables which act on the elastic and/or plastic behaviour of the material at the macroscopic level. First, we will deal with the scalar damage model and present several versions of this type of constitutive relation. After having recalled how the local integration of the constitutive relation can be performed in the same spirit as for plasticity-based models, we will deal with with more specific applications: cases where the evolution of damage is specified in an integrated form, where damage induced inelastic strains are introduced, and finally where crack closure effects are modelled.

The following section deals with damage induced anisotropy. A general framework for capturing directionality of damage is recalled. Coupling with plasticity and crack closure effects is discussed. The relation with smeared crack model is also presented. Non local damage is discussed in the fourth section. We present integral and gradient models. In particular, a general scheme for the implementation of the gradient model, implemented into the finite element code Code_Aster, is described.

The fifth section deals with several computational issues related to damage such as the implementation of the model within secant or tangent algorithms and convergence properties of the tangent implementation. Solution control techniques whenever bifurcation or snap-back occurs are recalled. The paper concludes with an example of 3D computation of a reinforced concrete bending beam.

2. Scalar Damage Model

2.1. Incremental Damage Model

When damage is assumed to be isotropic, it is considered that it produces a degradation of the elastic stiffness of the material through a variation of the Young's modulus:

$$\varepsilon_{ij} = \frac{1+v_0}{E_0(1-d)}\sigma_{ij} - \frac{v_0}{E_0(1-d)}[\sigma_{kk}\delta_{ij}] \qquad [1]$$

where v_0, E_0 and δ_{ij} are the Poisson's ratio and Young's modulus of the undamaged isotropic material and the Kronecker symbol respectively. The elastic (i.e. free) energy per unit mass of material is:

$$\rho\psi = \frac{1}{2}(1-d)\varepsilon_{ij}C^0_{ijkl}\varepsilon_{kl} \qquad [2]$$

where C^0 is the stiffness tensor of the undamaged material. This energy is assumed to be the state potential. The damage energy release rate is defined as the variable associated to the damage state variable in the state potential:

$$Y = -\rho \frac{\partial \psi}{\partial d} = \frac{1}{2} \varepsilon_{ij} C^0_{ijkl} \varepsilon_{kl}$$ [3]

with the rate of dissipated energy:

$$\dot{\phi} = -\frac{\partial \rho \psi}{\partial d} \dot{d}$$ [4]

For an isotropic damage model, this equation reduces to:

$$\dot{\phi} = Y \dot{d}$$ [5]

The damage energy release rate is always positive and thus the rate of damage must be positive in order to comply with the second principle of thermodynamics.

The rules governing the evolution of damage can be defined following the same principles as those in elasto-plasticity. In fact, the general formalism is that of generalised standard materials where the loading function is assumed to be a pseudo potential of dissipation (Lemaitre, 1992). The evolution of damage requires the definition of a loading function:

$$f(Y, d) = Y - Y_0 - Z$$ [6]

where Y_0 is a parameter which defines the threshold of damage, and Z is the hardening-softening controlling variable. The evolution law is prescribed according to the normality rule:

$$\dot{d} = \dot{\lambda} \frac{\partial f}{\partial Y}$$ [7]

with the Kuhn-Tucker conditions $\dot{\lambda} \geq 0, f \leq 0$, and $\dot{\lambda} f = 0$. The evolutionary equations are completed by the definition of a hardening function:

$$\dot{\lambda} = H \dot{Z}$$ [8]

where H can be regarded as the equivalent of a hardening-softening modulus in generalised plasticity. In most applications, this modulus is not constant. The major difference with plasticity is that the loading function and the evolution equation for damage are provided in an explicit form because these quantities are function of the total strain. For the integration of this constitutive relation, an Euler forward integration technique may be used. This method consists of an elastic predictor/damage corrector technique. The elastic predictor is:

$$\delta\sigma_{ij}^{e} = (1-d)C_{ijkl}^{0}\delta\varepsilon_{kl}$$ [9]

and the damage corrector is:

if $\frac{1}{2}(\varepsilon + \delta\varepsilon) : C^{0} : (\varepsilon + \delta\varepsilon) - Z > 0$ then

$$\sigma^{new} = \sigma + \delta\sigma^{e} - \delta\sigma^{dam}, \quad \delta\sigma^{dam} = \left\{\langle H(Z)\rangle\left(C^{0} : \varepsilon\right)\otimes\left(C^{0} : \varepsilon\right)\right\}\delta\varepsilon$$ [10]

else $\sigma^{new} = \sigma + \delta\sigma^{e}$

As opposed to plasticity, there is no risk for divergence when the loading point is mapped back onto the loading surface. The reason for this is that the loading surface is in fact a one-dimensional segment in the Y space. In fact, the finite element implementation of such a model is exactly the same as the finite element implementation of one dimensional plasticity because damage is a scalar, not a second order tensor like the plastic strains in the 3D case.

2.2. Damage Model with Crack Closure

Crack closure effects are of importance when the material is subjected to alternated loads, for instance reinforced concrete structures subjected to earthquakes. The fundamental property of the scalar damage model is that the time derivative of damage is constrained to be positive or zero (second principle of thermodynamics). During load reversals, however, micro cracks are closing progressively and the tangent stiffness of the material should increase while damage is constant (Figure 1).

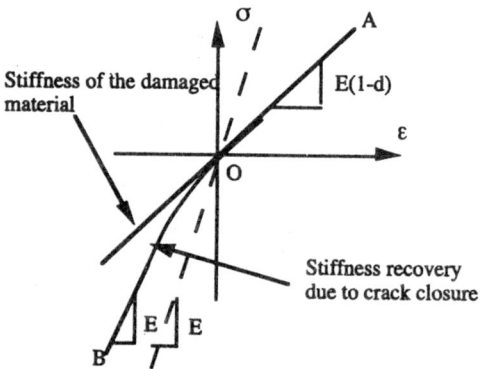

Material damaged in tension (path OA) and unloaded up to point B

Figure 1. *Schematic response of a material subjected to uniaxial compression after being damaged in tension*

Within isotropic damage modelling, one solution is to introduce two damage scalars, instead of one, in order to separate the mechanical effect of micro cracking as a function of the type of loading. La Borderie (1991) developed the corresponding constitutive relations and applied it to concrete. The elastic energy is:

$$\rho\psi = \frac{1}{2}[\frac{1}{E_0(1-d_t)} <\sigma>^+_{ij} <\sigma>^+_{ij} + \frac{1}{E_0(1-d_c)} <\sigma>^-_{ij} <\sigma>^-_{ij}$$

$$+\frac{v_0}{E_0}(\sigma_{ij}\sigma_{ij} - \sigma^2_{kk})] + \frac{\beta_1 d_t}{E_0(1-d_t)} f(\sigma) + \frac{\beta_2 d_c}{E_0(1-d_c)}\sigma_{kk}$$

[11]

where $\langle\sigma\rangle^+$ is the positive part of the stress, i.e. the stress tensor expressed in its principal directions with the positive principal stresses only, and $\langle\sigma\rangle^- = \sigma - \langle\sigma\rangle^+$.

Anelastic strains due to the growth of damage are also taken into account in this model with a function $f(\sigma)$ introduced in order to represent the progressive vanishing of anelastic strains due to damage in tension when the material is subjected to compression:

$$f(\sigma) = \sigma_{kk} \text{ when } \sigma_{kk} \in]0,\infty]f(\sigma)$$

$$f(\sigma) = (1+\frac{\sigma_{kk}}{2\sigma_c})\sigma_{kk} \text{ when } \sigma_{kk} \in]-\sigma_c,0]$$

[12]

$$f(\sigma) = -\frac{\sigma_c}{2}\sigma_{kk} \text{ when } \sigma_{kk} \in [-\infty,-\sigma_c]$$

where σ_c is the crack closure stress (material parameter). The stress-strain relation reads:

$$\varepsilon_{ij} = \frac{1}{E_0(1-d_t)} <\sigma>^+_{ij} + \frac{1}{E_0(1-d_c)} <\sigma>^-_{ij}$$

$$+\frac{v_0}{E_0}[\sigma_{ij} - \sigma_{kk}\delta_{ij}] + \frac{\beta_1 d_t}{(1-d_t)}\frac{\partial f(\sigma)}{\partial \sigma_{ij}} + \frac{\beta_2 d_c}{(1-d_c)}\delta_{ij}$$

[13]

Since the model has two damage variables, the evolutionary equations include two yield functions. These functions are constructed on the same principles as the function used for the one scalar model, except that there are two energy release rates associated to the two damage variables respectively:

$$Y_t = -\rho\frac{\partial\psi}{\partial d_t}, \quad Y_c = -\rho\frac{\partial\psi}{\partial d_c}$$

[14]

The two loading functions are:

$$f_t(Y_t,d_t) = Y_t - Y_{0t} - Z_t, \quad f_c(Y_c,d_c) = Y_c - Y_{0c} - Z_c$$

[15]

and an associated model can be constructed with two independent hardening functions similar to Eq. [8].

Note that this constitutive relation cannot be integrated explicitly. The stress decomposition into positive and negative parts is not known a priori. Therefore, the conditions of evolution of damage are not known for each integration step. Same as in classical plasticity, the integration of the constitutive relations is performed with an iterative scheme: the equations of the system are the state laws and the evolution laws recast as follows in order to obtain a well-conditioned algebraic system :

$$\begin{cases} \varepsilon_{ij} = \dfrac{1}{E_0(1-d_t)} <\sigma>^+_{ij} + \dfrac{1}{E_0(1-d_c)} <\sigma>^-_{ij} \\[2mm] + \dfrac{v_0}{E_0}[\sigma_{ij} - \sigma_{kk}\delta_{ij}] + \dfrac{\beta_1 d_t}{(1-d_t)} \dfrac{\partial f(\sigma)}{\partial \sigma_{ij}} + \dfrac{\beta_2 d_c}{(1-d_c)}\delta_{ij} \\[2mm] Y_t - Z_t = 0 \\[2mm] Y_c - Z_c = 0 \\[2mm] \dfrac{1}{(1-d_t)} = F_t(Y_t, Z_t) \\[2mm] \dfrac{1}{(1-d_c)} = F_c(Y_c, Z_c) \end{cases}$$ [16]

Here, F_t, F_c are calculated from the integrated equation of evolution of damage instead of an incremental one. It does not change the general picture of the integration process, however. The system is written in a matrix form:

$$[J]\bar{x} = 0 \quad \text{with } \bar{x}^T = \left\{\sigma \quad Y_t \quad Y_c \quad \frac{1}{1-d_t} \quad \frac{1}{1-d_c}\right\}$$ [17]

where matrix $[J]$, can be computed and programmed using formal calculus. A full Newton algorithm is implemented in order to solve this non-linear system. Note that there are two sorts of nonlinearities: the first one is due to the evolution of damage (i.e. whether damage grows or not), the second is due to crack closure effects. For constant values of the damage variables, the material is orthotropic in the coordinate system of the principal stresses.

2.3. Integrated Damage Model

In continuum damage mechanics, the evolution of damage is very often related to the state of strain. Moreover, it is defined in an explicit, integrated way, which is easier to handle. The damage loading function in Eq. [6] can be redefined as:

$$f(\tilde{\varepsilon},\kappa) = \tilde{\varepsilon} - \kappa \qquad [18]$$

where $\tilde{\varepsilon}$ is a positive equivalent measure of strain and κ is a threshold value. The equation $f = 0$ represents a loading surface in strain space. For the uni-axial tensile case, the equivalent uniaxial strain in Eq. [18] is straightforward. It is the axial strain if the lateral strains are neglected. However, for general states of stress, damage evolution should be related to some scalar quantity, function of the state of strain. There are, in this regard, several proposals. For example, an appropriate definition for metals is rooted in the elastic stored energy (Peerlings et al. 1998):

$$\tilde{\varepsilon} = \sqrt{\frac{1}{E} \varepsilon_{ij} C_{ijkl} \varepsilon_{kl}} \qquad [19]$$

which is depicted in Figure 2. It is the integrated version of the relation presented in section 2.1. For concrete, Mazars (1984) proposed the following form:

$$\tilde{\varepsilon} = \sqrt{\sum (< \varepsilon_i >)^2} \qquad [20]$$

where ε_i are the principal strains. A third possibility, which is also mentioned by Peerlings, is the modified von Mises definition. It is written as follows

$$\tilde{\varepsilon} = \frac{k-1}{2k(1-2v)} I_1 + \frac{1}{2k}\sqrt{\frac{(k-1)^2}{(1-2v)^2} I_1^2 - \frac{12k}{(1+v)^2} J_2} \qquad [21]$$

I_1 and J_2 are the first invariant of the strain tensor and the second invariant of the deviatoric strain tensor respectively. Only the modified von Mises criterion leads to a new material parameter, namely the factor k. The parameter k is the ratio between uni-axial compressive and uni-axial tensile strength. These criteria are plotted in Figure 2 (with $k = 10$). The loading surfaces (Eqs. 19-21) are closed contours around the origin. The dashed lines represent the constant uniaxial compression and uniaxial tension stress paths.

The evolution of damage has the same form as in the previous sections:

$$\text{if } f(\tilde{\varepsilon},\kappa) = 0 \text{ and } \dot{f}(\tilde{\varepsilon},\kappa) = 0 \text{ then } \begin{cases} d = h(\kappa) \\ \kappa = \tilde{\varepsilon} \end{cases} \text{ where } \dot{d} \geq 0$$

$$\text{otherwise } \begin{cases} \dot{d} = 0 \\ \dot{\kappa} = 0 \end{cases} \qquad [22]$$

The function $h(\kappa)$ is specific, depending on different models. For tension only, exponential softening can be used and

$$h(\kappa) = 1 - \frac{\kappa_0}{\kappa}(1 - \alpha + \alpha e^{-\eta(\kappa - \kappa_0)}) \qquad [23]$$

where κ_0, α, η are model parameters.

In order to capture the differences of mechanical responses of the material in tension and in compression, Mazars proposed to split the damage variable into two parts and used the equivalent strain defined in Eq. [20]:

$$d = \alpha_t d_t + \alpha_c d_c \tag{24}$$

where d_t and d_c are the damage variables in tension and compression, respectively.

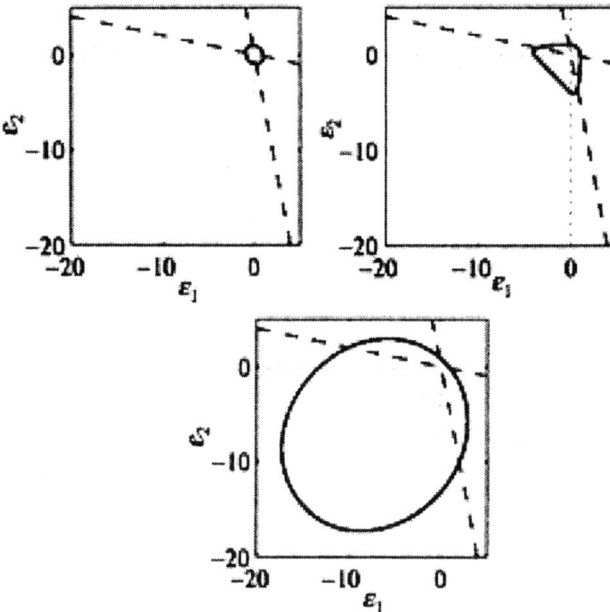

Figure 2. *Contour plots for $\tilde{\varepsilon}$ for the elastic stored energy (top left), Mazars definition (top right), and the modified von Mises expression (bottom), after Peerlings et al. (1998)*

They are combined with the weight coefficients α_t and α_c defined as function of the principal values of the strains ε_{ij}^t and ε_{ij}^c, due to positive and negative stresses (see Mazars, 1984).

$$\varepsilon_{ij}^t = (1-d)C_{ijkl}^{-1}\sigma_{kl}^t, \quad \varepsilon_{ij}^c = (1-d)C_{ijkl}^{-1}\sigma_{kl}^c \tag{25}$$

$$\alpha_t = \sum_{i=1}^{3}\left(\frac{\langle\varepsilon_i^t\rangle\langle\varepsilon_i\rangle_+}{\tilde{\varepsilon}^2}\right)^{\beta}, \quad \alpha_c = \sum_{i=1}^{3}\left(\frac{\langle\varepsilon_i^c\rangle\langle\varepsilon_i\rangle_+}{\tilde{\varepsilon}^2}\right)^{\beta} \tag{26}$$

In uniaxial tension $\alpha_t=1$ and $\alpha_c=0$. In uniaxial compression $\alpha_c=1$ and $\alpha_t=0$. Hence, d_t and d_c can be obtained separately from uniaxial tests. The evolution of damage is provided in an integrated form, as a function of the variable κ (Mazars, 1984):

$$d_t = 1 - \frac{\kappa_0(1 - A_t)}{\kappa} - \frac{A_t}{\exp(B_t(\kappa - \kappa_0))}$$

$$d_c = 1 - \frac{\kappa_0(1 - A_c)}{\kappa} - \frac{A_c}{\exp(B_c(\kappa - \kappa_0))}$$

[27]

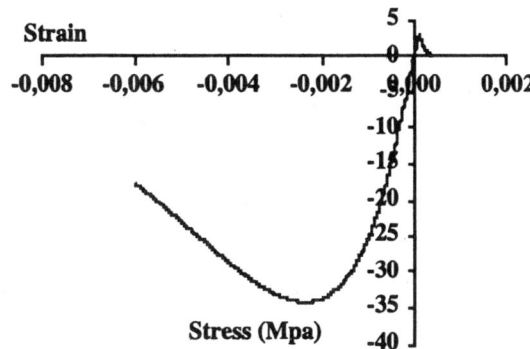

Figure 3. *Uniaxial response of the model by Mazars (1984)*

The purpose of exponent β is to reduce the effect of damage on the response of the material under shear compared to tension (Pijaudier-Cabot et al. 1991). Figure 3 shows the uniaxial response of the model in tension and compression with the following parameters: $E_0= 30000$ MPa, $v_0=0.2$, $\kappa_0=0.0001$, $A_t=1$, $B_t = 15000$, $A_c = 1.2$, $B_c=1500$, $\beta=1$.

3. Damage Induced Anisotropy

Microcracking is usually geometrically oriented as a result of the loading history on the material. In tension, microcracks are perpendicular to the tensile stress direction, in compression microcracks open parallel to the compressive stress direction. Although a scalar damage model, which does not account for directionality of damage, might be a sufficient approximation in usual applications, i.e. when tensile failure is expected with a quasi-radial loading path, damage induced anisotropy is required for more complex loading histories. The influence of crack closure is needed in the case of alternated loads: microcracks may close and the effect of damage on the material stiffness disappears. Finally, plastic strains are observed when the material unloads in compression. The following section describes a constitutive relation based of elasto-plastic damage which addresses these issues (Fichant et al. 1999).

3.1. *Principle*

The model is based on the approximation of the relationship between the overall stress (latter simply denoted as stress) and the effective stress in the material defined by the equation:

$$\sigma_{ij}^t = C_{ijkl}^0 \varepsilon_{kl}^e \quad \text{or} \quad \sigma_{ij}^t = C_{ijkl}^0 \left(C^{damaged}\right)_{klmn}^{-1} \sigma_{mn} \qquad [28]$$

where σ_{ij}^t is the effective stress component, ε_{kl}^e is the elastic strain, and $C_{ijkl}^{damaged}$ is the stiffness of the damaged material. We define the relationship between the stress and the effective stress along a finite set of directions of unit vectors $\bar{n}$ at each material point:

$$\sigma = \left(1 - d(n)\right)n_i \sigma_{ij}^t n_j, \quad \tau = \left(1 - d(n)\right)\sqrt{\sum_{i=1}^{3}\left(\sigma_{ij}^t n_j - \left(n_k \sigma_{kl} n_l\right)n_i\right)^2} \qquad [29]$$

σ and τ are the normal and tangential components of the stress vector respectively. $d(n)$ is a scalar valued quantity which introduce the effect of damage in each direction $\bar{n}$.

The basis of the model is the numerical interpolation of $d(n)$ (called damage surface) which is approximated by its knowledge *over a finite set of directions*. The stress is solution of the virtual work equation:

find σ_{ij} such that $\forall \varepsilon_{ij}^*$

$$\frac{4\pi}{3}\sigma_{ij}\varepsilon_{ij}^* = \int_S \left(\left[(1 - d(n))n_k \sigma_{kl}^t n_l n_i + (1 - d(n))\left(\sigma_{ij}^t n_j - n_k \sigma_{kl}^t n_l n_i\right)\right] \cdot \varepsilon_{ij}^* n_j\right)d\Omega \qquad [30]$$

Depending on the interpolation of the damage variable $d(n)$, several forms of damage induced anisotropy can be obtained. Here, it is defined by three scalars in three mutually orthogonal directions. It is the simplest approximation that yields anisotropy of the damaged stiffness of the material. The material is orthotropic with a possibility of rotation of the principal axes of orthotropy.

The stiffness degradation occurs mainly for tensile loads. Hence, the evolution of damage will be indexed on tensile strains. The evolution of damage is controlled by a loading surface f, which is similar to Eq. [6]:

$$f(n) = n_i \varepsilon_{ij}^e n_j - \varepsilon_d - \chi(n)$$ [31]

χ is an hardening softening variable which is interpolated in the same fashion as the damage surface. The initial threshold of damage is ε_d. The evolution of the damage surface is defined by an evolution equation inspired from that of an isotropic model:

if $f(n^*) = 0$ and $n_i^* d\varepsilon_{ij}^e n_j^* > 0$

then $\begin{cases} dd(n^*) = \left[\dfrac{\varepsilon_d\left(1 + a\left(n_i^* \varepsilon_{ij}^e n_j^*\right)\right)}{\left(n_i^* \varepsilon_{ij}^e n_j^*\right)^2} \exp(-a(n_i^* \varepsilon_{ij}^e n_j^* - \varepsilon_d) \right] n_i^* d\varepsilon_{ij}^e n_j^* \\[4mm] d\chi(n^*) = n_i^* d\varepsilon_{ij}^e n_j^* \end{cases}$ [32]

else $dd(\bar{n}^*) = 0$, $d\chi(n^*) = 0$

The model parameters are ε_d and a. Note that the vectors $\vec{n}^*$ are the three principal directions of the *incremental* strains whenever damage grows. After an incremental growth of damage, the new damage surface is the sum of two ellipsoidal surfaces: the one corresponding to the initial damage surface, and the ellipsoid corresponding to the incremental growth of damage.

3.2. Coupling with plasticity

We decompose the strain increment into elastic and plastic components:

$$d\varepsilon_{ij} = d\varepsilon_{ij}^e + d\varepsilon_{ij}^p$$ [33]

The evolution of the plastic strain is controlled by a yield function which is expressed in term of the effective stress in the undamaged material. We have implemented the yield function due to Nadai (1950). It is the combination of two Drucker-Prager functions F_1 and F_2 with the same hardening evolution:

$$F_i = \sqrt{\frac{2}{3} J_2'} + A_i \frac{I_1'}{3} - B_i w$$ [34]

where J_2' and I_1' are the second invariant of the deviatoric effective stress and the first invariant of the effective stress respectively. w is the hardening variable and (A_i, B_i) are four parameters ($i = 1, 2$) which were originally related to the ratios of the tensile strength to the compressive strength denoted γ and of the biaxial compressive strength to the uniaxial strength denoted β:

$$A_1 = \sqrt{2}\,\frac{1-\gamma}{1+\gamma}, \quad A_2 = \sqrt{2}\,\frac{\beta-1}{2\beta-1}, \quad B_1 = 2\sqrt{2}\,\frac{\gamma}{1+\gamma}, \quad B_2 = \sqrt{2}\,\frac{\beta}{2\beta-1} \qquad [35]$$

These two ratios will be kept constant in the model: $\beta = 1.16$ and $\gamma = 0.4$. The evolution of the plastic strains is associated to these surfaces. The hardening rule is given by:

$$w = qp^r + w_0 \qquad [36]$$

where q and r are model parameters, w_0 defines the initial reversible domain in the stress space, and p is the effective plastic strain.

3.3. Crack closure effects

A decomposition of the stress tensor into a positive and negative part is introduced again: $\sigma = \langle\sigma\rangle_+ + \langle\sigma\rangle_-$, where $<\sigma>_+$, and $<\sigma>_-$ are the positive and negative parts of the stress tensor. The relationship between the stress and the effective stress defined in Eq. [29] of the model is modified:

$$\sigma_{ij}n_j = \left(1 - d(n)\right)\langle\sigma\rangle^t_{+\,ij}n_j + \left(1 - d_c(n)\right)\langle\sigma\rangle^t_{-\,ij}n_j. \qquad [37]$$

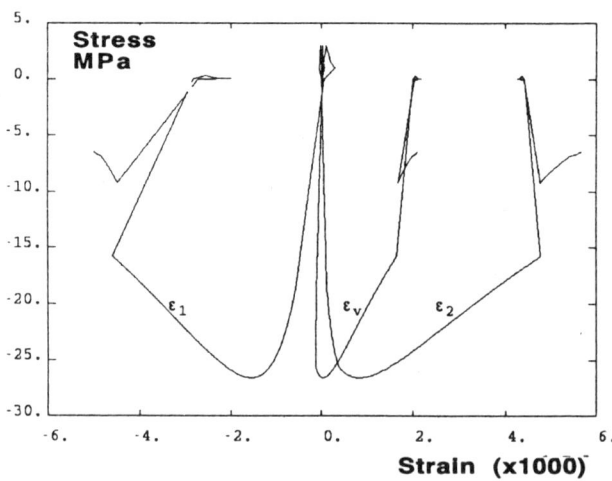

Figure 4. *Uniaxial tension-compression response of the anisotropic model (longitudinal (1), transverse (2) and volumetric (v) strains as functions of the compressive stress)*

$d_c(n)$ is a new camage surface which describes the influence of damage on the response of the material in compression. Since this new variable refers to the same physical state of degradation as in tension, $d_c(n)$ is directly deduced from $d(n)$. It is defined by the same interpolation as $d(n)$ and along each principal direction i, we have the relation:

$$d_c^i = \left(\frac{d_j(1 - \delta_{ij})}{2} \right)^{\alpha}, \ i \in [1,3]$$ [38]

α is a model parameter. The constitutive relations contain 6 parameters in addition to the Young's modulus of the material and the Poisson's ratio. Their determination benefits from the fact that in tension, plasticity is negligible. Once the evolution of damage in tension has been fitted, the remaining parameters are fitted from a compression test. Figure 4 shows a typical uniaxial compression-tension response of the model corresponding to a concrete with a tensile strength of 3 MPa and a compressive strength of 40 MPa.

3.4. Damage and smeared crack models

Historically, smeared crack models have been developed for modelling concrete fracture fifteen years before continuous damage models started to become popular. In smeared crack models the directionality of the material decohesion is a fundamental characteristic, which was not included into the first damage models. Therefore, it is natural to relate damage models to smeared crack models (de Borst and Gutierrez, 1999). In the latter model, the material response is defined in a local coordinate system (n, s). Direction n is the normal to the plane in which the greatest positive normal strain denoted as ε_{nn} is found. In the (n,s) system, the secant stress-strain relation is:

$$\sigma_{n,s} = D_{ns}^s \varepsilon_{n,s}$$ [39]

with $\sigma_{n,s} = \begin{bmatrix} \sigma_{nn} & \sigma_{ss} & \sigma_{ns} \end{bmatrix}^T$, $\varepsilon_{n,s} = \begin{bmatrix} \varepsilon_{nn} & \varepsilon_{ss} & \varepsilon_{ns} \end{bmatrix}^T$, and the secant stiffness D_{ns}^s is

$$D_{ns}^s = \begin{bmatrix} \dfrac{(1 - d_1)E}{1 - (1 - d_1)v^2} & \dfrac{(1 - d_1)vE}{1 - (1 - d_1)v^2} & 0 \\ \dfrac{(1 - d_1)vE}{1 - (1 - d_1)v^2} & \dfrac{E}{1 - (1 - d_1)v^2} & 0 \\ 0 & 0 & (1 - d_2)G \end{bmatrix}$$ [40]

where d_1 and d_2 are two damage parameters. $(1 - d_2)$ is the degradation of the shear stiffness G and can be related to the shear retention factor in traditional smeared crack models. The evolution of these two damage parameters is defined with the help of a loading function in the (n,s) coordinate system: $f(\varepsilon_{nn},\kappa) = \varepsilon_{nn} - \kappa$ and the appropriate Kuhn – Tucker conditions. d_1 and d_2 are functions of the history variable κ, same as in the other damage models. If we introduce ϕ as the angle between the (x,y) coordinate system and the (n,s) coordinate system, the strain and stress components in the two systems can be related as follow:

$$\varepsilon_{n,s} = T_\varepsilon(\phi)\varepsilon_{x,y} \text{ and } \sigma_{n,s} = T_\sigma(\phi)\sigma_{x,y} \qquad [41]$$

where $T_\varepsilon(\phi)$ and $T_\sigma(\phi)$ are the appropriate transformation matrices. The secant relation in Eq. [42] becomes:

$$\sigma_{x,y} = T_\sigma^{-1}(\phi)D_{ns}^s T_\varepsilon(\phi)\varepsilon_{x,y} \qquad [42]$$

This equation, and the loading function $f(\varepsilon_{nn},\kappa)$, incorporate the fixed smeared crack model and the rotating crack model as well. The difference is that in the fixed crack model the angle ϕ is constant whereas in the rotating crack model, the coaxiality requirement enforces that the n-direction is always the major principal strain direction.

As we can see, smeared crack models are indeed damage models, with two damage coefficients and some rule upon which the principal directions of damage may change. In the fixed crack model, the principal directions of damage are fixed; in the rotating crack model, they rotate according to the condition of coaxiality. There is a similarity between the above formulation of the smeared crack model and the microplane-based damage model described in section 3. The integral which defines the overall stress in the microplane model, is the same as in the model by Fichant [Eq. 30]. This integral can be transformed in order to arrive to a format that is very similar to the multiple fixed crack model:

$$\sigma_{ij} = \sum_n \omega^n \left(\left[\begin{array}{l} (1 - d(\bar{n}))E_N n_k \varepsilon_{kl} n_l n_i \\ +(1 - D(\bar{n}))E_T\left(\varepsilon_{ij}n_j - n_k\varepsilon_{kl}n_l n_i\right) \end{array} \right] \cdot n_j \right) \qquad [43]$$

where ω^n is a weighting factor.

4. Non Local Damage

We turn now attention to the non local generalisation of damage models. It is now established that non locality, in a gradient or integral format, is mandatory for a proper, consistent, modelling of fracture (see e.g. de Borst et al. 1993, or the contribution by M. Jirasek in this volume). It avoids the difficulties encountered upon material softening and strain localisation. Within a single approach, it encompasses both crack initiation (for which continuum models are very well fitted) and crack propagation (for which discrete fracture approaches have been developed).

4.1. Integral Model

Consider for instance the scalar damage model in which evolution of damage is controlled by the equivalent strain $\tilde{\varepsilon}$ introduced by Mazars. The principle of nonlocal continuum models with local strains is to replace the equivalent strain $\tilde{\varepsilon}$ with its average (Pijaudier-Cabot and Bazant, 1987):

$$\bar{\varepsilon}(x) = \frac{1}{V_r(x)} \int_\Omega \psi(s)\tilde{\varepsilon}(s+x)ds \text{ with } V_r(x) = \int_\Omega \psi(s)ds \qquad [44]$$

where Ω is the volume of the structure, $V_r(x)$ is the representative volume at point x, and $\psi(s)$ is the weight function, for instance:

$$\psi(s) = \exp\left(-\frac{4\|s\|^2}{l_c^2}\right) \qquad [45]$$

l_c is the internal length of the non local continuum. $\bar{\varepsilon}$ replaces the equivalent strain in the evolution of damage. In particular, the loading function becomes $f(\bar{\varepsilon},\kappa) = \bar{\varepsilon} - \kappa$. As we will see in section 5, it should be noticed that this model is easy to implement in the context of explicit, total strain models. Its extension to plasticity and to implicit incremental relations is awkward. The local tangent stiffness operator relating incremental strains to incremental stresses becomes non symmetric, and more importantly its bandwidth can be very large due to non local interactions. This is one of the reasons why gradient damage models have become popular over the past few years.

4.2. Gradient damage model

A simple method to transform the above non local model to a gradient model is to expand the equivalent strain into Taylor series truncated for instance to the second order:

$$\bar{\varepsilon}(x+s) = \tilde{\varepsilon}(x) + \frac{\partial\tilde{\varepsilon}(x)}{\partial x}s + \frac{\partial^2\tilde{\varepsilon}(x)}{\partial x^2}\frac{s^2}{2!} + \ldots \qquad [46]$$

Substitution in Eq. [44] and integration with respect to variable s yields:

$$\bar{\varepsilon}(x) = \tilde{\varepsilon}(x) + c^2\nabla^2\tilde{\varepsilon}(x) \qquad [47]$$

where c is a parameter which depends on the type of weight function in Eq. [47]. Its dimension is m^2 and it can be regarded as the square of an internal length. Substitution of the new expression of the non local equivalent strain in the non local damage model presented above yields a gradient damage model. Computationally, this model is still delicate to implement because it requires higher continuity in the interpolation of the displacement field. This difficulty can be solved if an implicit format of the gradient damage model is used. Eq. [47] is replaced with

$$\bar{\varepsilon}(x) - c^2 \nabla^2 \bar{\varepsilon}(x) = \tilde{\varepsilon}(x) \qquad [48]$$

Here, the definition of the non local equivalent strain is implicit. It is the solution of a Fredholm equation. As shown by Peerlings et al. (1996), the implicit form is in fact an exact representation of the integral relation devised by Pijaudier-Cabot and Bazant (1987), provided an exponential weight function is used.

4.3. Extension of the Gradient Approach

In the anisotropic damage models described in section 3, the evolution of damage is directional. The evolution of damage is also directional in the microplane-based models and in the smeared crack models. The extension of the gradient damage model to anisotropy, or to any strain-based damage model, can be performed in a very systematic way (Godard, 2001). The Fredholm equation is written for each strain component ε_{nn}:

$$\bar{\varepsilon}_{nn} - c^2 \nabla^2 (\bar{\varepsilon}_{nn}) = \varepsilon_{nn} \qquad [49]$$

Therefore, a non local (Gradient type) strain tensor is computed within each iteration. It is from this tensor that an equivalent strain can be computed, or that directional damage growth can be devised depending on the type of damage model that is implemented. This extended form of a gradient model has been implemented in the general purpose finite element code Code_Aster at EDF.

5. Some Computational Issues

5.1. Secant Stiffness Algorithm

Since the major effect of damage is a secant stiffness reduction when micro-cracking progresses, a non-linear computational procedure based on the secant material stiffness seems quite appropriate. Total displacement v.s. total force relations can be used at each load increment:

$$K^{sec} \vec{u} = \vec{f} \qquad [50]$$

K^{sec} is the secant stiffness matrix of the discrete solid which is analysed:

$$K^{sec} = \int_v (1-d)B^T C^0 B dv$$

[51]

with $\varepsilon = B\bar{u}$

where $\bar{u}$ is the vector of nodal displacements. During each iteration, a new displacement field is computed. Convergence is obtained when the residual forces are less than a maximum tolerance. These residual forces at iteration n are:

$$\delta \bar{f}_n = K_n^{sec} \bar{u}_n - \bar{f}, \quad \text{with} \quad K_n^{sec} = \int_v (1-d_n)B^T C^0 B dv$$

[52]

In most applications, the norm of the local residual forces at each node and the norm of the residual forces vector over the entire structure should be less than the maximum tolerance. These residuals are non dimensionalised (e.g. by dividing by the applied load vector).

This type of algorithm is also widely used when an integral non local damage model is implemented. The major reason is that the tangent stiffness operator (in the case of loading) is far from being straightforward because of non locality. It is possible to obtain an expression of the rate constitutive relations where we can see that the local tangent operator is non local, i.e. does not depend on the behaviour of a single material point:

$$\dot{\sigma}(x) = \left\{ (1-d_0)C^0 \dot{\varepsilon}(x) - \left(\frac{\partial h(\kappa)/\partial \kappa}{V_r(x)} C^0 : \varepsilon_0(x) \int_V \psi(s)\, \dot{\bar{\varepsilon}}(x+s)dv \right) \right\}$$

[53]

One central feature of this model is that the rate of damage and the rate of stress are explicit functions of the rate of strain. Compared to non local plasticity where the incremental strain is split into a plastic and an elastic part which are unknown in advance, the consistency condition is here an integral relation whose kernel is constant and known once for all in an integration step. This property makes the time integration of the constitutive relation more simple than in non local plasticity. It is also a reason why gradient plasticity (de Borst et al. 1993) is generally preferred to non local (integral) plasticity because the consistency condition in non local plasticity is an implicit integral relation which is complex to solve iteratively.

The integral relation due to the non local term is discretised according to the finite element mesh used for the analysis and a usual quadrature rule is employed for its evaluation. In finite element calculations, the weight function is chopped off: the weights that are less than 0.001 are set to zero. The actual volume of integration does not span over the entire volume of the solid and the calculation of the integrals requires less computer time and memory as the number of neighbouring integration points is reduced. Nevertheless, the bandwidth of the tangent stiffness matrix is much larger with the non local model than with the local model. Averaging expands the number of non zero terms simply because non local terms correspond to long range interactions. The tangent stiffness matrix of the finite element model is, however, non symmetric for two reasons sketched in Figure 5:

1. Non locality is an interaction that exists only when damage grows. Consider two points A and B sufficiently close to each other so that the non local interaction is strong. Damage grows at point A and remains constant at point B. The growth of damage in A is non local and depends on the incremental strain at point B. The behaviour of point B is local and does not depend on the incremental strain at point A. The influence functions between points A and B are not mutually equal. Therefore the corresponding terms placed symmetrically in the stiffness matrix are not equal.

2. The volume $V_r(x)$ in Eq. [53] is not constant. In particular, this quantity which normalises the averaging decreases near free boundaries. The influence of a point A located near a free surface on a point B located farther from the free surface is not the same as the influence of B on A because the averages are different.

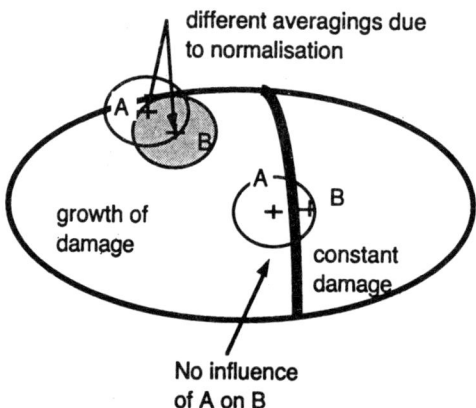

Figure 5. *Sources of non symmetry of the tangent stiffness matrix*

With the secant algorithm, the non local model is relatively easy to implement since the equilibrium equations remain standard. Their weak form is the same as for the classical, local, damage model.

It should be kept in mind, however, that the secant stiffness algorithm presents serious difficulties; mainly there is no proof of convergence of such an algorithm. In large scale finite element computations, a standard Newton – Raphson algorithm is highly preferable, as convergence is more robust. Such an algorithm relies on the derivation of the consistent tangent stiffness of the material response.

5.2. Consistent Tangent Stiffness and Convergence

In this section, we are going to focus on the scalar damage model, coupled to the gradient approach. The implementation of damage induced anisotropy following the same procedure still remains to be performed.

Consider again the discrete equations of equilibrium in a total displacement format:

$$\int_v (1 - d)B^T C^0 Bdv\bar{u} = \bar{F} \qquad [54]$$

where $\bar{u}$ and $\bar{F}$ are the unknown nodal displacement and external forces respectively. The tangent stiffness is formally an unknown in this equation as it depends on the displacements (through the unknown distribution of damage). The fundamental principle of the Newton-Raphson method is the construction of a series of approximation of the residual $R(\bar{u})$

$$R(\bar{u}) = \int_v (1 - d)B^T C^0 Bdv\bar{u} - \bar{F} \qquad [55]$$

which converges to zero for a finite number of iterations, and for a fixed value of the external loads. For this purpose, the residual at iteration $i+1$ is computed from that at iteration i according to a first order Taylor expansion:

$$R(\bar{u}_{i+1}) = R(\bar{u}_i) + \left(\frac{\partial R}{\partial \bar{u}}\right)_i \delta\bar{u}_i \qquad [56]$$

with
$$\bar{u}_{i+1} = \bar{u}_i + \delta\bar{u}_i \qquad [57]$$

The corrective term in the displacement field $\delta\bar{u}_i$ is evaluated as the solution of the system:

$$\int_v B^T \left(\frac{\partial \sigma}{\partial \varepsilon}\right)_i Bdv\delta\bar{u}_i = -R(\bar{u}_i) \qquad [58]$$

In this equation, the matrix that is on the left hand-side term is the consistent tangent stiffness matrix. Its evaluation requires an exact derivation of the constitutive relation. The advantage is that with this matrix, quadratic convergence of the algorithm is observed.

In the modified Newton-Raphson scheme, the tangent matrix is substituted with the elastic matrix:

$$\int_v B^T C^0 Bdv\delta\bar{u}_i = -R(\bar{u}_i)$$ [59]

The price to pay to this simplification is that convergence is not quadratic. It is linear. On the other side, the evaluation of this elastic matrix is less time consuming. Recall that the tangent stiffness matrix needs to be updated at each time step.

In the gradient damage model, the equilibrium equations and the Fredholm equations are solved as a coupled problem. The non local strain tensor can be discretised according to the same finite element grid as the displacements, with the same interpolation fuction N. Eq. [58] becomes:

$$\begin{bmatrix} K_{aa} & K_{ae} \\ K_{ea} & K_{ee} \end{bmatrix} \begin{pmatrix} \delta\bar{u}_i \\ \delta\bar{\varepsilon}_i \end{pmatrix} = -\left(\begin{array}{c} R(\bar{u}_i) \\ K_{ee}\bar{\varepsilon}_i - \int_v N^T \varepsilon_i dv \end{array} \right)$$ [60]

with the matrices K defined as

$$K_{aa} = \int_v (1-d)B^T C^0 Bdv, \quad K_{ae} = \int_v B^T \frac{\partial\sigma}{\partial\bar{\varepsilon}} Ndv$$

$$K_{ea} = \int_v -N^T Bdv, \quad K_{ee} = \int_v \left(c^2 B^T B + N^T N \right) dv$$ [61]

In order to avoid stress oscillations, the non local strain tensor can be also discretised with interpolation functions that are linear while the interpolation of displacements is quadratic. This is what has been implemented in the FE software Code_ASTER. The iterations are carried out until the relative residual forces are less than a given tolerance:

$$\left\| \frac{R(\bar{u}_i)}{\bar{F}} \right\| \le Tolerance$$ [62]

The difficulty in this algorithm is the proper calculation of the tangent stiffness at the material level. Very often, small errors, or approximations, in the derivation result in a loss of the quadratic convergence of the Newton-Raphson scheme. In particular cases such as the damage model proposed by Mazars and presented Eqs. [20, 22-27], the tangent operator is very difficult to derive properly. The reason is that the differenciation of the factors α_t, α_c is very uneasy. In the calculations presented in this paper, the variation of these coefficients is neglected. Hence quadratic convergence is expected for radial loading only (constant α_t, α_c).

Figure 6 shows the evolution of the relative residual during the iterations for several initial states of damage. The computation corresponds to uniaxial compression. The Newton-Raphson scheme [Eq. 58] and the modified scheme Eq. [59] have been compared on this plot. Quadratic convergence yields an evolution of the residual which follows a parabola approximately whereas linear convergence results into a straight line on this plot. It is clear that the modified scheme yields a slowest convergence, resulting into a larger number of iteration for the same tolerance.

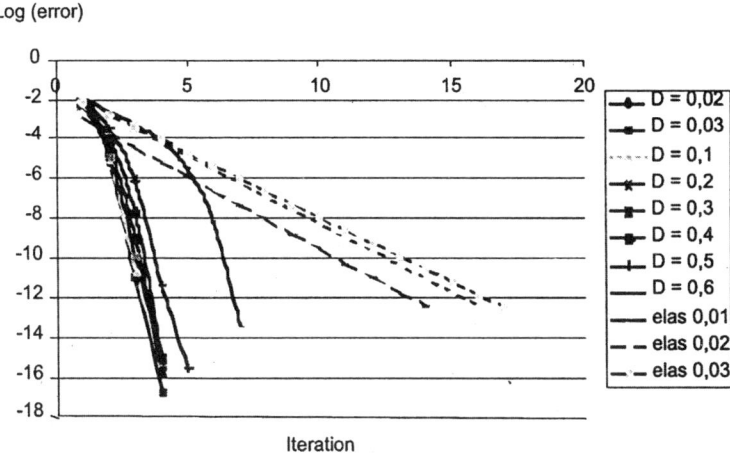

Figure 6. *Evolution of the relative residuals for the Newton-Raphson and modified Newton-Raphson schemes*

5.3. Solution Control

Upon localisation of damage and strain, i.e. in the course of the failure process, uniqueness of the equilibrium equations can be lost. Bifurcation can occur in the continuum solution and should be expected for the discrete approximation as well, and stability might be lost too. Therefore, finite element codes must be equipped with solution checkings capable of detecting loss of uniqueness and loss of stability. If bifurcation occurs, the solution should be controlled so that it follows the stable path. If stability is lost, there should be the possibility for switching to indirect displacement control so that the finite element solution can be traced during instability. In the following we shall consider the rate equation of equilibrium of the discrete problem written generically as:

$$\int_v B^T .\dot\sigma dv = K\dot{\bar u} = \dot{\bar F} \qquad [63]$$

where K is the tangent stiffness matrix of the discrete solid. Uniqueness of the solution can be assessed by considering the equations of equilibrium, Eq. [63]. For the sake of simplicity, attention is restricted to the case where the loading is not stationary as bifurcation points and limit points rarely coincide. If the solution of the rate boundary value problem is non unique, the tangent stiffness matrix should be singular :

$$\det(K) = 0 \qquad [64]$$

However, K is not a single valued matrix and depends on the loading conditions. Rigorously, all the possible (loading-unloading) combinations should be investigated. In most applications, K is calculated for the incrementally linear comparison solid only. It means that, locally, the tangent modulus computed at each material point correponds to loading if the consistency condition is met at the time step of the calculation considered. Our experience shows that the loss of uniqueness for the comparison solid occurs before it can be detected for any other loading-unloading configurations.

At a state of equilibrium under dead load, the equilibrium of a discrete system is critical if there exists a kinematically admissible field $\dot{u}$ such that:

$$\dot{u} K \dot{u} = 0 \qquad [65]$$

Again K is assumed to be single valued. This condition is equivalent to:

$$\begin{cases} \det(K) = 0 \\ \dot{u} \text{ orthogonal to } K\dot{u} \end{cases} \qquad [66]$$

The loss of stability does not necessarily occur at a bifurcation point because the tangent stiffness operator is not symmetric, and theoretically it may be possible to find a velocity field such that the second condition in Eq. [66] is satisfied. However it is easy to show (Pijaudier-Cabot and Huerta, 1991) that Eq. [66] is equivalent to:

$$\det(S) = 0 \qquad [67]$$

where S is the symmetric part of K.

During the calculation and once the conditions for stability or uniqueness are not satisfied, it is necessary to investigate whether there exists a stable response of the structure and what are the different possible paths in the post-bifurcation regime. As the loading progresses, the loss of stability is encountered first. If the criterion for uniqueness is not met, i.e. if K is non singular, then the solution is unique and the equilibrium of the structure is critical if the vanishing eigenvector of S is collinear to the solution of the problem Eq. [63].

If this is the case, load control or displacement control techniques fail to trace the snap-back portion of the response curve of the structure, and other algorithms should be implemented to circumvent the singularities beyond critical points. In fact, most of the continuation techniques are based on the introduction of another variable which governs the load factor and induces an extra-equation. Some of the available methods are indirect displacement control (de Borst, 1986) or the well known arc-length control method (Riks, 1979; Crisfield, 1981).

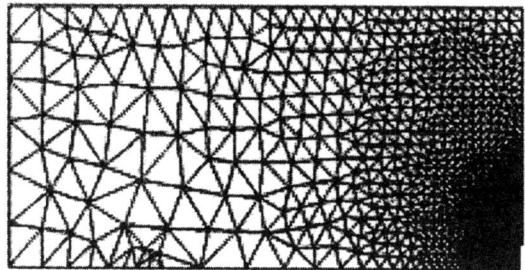

Figure 7. *Finite element mesh for three point bending beams. The load is applied at the top-right corner (axis of symmetry) and vertical displacements are fixed at the bottom left corner*

As an example, let us consider a notched bending beam. Figure 7 shows the finite element mesh (half of it due to the symmetry) that has been implemented. Figure 8 shows a comparison of the responses computed for the non local gradient damage model and for the non local integral damage model. In both cases, an indirect control displacement procedure is needed. Indeed, a displacement controlled computation would yield an unstable response due to snap-back.

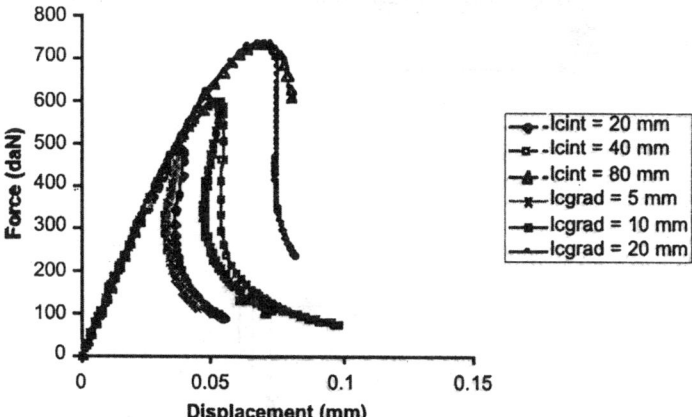

Figure 8. *Non Local computations of notched three point bending beams with the gradient (lcgrad) and integral (lcint) models for several internal lengths*

Finally, the case where the criterion for uniqueness is not met must be investigated. Again we know that there is at least one solution denoted as $\dot{\vec{u}}^*$ to the rate equilibrium problem. Let us call this solution the fundamental solution. The objective now is to perturb the fundamental solution adequately in order to obtain the solution over the stable path. This perturbation is realised by means of the right hand eigenvector $\vec{v}$ associated to the vanishing eigenvalue of the tangent stiffness matrix K because it can be demonstrated (de Borst, 1986 and 1988) that for any arbitrary scalar β, the velocity field $\dot{\vec{u}}^* + \beta\vec{v}$ is also a solution of the problem:

$$K(\dot{\vec{u}}^* + \beta\vec{v}) = \dot{\vec{F}} \qquad [68]$$

For more details on this specific subject, see Pijaudier-Cabot and Huerta (1991) and de Borst (1988).

6. Conclusion - Example of Computation

As a conclusion, the 3D computation of a reinforced concrete bending beam is presented in this section. This test case belongs to the benchmark study that has been carried out by EDF (Ghavamian, 1999).

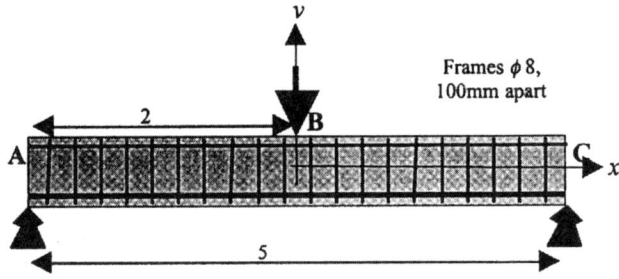

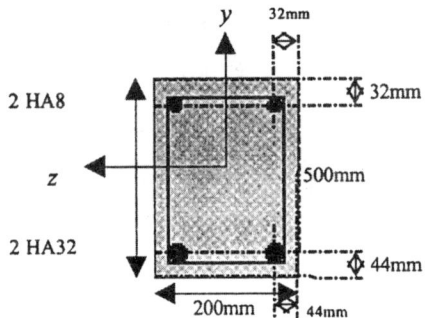

Figure 9. *Three point bending beam*

The beam and the load system are shown in Figure 9. The computation has been carried out with a scalar damage model which derives from the damage induced anisotropic model presented in section 3. The equations are the same, except that the interpolation of damage is carried out over a single direction instead of three mutually orhtogonal ones. The finite element mesh is made of brick elements for concrete and elasto-plastic bar elements for the longitudinal reinforcements and for the stirrups. Figure 10 shows the load deflection curve.

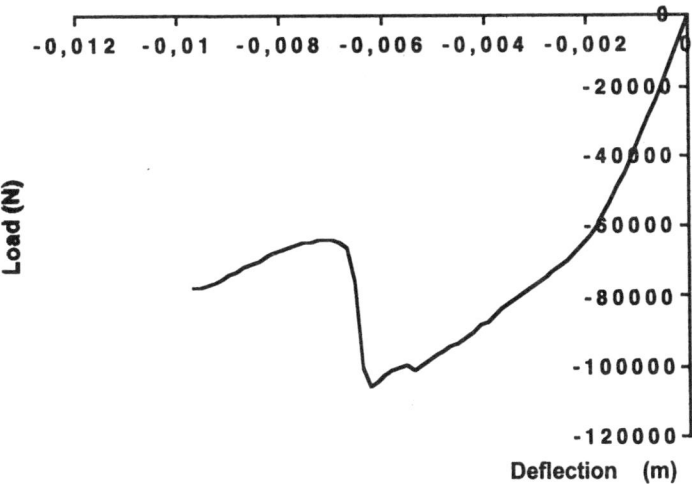

Figure 10. *Load deflection curve*

Figures 11 and 12 show the distribution of damage in two cross sections of the beam. The first one is longitudinal and the second one is a transverse cross section at mid-span.

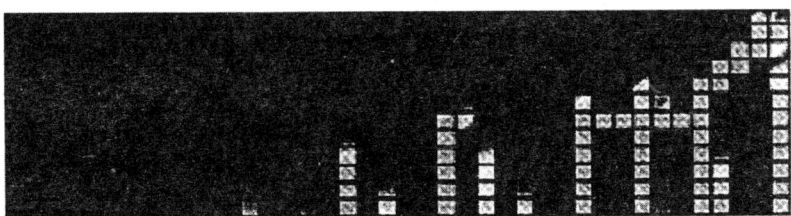

Figure 11. *Distribution of damage in a longitudinal cross section. The load is applied at the top-right corner (axis of symmetry) and vertical displacements are fixed at the bottom left corner*

In these figures, the dark grey colour corresponds to zero damage and the white colour corresponds to complete damage. Overall, the distribution of damage is very much in agreement with experiments, which show regularly spaced cracking.

Figure 12. *Distribution of damage in a transverse cross section at mid-span*

As seen in these figures, 3D effects are captured in addition to the map of regularly spaced vertical cracks and diagonal cracks. The computation has been stopped when the applied vertical displacement reached –1 cm.

It is on this type of 3D finite element computations, which correspond to major industrial applications, that several issues involving computational damage are faced. One can mention the pertinence of the constitutive relation, of course, but also with respect to "non mechanical" issues involved in safety analyses (occurrence of leaks, ageing, …). The robustness of the finite element implementation is another one, in which the simple question "is a solution expected from the finite element calculation?" is far from being trivial in large scale analyses. Finally, the quality "at large" of finite element simulation of failure still remains an outstanding issue which should be addressed if this type of modelling is intended to be more widely used in engineering practice.

Acknowledgements
Partnership of Electricité de France with R&DO is gratefully acknowledged. The authors would like to thank Prof. A. Huerta (UPC Barcelona, Spain) for helpful discussions about the implementation of the gradient non local damage model.

7. References

Crisfield M.A. (1981), "A Fast Incremental / Iterative Solution Procedure that Handles Snap-Through", *Comp. and Struct.*, Vol. 13, pp. 55-62.

de Borst R., Sluys L.J., Muhlhaus H.B., and Pamin J. (1993), "Fundamental Issues in Finite Element Analyses of Straain Localisation", *Engrg. Comput.*, Vol. 10, pp. 99-121.

de Borst R. (1988), "Bifurcation in Finite Element Models with Non-Associated Flow Law", *Int. J. for Num. and Analytical Meth. in Geomechanics*, Vol. 12, pp. 99-116.

de Borst R. (1986), Non Linear Analysis of Frictional Materials, Dissertation, TU-Delft.

de Borst R. and Gutierrez M. (1999), "A unified framework for concrete damage and fracture models including size effects", *Int. J. Fracture*, Vol. 95, pp. 261-277.

Fichant S., La Borderie C., and Pijaudier-Cabot G. (1999), "Isotropic and Anisotropic Descriptions of Damage in Concrete Structures", *Int. J. Mechanics of Cohesive Frictional Materials*, Vol. 4, pp. 339-359.

Ghavamian S. (1999), Evaluation tests on models of non-linear behaviour of cracking concrete, using three dimensional modelling, Benchmark EDF/Division R&D, CR – MMN 99/232.

Godard V. (2001), Modélisation mécanique non locale des matériaux fragiles : délocalisation de la déformation, Stage de DEA de Mécanique, Université Pierre et Marie Curie.

La Borderie C. (1991), Phénomènes unilatéraux dans un matériau endommageable, Thèse de Doctorat de l'Université Paris 6.

Lemaitre J. (1992), *A Course on Damage Mechanics*, Springer-Verlag.

Mazars J. (1984), Application de la mécanique de l'endommagement au comportement non linéaire et à la rupture du béton de structure, Thèse de Doctorat ès Sciences, Université Paris 6.

Nadai A. (1950), *Theory of Flow and Fracture of Solids*, Vol. 1, second edition, Mc Graw Hill, New York, p. 572.

Peerlings R.H., de Borst R., Brekelmans W.A.M., de Vree J.H.P. (1996), "Gradient enhanced damage for quasi-brittle materials", *Int. J. Num. Meth. Engrg.*, Vol. 39, pp. 3391-3403.

Peerlings R.H., de Borst R., Brekelmans W.A.M, Geers M.G.D. (1998), "Gradient enhanced modelling of concrete fracture", *Int. J. Mechanics of Cohesive Frictional Materials*, Vol. 3, pp. 323-343.

Pijaudier-Cabot G. and Bazant Z.P. (1987), "Nonlocal Damage Theory", *J. of Engrg. Mech., ASCE*, Vol. 113, pp.1512-1533.

Pijaudier-Cabot G., Mazars J., and Pulikowski J. (1991), "Steel-Concrete Bond Analysis with Nonlocal Continuous Damage", *J. of Structural Engineering, ASCE*, Vol. 117, No.3, pp. 862-882.

Pijaudier-Cabot G., and Huerta A. (1991), "Finite Element Analysis of Bifurcation in Nonlocal Strain Softening Solids", *Comp. Meth. in Applied Mech. and Engrg.*, Vol. 90, pp. 905-919.

Riks E. (1979), "An Incremental Approach to the Solution of Snapping and Buckling Problems", *Int. J. Solids and Struct.*, Vol. 15, pp. 529-551.

Chapter 6

Non Linear Problems: An Introduction

Pablo Mira and Manuel Pastor

Centro de Estudios y Experimentación de Obras Públicas, CEDEX, Ministerio de Fomento, Madrid, Spain, and ETS Ingenieros de Caminos, Universidad Politécnica de Madrid, Spain

1. Introduction

Solid mechanics problems in general, and specifically in the case of geomaterials, are very often significantly non linear. Although linear numerical models may be used as rough approximations to the problem solution, non linearities are very often too significant to be neglected. Under those circumstances it is therefore difficult to remain satisfied with a solution based on a linear model. To solve these problems special non linear strategies are necessary. Since different type of non linearities require different strategies and there is no such thing as a generally valid non linear algorithm it is advisable to include several types of non linear algorithms in the model covering the most usual type of non linear situations. There are several very good texts that cover this topic extensively, such as Bathe [BAT 96], Belytschko, Liu and Moran [BEL 00], Crisfield [CRI 91] and [CRI 97], Simo and Hughes [SIM 98]and Zienkiewicz and Taylor [ZIE 89].

The non linearities in a problem may have different causes. A first attempt to classify these causes would divide them into two big groups : geometrical and material non linearities. Geometrical non linearities arise when some of the geometrical parameters of the model change during the loading process. A typical example of this would be a buckling problem where small displacement theory ceases to be valid and large displacement theory is required. Another example of geometrical non linearity is a contact problem where boundary conditions change during the loading process. On the other hand material non linearities arise when some of the constitutive parameters of the model change during the loading process. Typical examples of this include non linear elastic, elastoplastic or viscoplastic material behaviour, damage models, etc.

Numerical solution of non linear problems requires as expected many more operations than linear problems. Special numerical techniques are necessary not only to reach a solution but also to do so in an efficient manner.

Traditionally, non linear finite element problems have been solved using the Newton-Raphson method in any of its different versions. It is therefore the basic tool that virtually every non linear finite element program should include. However, there are many different ways to implement this algorithm and often the "typical" or "standard" versions of the Newton-Raphson method are not sufficient to solve certain problems. Although which techniques may be referred to as "standard" and which may not is a subjective matter an attempt to do so will be made in this book. The present chapter will be devoted to so called standard techniques such as the following versions of the Newton-Raphson method: load control, displacement control, modified Newton Raphson methods and Quasi Newton methods. The next chapter will be devoted to advanced or non standard techniques such as arc-length control and line searches.

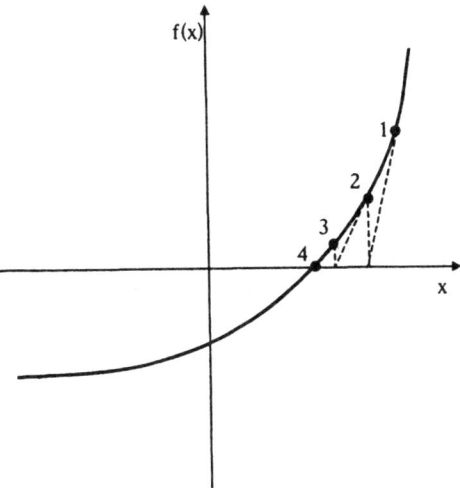

Figure 1. *One dimensional Newton-Raphson method*

2. The Newton-Raphson method

The Newton-Raphson method, also called Newton's method, is a classical one dimensional root-finding algorithm [PRE 92]. Assuming the problem is stated as finding the value or values of x such that the following equation is satisfied:

$$f(x) = 0$$

the main distinguishing feature of this method from other root-finding methods is the fact that it requires the evaluation of both the function $f(x)$ and the derivative $f'(x)$ in each iteration. The method consists in geometrical terms of extending the tangent line at a current point x_i until it crosses zero, then setting the next guess x_{i+1} to the abscissa of that zero crossing as sketched in Figure 1:

$$x_{i+1} = x_i - \frac{f(x_i)}{f'(x_i)} \tag{1}$$

Algebraically the method derives from the Taylor expansion series in the neighborhood of a point:

$$f(x + \delta) = f(x) + f'(x)\delta + \frac{1}{2}f''(x)\delta^2 + \ldots\ldots \tag{2}$$

Given x_i such that $f(x_i) \neq 0$ and using a two term expansion based on (2), a value $x_{i+1} = x_i + \delta$ is looked for such that:

$$f(x_i + \delta) = f(x_i) + f'(x_i)\delta = 0 \tag{3}$$

Solving for δ:

$$\delta = -\frac{f(x_i)}{f'(x_i)}$$

Correcting x_i with the value δ of to obtain x_{i+1} would lead to expression (1).

It is also possible to use this type of analysis to obtain an estimate of the iterative error ε. Expressing x_{i+1} x_i in terms of the root x and iterative errors ε_{i+1} and ε_i:

$$x_{i+1} = x + \varepsilon_{i+1}$$

$$x_i = x + \varepsilon_i$$

and substituting these expressions into equation (1) a relationship between consecutive iterative errors is obtained:

$$\varepsilon_{i+1} = \varepsilon_i - \frac{f(x_i)}{f'(x_i)} \tag{4}$$

Estimating $f(x_i)$ and $f'(x_i)$ through Taylor series expansions in the neighbourhood of the root x and taking into account that $f(x) = 0$ we obtain:

$$f(x_i) = f(x + \varepsilon_i) = f(x) + f'(x)\varepsilon_i + \frac{1}{2}f''(x)\varepsilon_i^2 + \ldots\ldots \simeq f'(x)\varepsilon_i + \frac{1}{2}f''(x)\varepsilon_i^2$$

$$f'(x_i) = f'(x + \varepsilon_i) = f'(x) + f''(x)\varepsilon_i + \ldots\ldots \simeq f'(x) + f''(x)\varepsilon_i$$

Substituting these expressions into equation (1):

$$
\begin{aligned}
\varepsilon_{i+1} &= \varepsilon_i - \frac{f(x_i)}{f'(x_i)} = \varepsilon_i - \frac{f'(x)\varepsilon_i + \frac{1}{2}f''(x)\varepsilon_i^2}{f'(x) + f''(x)\varepsilon_i} \\
&= \frac{\varepsilon_i(f'(x) + f''(x)\varepsilon_i) - f'(x)\varepsilon_i - \frac{1}{2}f''(x)\varepsilon_i^2}{f'(x) + f''(x)\varepsilon_i} \\
&= -\frac{f''(x)\varepsilon_i^2}{2(f'(x) + f''(x)\varepsilon_i)} \simeq -\varepsilon_i^2 \frac{f''(x)}{2f'(x)}
\end{aligned}
\tag{5}
$$

Equation (5) is a recursive expression between consecutive iterative errors saying that the "new" iterative error ε_{i+1} is proportional to the square of the "old" one. In other words, the Newton-Raphson procedure converges *quadratically*. This is a very powerful feature of this procedure, but in order to preserve it we have to make sure we use a good estimate of $f'(x_i)$. Frequently, rough estimates of $f'(x_i)$ are used and quadratic convergence is lost.

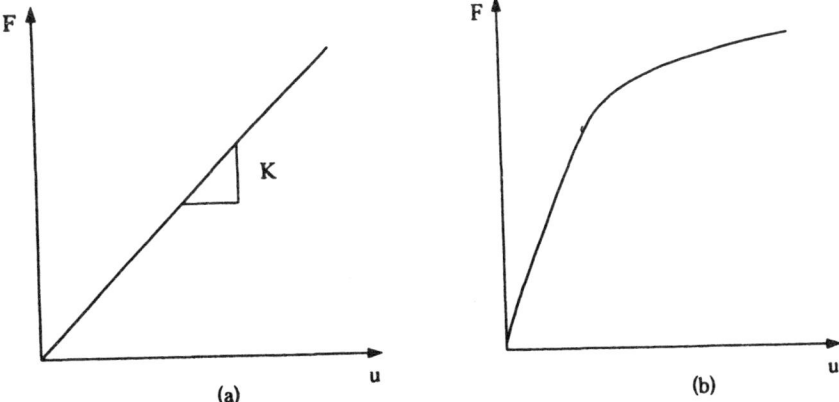

Figure 2. *(a) Linear problem   (b) Non linear problem*

2.1. *The Newton-Raphson method in a finite element context*

The use of the Newton-Raphson procedure is not limited to one dimensional problems as the one presented at the beginning of this section. The procedure is perfectly applicable to multidimensional problems such as the ones arising from the application of the finite element method.

Let us consider a solid mechanics problem defined in a domain Ω to be solved through a displacement formulation of the finite element method based on small deformation theory. Let us assume also that the problem is non linear, that is, the relationship between force and displacement is, in graphical terms of the type sketched in Figure 2.b as opposed to a linear relationship of the type presented in Figure 2.a. In Figure 2.b, as the loading process advances and additional gauss points in different elements enter the non linear range the curve becomes less and less steep. In more rigorous and mathematical terms, the objective of the problem could be stated as finding a displacement vector u such that the following set of equations is satisfied:

$$A_{e=1,nelem}\left[\int_{\Omega e} \mathbf{B}^T\sigma(\epsilon(\mathbf{u}))d\Omega_e\right] -\mathbf{f}^{ext} = 0 \tag{6}$$

where :

$A_{e=1,nelem}=$ Assembly operator to obtain global variables from element variables.

$\mathbf{f}^{ext} =$ Global external force vector

$\int_{\Omega e} \mathbf{B}^T\sigma(\epsilon(\mathbf{u}))d\Omega_e =$ Element internal force vector

$\mathbf{B}$ = Matrix relating deformations and displacement in the following way: $\epsilon = \mathbf{Bu}$

σ = Stress vector. The relationship between σ and ϵ is non linear

Equation (6) may be also expressed in a more compact form as :

$$\Psi(\mathbf{u}) = 0$$

with $\Psi(\mathbf{u})$ = Residual force vector = $A_{e=1,nelem}\left[\int_{\Omega_e} \mathbf{B}^T\sigma(\epsilon(\mathbf{u}))d\Omega_e\right] - \mathbf{f}^{ext}$

The problem, as stated is a non linear problem in u and as such, an iterative procedure should be used to solve it. Using the Newton-Raphson to do so a displacement state $\mathbf{u}_o$ may be assumed to exist such that $\Psi_o = \Psi(\mathbf{u}_o) \neq 0$. A new displacement state $\mathbf{u}_n = \mathbf{u}_o + \delta\mathbf{u}$ is looked for such that $\Psi_n = \Psi(\mathbf{u}_n) = 0$. Using a two term Taylor expansion as in the one dimensional case:

$$\Psi_n = \Psi_o + \frac{\partial\Psi}{\partial\mathbf{u}}\delta\mathbf{u} = \Psi_o + \mathbf{K}_t\delta\mathbf{u} = 0 \tag{7}$$

where :

$$
\begin{aligned}
\frac{\partial\Psi}{\partial\mathbf{u}} &= \frac{\partial(A_{e=1,nelem}\left[\int_{\Omega_e}\mathbf{B}^T\sigma(\epsilon)d\Omega_e\right])}{\partial\mathbf{u}} = A_{e=1,nelem}\left[\int_{\Omega_e}\mathbf{B}^T\frac{\partial\sigma(\epsilon)}{\partial\mathbf{u}}d\Omega_e\right] \\
&= A_{e=1,nelem}\left[\int_{\Omega_e}\mathbf{B}^T\frac{\partial\sigma(\epsilon)}{\partial\epsilon}\frac{\partial\epsilon}{\partial\mathbf{u}}d\Omega_e\right] = A_{e=1,nelem}\left[\int_{\Omega_e}\mathbf{B}^T\mathbf{D}^{ep}\mathbf{B}d\Omega_e\right] = \mathbf{K}_t
\end{aligned}
$$

Expression (7) would lead to the following linear equation set:

$$\mathbf{K}_t\delta\mathbf{u} = -\Psi_o$$

Solving for $\delta\mathbf{u}$:

$$\delta\mathbf{u} = -\mathbf{K}_t^{-1}\Psi_o$$

In this format the problem is load controled, that is external forces $\mathbf{f}^{ext}$ are fixed and displacements u are the unknowns. In this context, the Newton-Raphson procedure is usually applied in an incremental fashion, that is, the load vector $\mathbf{f}^{ext}$ is divided into increments. The iterative procedure sketched in Figure 3 is applied at each increment, in the following way:

1) Let us assume a displacement state $\mathbf{u}_A$ exists such that $\Psi_A = 0$.This means the internal forces $\mathbf{f}_A^{int}$ are in equilibrium with external forces $\mathbf{f}_A^{ext}$. Presumably

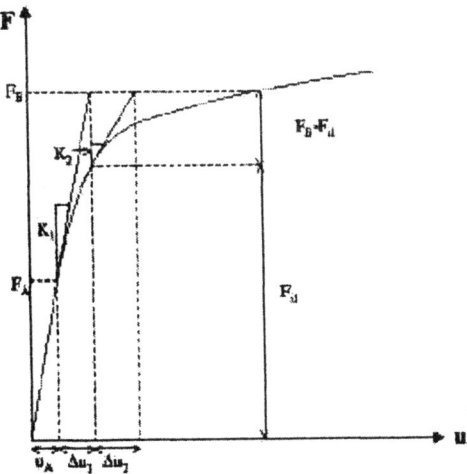

Figure 3. *The Newton-Raphson method*

this state was reached by a Newton-Raphson procedure which converged after several iterations. The convergence condition was:

$$\frac{\|\Psi_A\|}{\|f_A^{ext}\|} = \frac{\|f_A^{ext} - f_A^{int}\|}{\|f_A^{ext}\|} < tol \tag{8}$$

where $\|\|$ is the vector norm operator and *tol* is a tolerance factor previously defined by the analyst.

2)   We are now looking for displacements u_B associated to a load vector f_B^{ext}. The next load increment to be applied is therefore $f_B^{ext} - f_A^{ext}$. Since we know u_A we are really looking for a displacement corrector Δu such that:

$$u_B = u_A + \Delta u$$

Our first predictor for Δu will come from solving the following linear equation system:

$$f_B^{ext} - f_A^{ext} = K_1 \delta u_1$$

where K_1 is the stiffness matrix at A, that is $K_1 = K_A$

3)   Once δu_1 has been obtained, and the displacement vector updated $u_1 = u_A + \delta u_1$ the internal force vector $f_1^{int}(u_1)$ is evaluated. To do so stresses at all Gauss points will have to be integrated to compute the following integral:

$$f_1^{int}(u_1) = \int B^T \sigma(\epsilon(u_1)) d\Omega$$

4)   The next step is to evaluate the residual force vector $\mathbf{G}_1$ to check whether the convergence test is satisfied or not:

$$\frac{\|\mathbf{\Psi}_1\|}{\|\mathbf{f}_B^{ext}\|} = \frac{\left\|\mathbf{f}_B^{ext} - \mathbf{f}_1^{int}\right\|}{\|\mathbf{f}_B^{ext}\|} < tol \tag{9}$$

If the previous condition is satisfied the iterative process ends at this point.

5)   If the convergence test is not satisfied a new stiffness matrix $\mathbf{K}_2$ will have to be obtained based on the new displacement state $\mathbf{u}_1$ to solve the following system of equations:

$$-\mathbf{\Psi}_1 = \mathbf{K}_2\delta\mathbf{u}_2$$

In general $\mathbf{K}_2$ will be different from $\mathbf{K}_1$ since most likely there will be Gauss points that were elastic for $\mathbf{u}_A$ but have become plastic for $\mathbf{u}_1$. The iterative procedure will continue until at the ith iteration $\mathbf{f}_i^{int}$ is evaluated and the convergence condition $\frac{\left\|\mathbf{f}_B^{ext} - \mathbf{f}_i^{int}\right\|}{\|\mathbf{f}_B^{ext}\|}$ satisfied.

Norm operators used to evaluate the convergence condition can be defined in different ways. Typically the square root of the sum of squares is used:

$$\|\mathbf{v}\| = \sqrt{v_i^2}$$

but other definitions may be used such as:

$$\|\mathbf{v}\| = Max(v_i)$$

In the present version of the Newton-Raphson algorithm the convergence criterion was based on the norm of the residual force vector, but other criterions may be used. One of the most frequently used criteria, apart from the residual force one, is the iterative displacement correction:

$$\frac{\|\delta\mathbf{u}_i\|}{\|\Delta\mathbf{u}_i\|} < tol$$

where $\delta\mathbf{u}_i$ is the iterative displacement correction and $\Delta\mathbf{u}_i$ is the incremental displacement correction, both corresponding to the ith iteration:

$$\Delta\mathbf{u}_i = \Delta\mathbf{u}_{i-1} + \delta\mathbf{u}_i$$

$$\mathbf{u}_i = \mathbf{u}_A + \Delta\mathbf{u}_i$$

Finally, another frequently used convergence criterion is the residual energy:

$$\frac{\Delta\mathbf{u}_i \cdot \mathbf{\Psi}_i}{\Delta\mathbf{u}_i \cdot \mathbf{f}^{ext}} < tol$$

As in the previous criterion the $_i$ subindex refers to the ith iteration

Although the most usual criterion is the residual force, it is advisable to use an additional criterion, typically the iterative displacement, and stop the iterative process when both criteria are satisfied.

The version of the Newton-Raphson method presented here corresponded to a single field static problem. The method can of course be applied in a more general context to multifield or time dependent problems. For example non linear equations a general problem formulated in displacements and pressures would be stated as:

$$\Psi(\mathbf{u}, \mathbf{p}) = \left\{ \begin{array}{c} \Psi_u(\mathbf{u}, \mathbf{p}) \\ \Psi_p(\mathbf{u}, \mathbf{p}) \end{array} \right\} = 0$$

or in a more compact manner:

$$\Psi(\mathbf{x}) = 0$$

with:

$$\mathbf{x} = \left\{ \begin{array}{c} \mathbf{u} \\ \mathbf{p} \end{array} \right\}$$

The general Newton-Raphson equation would now be:

$$\Psi_n = \Psi_o + \frac{\partial \Psi}{\partial \mathbf{x}} \delta \mathbf{x} = \Psi_o + \mathbf{J} \delta \mathbf{x} = 0 \tag{10}$$

with:

$$\delta \mathbf{x} = \left\{ \begin{array}{c} \delta \mathbf{u} \\ \delta \mathbf{p} \end{array} \right\} \qquad \mathbf{J} = \left[\begin{array}{cc} \frac{\partial \Psi_u}{\partial \mathbf{u}} & \frac{\partial \Psi_u}{\partial \mathbf{p}} \\ \frac{\partial \Psi_p}{\partial \mathbf{u}} & \frac{\partial \Psi_p}{\partial \mathbf{p}} \end{array} \right]$$

If the problem were both mixed and time dependent, with first time derivatives, we would have:

$$\Psi_{n+1}(\mathbf{u}_{n+1}, \dot{\mathbf{u}}_{n+1}, \mathbf{p}_{n+1}, \dot{\mathbf{p}}_{n+1}) = \left\{ \begin{array}{c} \Psi_u(\mathbf{u}_{n+1}, \dot{\mathbf{u}}_{n+1}, \mathbf{p}_{n+1}, \dot{\mathbf{p}}_{n+1}) \\ \Psi_p(\mathbf{u}_{n+1}, \dot{\mathbf{u}}_{n+1}, \mathbf{p}_{n+1}, \dot{\mathbf{p}}_{n+1}) \end{array} \right\} = 0$$

Assuming a time integration scheme of the following type:

$$\mathbf{u}_{n+1} = \mathbf{u}_{n+1}(\mathbf{u}_n, \dot{\mathbf{u}}_n, \Delta\dot{\mathbf{u}}_n)$$
$$\dot{\mathbf{u}}_{n+1} = \dot{\mathbf{u}}_{n+1}(\dot{\mathbf{u}}_n, \Delta\dot{\mathbf{u}}_n)$$
$$\mathbf{p}_{n+1} = \mathbf{p}_{n+1}(\mathbf{p}_n, \dot{\mathbf{p}}_n, \Delta\dot{\mathbf{p}}_n)$$
$$\dot{\mathbf{p}}_{n+1} = \dot{\mathbf{p}}_{n+1}(\dot{\mathbf{p}}_n, \Delta\dot{\mathbf{p}}_n)$$

the Newton-Raphson equation would now be:

$$\Psi_{n+1,n} = \Psi_{n+1,o} + \frac{\partial \Psi}{\partial \mathbf{x}} \delta \mathbf{x} = \Psi_o + \mathbf{J} \delta \mathbf{x} = 0 \tag{11}$$

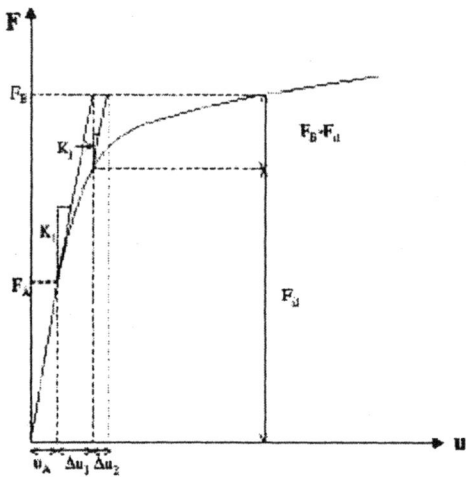

Figure 4. *Modified Newton-Raphson method*

with:

$$\delta \mathbf{x} = \left\{ \begin{array}{c} \delta(\Delta \dot{\mathbf{u}}_n) \\ \delta(\Delta \dot{\mathbf{p}}_n) \end{array} \right\} \qquad \mathbf{J} = \left[\begin{array}{cc} \frac{\partial \Psi_u}{\delta(\Delta \dot{\mathbf{u}}_n)} & \frac{\partial \Psi_u}{\delta(\Delta \dot{\mathbf{p}}_n)} \\ \frac{\partial \Psi_p}{\delta(\Delta \dot{\mathbf{u}}_n)} & \frac{\partial \Psi_p}{\delta(\Delta \dot{\mathbf{p}}_n)} \end{array} \right]$$

2.2. Modified Newton-Raphson

An alternative version of the method presented in the previous section is the method known as modified Newton-Raphson. The difference between this version and the previous one is that the stiffness matrix is only updated at the beginning of each increment or at least much less frequently than once every iteration. A single update per increment would imply in graphical terms that tangents for consecutive iterations are parallel as seen in the sketch of the algorithm presented in Figure 4. This updating strategy saves many numerical operations both at the element and global levels. Assuming a classical Gauss LU decomposition solver is being used a full Newton-Raphson would require a decomposition and a backsubstitution every iteration while the modified version would only require a single decomposition at the beginning of each increment and a back substitution at every iteration. This refers only to global level operations. If in addition to these, we consider the operations required to update the element stiffness matrices the computational cost per iteration is reduced very significantly.

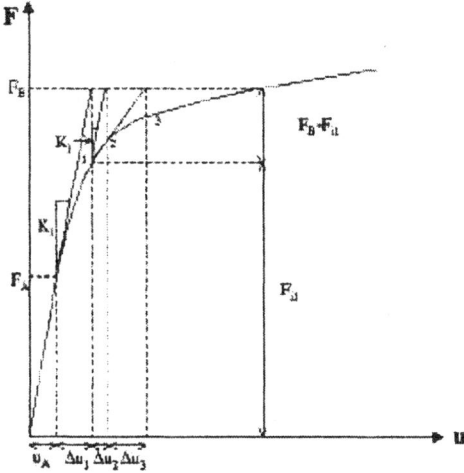

Figure 5. *Quasi Newton methods*

On the other hand, assuming the tolerance factor is the same, the convergence speed would be slower, so the number of iterations required to converge for the modified Newton-Raphson is higher than for the full version.

2.3. *Quasi Newton methods*

Another very well known family of algorithms arising from the Newton-Raphson method is that of the secant or Quasi-Newton methods. In this type of algorithm the displacement of each iteration is obtained based on a secant approximation of the displacements from the two previous iterations as seen in Figure 5. From the algorithmic point of view this implies that the new stiffness matrix should satisfy the so called secant condition:

$$\Psi(u_i) - \Psi(u_{i-1}) = K_i(u_i - u_{i-1}) \tag{12}$$

or in a more compact manner:

$$r_i = K_i \delta_i$$

where:

$$r_i = \Psi(u_i) - \Psi(u_{i-1}) \qquad \delta_i = u_i - u_{i-1}$$

The stiffness matrix of each iteration is computed in a recursive fashion from the previous iteration's stiffness matrix. There are many ways to do this. The simplest way is through a rank one update such as:

$$\mathbf{K}_i = \mathbf{K}_{i-1} + \frac{(\mathbf{r}_i - \mathbf{K}_{i-1}\delta_i)\mathbf{v}^T}{\mathbf{v}^T\delta_i} \qquad (13)$$

where $\mathbf{v}^T\delta_i \neq 0$. It is straightforward to check that expression (13) satisfies the secant condition (12). A first version of this algorithm is due to Broyden [BRO 65] taking $\mathbf{v} = \delta_i$ thus obtaining:

$$\mathbf{K}_i = \mathbf{K}_{i-1} + \frac{(\mathbf{r}_i - \mathbf{K}_{i-1}\delta_i)\delta_i^T}{\delta_i^T\delta_i}$$

A disadvantage of this method is that assuming $\mathbf{K}_{i-1}$ was symmetric, this type of update for $\mathbf{K}_i$ would not preserve $\mathbf{K}_{i-1}$'s symmetry. Davidon's version [GER 80] of the algorithm does preserve matrix symmetry by making $\mathbf{v} = \mathbf{r}_i - \mathbf{K}_{i-1}\delta_i$, thus obtaining:

$$\mathbf{K}_i = \mathbf{K}_{i-1} + \frac{(\mathbf{r}_i - \mathbf{K}_{i-1}\delta_i)(\mathbf{r}_i - \mathbf{K}_{i-1}\delta_i)^T}{(\mathbf{r}_i - \mathbf{K}_{i-1}\delta_i)^T\delta_i}$$

It is also possible to make Rank-two updates such as the one due to Davidon, Fletcher and Powell usually known as DFP [DEN 77] and the one due to Broyden, Fletcher, Goldfarb and Shanno usually known as BFGS [DEN 77], both of which preserve symmetry and positive definiteness.

An important advantage of quasi Newton methods is that apart from the computational cost saved on the stiffness matrix updating, as just shown, it is also possible to save on system solving through the so called inverse update. Through this scheme it is only necessary to perform matrix decomposition on the first iteration. The concept of inverse update is based on the classical matrix algebra formula known as Sherman and Morrison's:

$$(\mathbf{A} + \mathbf{b}\mathbf{c}^T)^{-1} = \mathbf{A} - \frac{\mathbf{A}^{-1}\mathbf{b}\mathbf{c}^T\mathbf{A}^{-1}}{1 + \mathbf{c}^T\mathbf{A}^{-1}\mathbf{b}}$$

where $\mathbf{b}$ and $\mathbf{c}$ are arbitrary vectors. The inverse update for Davidon's method based on Sherman and Morrison's formula is:

$$\mathbf{K}_i^{-1} = \mathbf{K}_{i-1}^{-1} + \frac{(\delta_i - \mathbf{K}_{i-1}^{-1}\mathbf{r}_i)(\delta_i - \mathbf{K}_{i-1}^{-1}\mathbf{r}_i)^T}{(\delta_i - \mathbf{K}_{i-1}^{-1}\mathbf{r}_i)^T\mathbf{r}_i}$$

or in more compact fashion:

$$\mathbf{K}_i^{-1} = \mathbf{K}_{i-1}^{-1} + a_i\mathbf{w}\mathbf{w}^T \qquad (14)$$

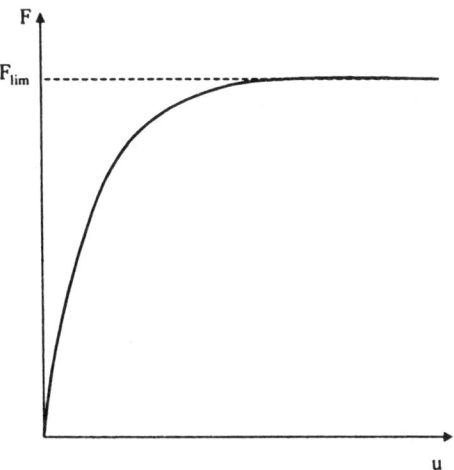

Figure 6. *Limit load problem*

where:

$$\mathbf{w} = \boldsymbol{\delta}_i - \mathbf{K}_{i-1}^{-1}\mathbf{r}_i \qquad a_i = \frac{1}{\mathbf{w}^T\mathbf{r}_i}$$

Inverse updating does not preserve matrix sparsity. It is therefore necessary, if advantage is to be taken of this feature, to solve the equation system by matrix-vector multiplication using expression 14 in a recursive fashion down to the first iteration:

$$\mathbf{K}_i^{-1} = \mathbf{K}_0^{-1} + \sum_{k=1}^{i} a_k \mathbf{v}_k \mathbf{v}_k^T$$

where:

$$\mathbf{v}_i = \boldsymbol{\delta}_i - \mathbf{K}_0^{-1}\mathbf{r}_i - \sum_{k=1}^{i-1} a_k \mathbf{v}_k \mathbf{v}_k^T \mathbf{r}_i$$

A large number of quasi-Newton updates may cause an ill conditioned iteration matrix. It is therefore recommended to restart the iteration procedure either using the initial stiffness matrix $\mathbf{K}_0$ or obtaining a new stiffness matrix [GER 80]. Details on rank two inverse updates may be obtained in [GER 80] and [DEN 77].

2.4. *Displacement control*

Non linear constitutive models such as those associated with elastic-perfectly plastic behaviour will frequently give rise to force-displacement diagrams such as the one in Figure 6. In this case the load control presented at the beginning of section 2.1 would obviously not converge. In such cases a displacement control would solve the problem. The first step would be to partition the problem into a first part including active degrees of freedom and a second part corresponding to restricted degrees of freedom. Thus, the global vectors would be of the following type:

$$\boldsymbol{\Psi}(\mathbf{u}^1, \mathbf{u}^2) = \left\{ \begin{array}{c} \boldsymbol{\Psi}^1(\mathbf{u}^1, \mathbf{u}^2) \\ \boldsymbol{\Psi}^2(\mathbf{u}^1, \mathbf{u}^2) \end{array} \right\} \quad \mathbf{u} = \left\{ \begin{array}{c} \mathbf{u}^1 \\ \mathbf{u}^2 \end{array} \right\} \quad \mathbf{f}^{ext} = \left\{ \begin{array}{c} 0 \\ \mathbf{f}^{ext,2} \end{array} \right\}$$

This means that $\mathbf{u}^1$ and $\mathbf{f}^{ext,2}$ would be the unknowns and $\mathbf{u}^2$ would be given.

$$A^1_{e=1,nelem} \left[\int \mathbf{B}^T \sigma(\epsilon(\mathbf{u}^1, \mathbf{u}^2)) d\Omega_e \right] \quad = \quad 0 \tag{15}$$

$$A^2_{e=1,nelem} \left[\int \mathbf{B}^T \sigma(\epsilon(\mathbf{u}^1, \mathbf{u}^2)) d\Omega_e \right] - \mathbf{f}^{ext,2} \quad = \quad 0 \tag{16}$$

where $A^1_{e=1,nelem}$ is the assembly operator associated to the active degrees of freedom and $A^2_{e=1,nelem}$ is the one corresponding to the restricted degrees of freedom. The jacobian of this equation set would be:

$$\mathbf{J} = \left[\begin{array}{cc} \frac{\partial \boldsymbol{\Psi}^1}{\partial \mathbf{u}^1} & \frac{\partial \boldsymbol{\Psi}^1}{\partial \mathbf{u}^2} \\ \frac{\partial \boldsymbol{\Psi}^2}{\partial \mathbf{u}^1} & \frac{\partial \boldsymbol{\Psi}^2}{\partial \mathbf{u}^2} \end{array} \right] = \left[\begin{array}{cc} \mathbf{K}^{11} & \mathbf{K}^{12} \\ \mathbf{K}^{21} & \mathbf{K}^{22} \end{array} \right]$$

1) Let us assume a displacement state $\mathbf{u}_A$ exists such that $\boldsymbol{\Psi}_A = 0$. This means the internal forces $\mathbf{f}_A^{int}$ are in equilibrium with external forces $\mathbf{f}_A^{ext}$. Presumably this state was reached by a Newton-Raphson procedure which converged after several iterations.

2) We are now looking for displacements $\mathbf{u}_B^1 = \mathbf{u}_A^1 + \Delta\mathbf{u}^1$ associated with a given displacement vector $\mathbf{u}_B^2 = \mathbf{u}_A^2 + \Delta\mathbf{u}^2$. Displacement increment $\Delta\mathbf{u}^2$ is given, $\Delta\mathbf{u}^1$ is unknown and it should be such that:

$$\boldsymbol{\Psi}(\mathbf{u}_B^1, \mathbf{u}_B^2) = 0$$

Since $\mathbf{f}^{ext,2}$ are reactions, and are also unknowns, the solution strategy should be to solve the first $\boldsymbol{\Psi}^1(\mathbf{u}_B^1, \mathbf{u}_B^2) = 0$ obtaining $\Delta\mathbf{u}^1$ and substituting $\Delta\mathbf{u}^1$ and $\Delta\mathbf{u}^2$ into $\boldsymbol{\Psi}^2(\mathbf{u}_B^1, \mathbf{u}_B^2) = 0$ would automatically produce $\mathbf{f}^{ext,2}$. Our first predictor for $\Delta\mathbf{u}^1$ shall be called $\Delta\mathbf{u}_1^1$ and will come from solving the following equation system:

$$\boldsymbol{\Psi}_1^1 = \boldsymbol{\Psi}_A^1 + \left[\begin{array}{cc} \mathbf{K}^{11} & \mathbf{K}^{12} \end{array} \right] \left\{ \begin{array}{c} \Delta\mathbf{u}_1^1 \\ \Delta\mathbf{u}_1^2 \end{array} \right\} = 0$$

In other words:

$$K_1^{11}\Delta u_1^1 = -K_1^{12}\Delta u_1^2$$

where K_1^{11} and K_1^{12} are stiffness matrices at stress state A

3) Once Δu_1^1 has been obtained, and displacement vectors updated through $u_1^1 = u_A^1 + \Delta u_1^1$ and $u_1^2 = u_A^2 + \Delta u_1^2$, internal force vectors $f_1^{int,1}(u_1^1, u_1^2)$ and $f_1^{int,2}(u_1^1, u_1^2)$ are evaluated. To do so stresses at all Gauss points will have to be integrated to compute the following integrals:

$$f_1^{int,1}(u_1^1, u_1^2) = A_{e=1,nelem}^1 \left[\int_{\Omega_e} B^T \sigma(\epsilon(u_1^1, u_1^2))d\Omega_e \right]$$

$$f_1^{int,2}(u_1^1, u_1^2) = A_{e=1,nelem}^2 \left[\int_{\Omega_e} B^T \sigma(\epsilon(u_1^1, u_1^2))d\Omega_e \right]$$

Since $f^{ext,2}$ are reactions it will also be necessary to update them:

$$f_1^{ext,2} = f_1^{int,2}(u_1^1, u_1^2) \tag{17}$$

4) The next step is check whether the convergence test is satisfied or not:

$$\frac{\|\Psi_1\|}{\|f_1^{ext}\|} = \frac{\|f_1^{ext} - f_1^{int}\|}{\|f_1^{ext}\|} < tol \tag{18}$$

It is interesting to remark that only Ψ_1^1 contributes to Ψ_1 since Ψ_1^2 is 0 due to equation (17) and only reactions stored in $f_1^{ext,2}$ contribute to f_1^{ext}. If the convergence condition is satisfied the iterative process ends at this point.

5) If the convergence test is not satisfied a new stiffness matrix K_2^{11} will have to be obtained based on the new displacement state u_1 to solve the following system of equations:

$$-\Psi_1^1 = K_2^{11}\delta u_2^1$$

The iterative procedure will continue until at the ith iteration f_i^{int} is evaluated and the convergence condition satisfied

2.5. Implementation of the Newton-Raphson method in a finite element code

Implementation of the Newton-Raphson method in a simple finite element code would require an algorithmic program structure of the type shown in Figure 7. Of course, there are different ways to do this, and the structure of a non linear finite

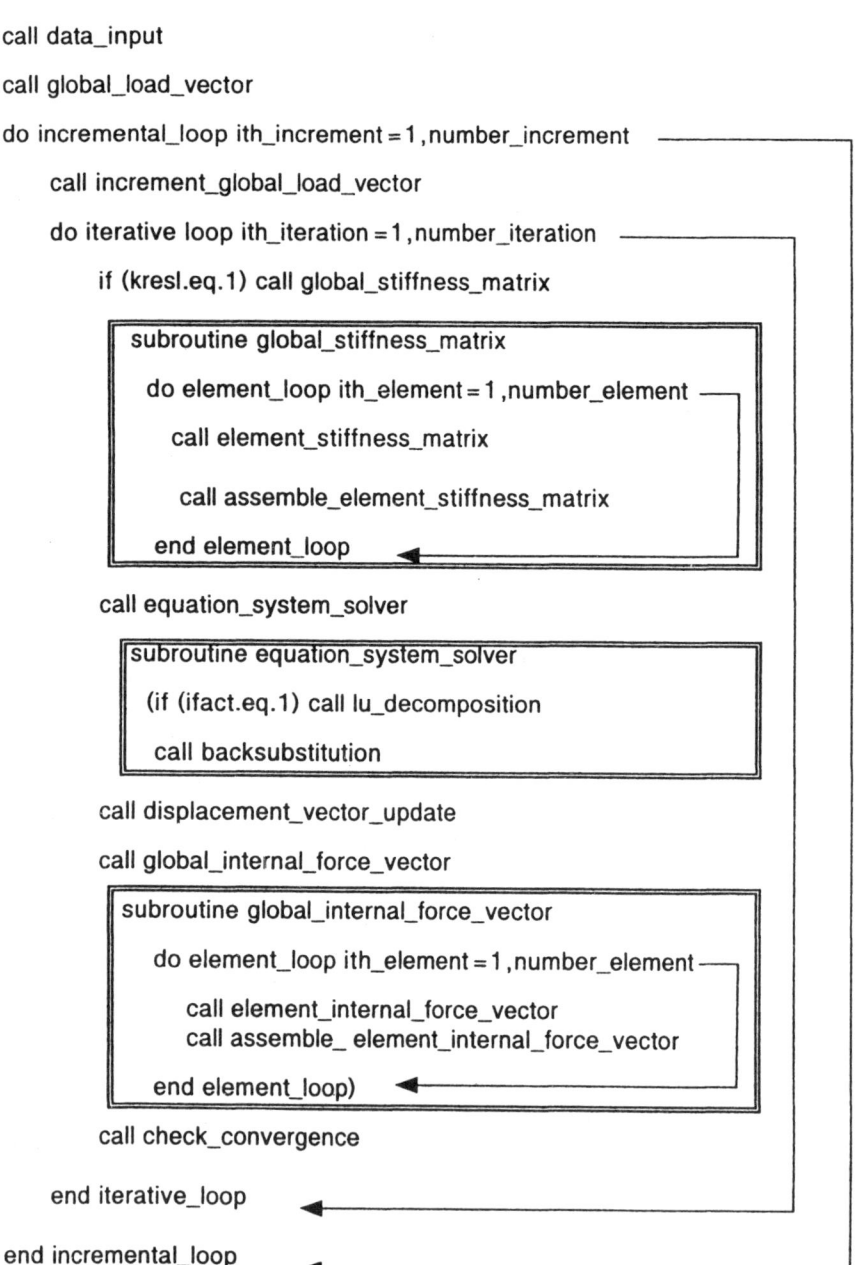

```
call data_input

call global_load_vector

do incremental_loop ith_increment = 1,number_increment ───────────────┐
                                                                        │
    call increment_global_load_vector                                   │
                                                                        │
    do iterative loop ith_iteration = 1,number_iteration ──────────┐   │
                                                                    │   │
        if (kresl.eq.1) call global_stiffness_matrix               │   │
                                                                    │   │
        ┌──────────────────────────────────────────────────────┐  │   │
        │ subroutine global_stiffness_matrix                    │  │   │
        │                                                       │  │   │
        │   do element_loop ith_element = 1,number_element ──┐  │  │   │
        │     call element_stiffness_matrix                  │  │  │   │
        │                                                    │  │  │   │
        │     call assemble_element_stiffness_matrix         │  │  │   │
        │   end element_loop  ◄──────────────────────────────┘  │  │   │
        └──────────────────────────────────────────────────────┘  │   │
        call equation_system_solver                                │   │
        ┌──────────────────────────────────────────────────────┐  │   │
        │ subroutine equation_system_solver                     │  │   │
        │   (if (ifact.eq.1) call lu_decomposition              │  │   │
        │   call backsubstitution                               │  │   │
        └──────────────────────────────────────────────────────┘  │   │
        call displacement_vector_update                            │   │
        call global_internal_force_vector                          │   │
        ┌──────────────────────────────────────────────────────┐  │   │
        │ subroutine global_internal_force_vector               │  │   │
        │   do element_loop ith_element = 1,number_element ──┐  │  │   │
        │     call element_internal_force_vector             │  │  │   │
        │     call assemble_ element_internal_force_vector   │  │  │   │
        │   end element_loop)  ◄─────────────────────────────┘  │  │   │
        └──────────────────────────────────────────────────────┘  │   │
        call check_convergence                                     │   │
    end iterative_loop  ◄──────────────────────────────────────────┘   │
                                                                        │
end incremental_loop  ◄─────────────────────────────────────────────────┘
```

Figure 7. *Non linear finite element program structure based on the Newton-Raphson method*

element program does not have to reproduce exactly the one shown in Figure 7. But the main ingredients are the ones shown there, and in any case the one presented there is a very typical non linear finite element program structure.

One of the main aspects of the program structure is a double loop consisting of an iterative loop nested inside an incremental loop. The incremental loop refers to loading increments in which the total external load is divided. The iterative loop is strictly speaking the Newton-Raphson procedure.

It shall be assumed that the linear equation solving algorithm used is of the direct type, and more specifically a sparse Gauss LU decomposition, which is very frequently used in the finite element community. In this context, a complete equation solving procedure typically includes matrix decomposition and backsubstitution. As a consequence of this assumption, the stiffness matrices and internal force vectors at the element level have to be assembled into the corresponding global matrices and vectors. From the point of view of computational cost the iterative loop is divided into three main parts:

1) Computation of the global stiffness matrix. This task includes computation of tangent stiffness at each gauss point, element level integration of the stiffness matrix, and assemblage into global stiffness matrix

2) Linear equation system solving.

3) Computation of internal force vector. This task includes stress integration at each gauss point, element level integration of internal force vector and assemblage into global internal force vector.

Control variables are introduced in the iterative loop to decide whether a new stiffness matrix should be computed and assembled and to decide whether both matrix decomposition and back substitution or just the latter of these tasks should be performed. These decisions depend on whether the present iteration is the first one or not, and on the algorithm used is a full or a modified Newton-Raphson. If the algorithm were a quasi Newton a few changes would have to be made in the stiffness matrix updating and linear equation solving tasks.

Finally, the iterative loop includes a convergence test to check whether convergence has beeen reached or on the contrary the iterative process should continue.

3. References

[BAT 83] K. J. BATHE, E.N. DVORKIN, " On the automatic solution of non linear finite element equations, *Computers and Structures*, vol. 17, 1983, p. 871-879.

[BAT 96] K. J. BATHE, *Finite Element Procedures* , Prentice Hall, 1996

[BEL 00] TED BELYTSCHKO, WING KAM LIU, BRIAN MORAN, *Non-linear Finite Elements for Continua and Structures*, Wiley, 2000

[BRO 65] C. G. BROYDEN, *Mathematics of Computation*, vol. 19,1965, p. 577-593

[PRE 92] WILLIAM H. PRESS, SAUL A. TEUKOLSKY, WILLIAM T. VETTERLING, BRIAN P. FLANNERY, *Numerical recipes in FORTRAN*, Cambridge University Press,2nd ed,1992

[CRI 91] M. A. CRISFIELD, *Non-linear Finite Element Analysis of Solids and Structures. Volume 1 : Essentials*, John Wiley and Sons,1991

[CRI 97] M. A. CRISFIELD, *Non-linear Finite Element Analysis of Solids and Structures. Volume 2 : Advanced Topics*, John Wiley and Sons,1997

[DEN 77] J. E. DENNIS, J. MORE, " Quasi Newton methods: motivation and theory, *SIAM Review*, vol. 19,1977, p. 46-89.

[GER 80] M. GERADIN, S. IDELSOHN, M. HOGGE, " Non linear structural dynamics via Newton and quasi Newton methods, *Nuclear Engineering Design*, vol. 58,1980, p. 339-348.

[HIN 80] D. R. J. OWEN, E. HINTON, *Finite Elements in Plasticity*, Pineridge Press Limited,1980

[SIM 98] J. SIMO, T. HUGHES, *Computational inelasticity*, Springer, 1998

[ZIE 89] O. C. ZIENKIEWICZ, R. L. TAYLOR, *The Finite Element Method*, McGraw-Hill,4th ed,vols 1 and 2,1989

Chapter 7

Small Strain vs Large Strain Formulation in Computational Mechanics

René Chambon
Laboratoire Sols Solide Structures, UJF, INPG, CNRS, Domaine Universitaire, Grenoble, France

1. Introduction

This paper deals with a comparison between the so-called large strain and small strain formulation of continuum mechanics. Let us first discuss the limitations of this work. The problems studied are only quasi static monophasic problems. The constitutive equations are not covered. It is only assumed that the Cauchy stress can be obtained as a function of the kinematical history. Only mechanical effects are considered.

Conservation equations are then only mass conservation and conservation of momentum which (for quasi static problems) becomes only a balance equation. Moreover, as our results have to be used in finite element computations, only the weak form of the balance equations is written here. In fact, when dealing with computational mechanics, it is our opinion that the strong forms of balance equations are not useful and induce useless complicated developments.

It is a classical exercise to deduce the weak form from the strong form of the balance equation (and vice versa) which can be seen in any text book of continuum mechanics. Let us recall first that the weak form of the balance equation is only the virtual power (or work) principle following the early works of J. Bernoulli, J. D'Alembert and first used extensively in his famous book by Lagrange [LAG 88]. To well understand the use of virtual power principle in continuum mechanics, the reader is recommended to read the papers of Germain [GER 73a], [GER 73b].

In section 2 the basic problem is recalled. Then in section 3, the same equations are written in another configuration called initial for simplicity. These equations are obtained from the previous ones through a simple change of variables. It is easy to exhibit classical measures of some virtual rate of deformation. It is then possible to study the corresponding stress measures. This allows us to define in a natural manner the first Piola-Kirchhoff (or nominal) stress tensor. Then with some small changes it is possible to show how the (second) Piola-Kirchhoff can be exhibited. Some remarks about boundary conditions conclude this section. The following section summarizes the previous results.

Section 5 is the main one, it exhibits the terms neglected when the change of configuration is not taken into account and subsequently indicates when the small strain assumption is suitable to solve continuum mechanics problems. It achieves the goal of this paper because it gives practical criteria needed to justify the use of the small strain assumption and subsequently discriminates cases for which such an assumption is dangerous. A subsection deals with some aspects related to constitutive equations.

It is finally concluded in section 6 that instead of justifying the use of the so called small strain assumption by inspecting the magnitude of the strain, it is possible to discard the non linear geometric terms only if the stresses are large enough with respect to the moduli used in the constitutive equation. This means that for a step close to plastic states the neglected terms can be greater than the ones remaining in the computation.

It is our opinion that this has been ignored for years though it was known for instance by Biot [BIO 65].

The approach of this introduction is somewhat different from the ones usually seen in continuum mechanics of in finite element textbooks. So the reader has to complete his knowledge by reading the presentations given in books dealing with non linear finite elements e.g. Belytchko [BEL 00], Bonet [BON 97], Crisfield [CRI 97], Simo [SIM 00] and Zienkiewicz [ZIE 00], just to mention the most recent ones. Another important reference to this work is the book by Malvern [MAL 69].

Let us finally give the principles of our notations. In the following, rectangular cartesian coordinates are used. Components are denoted by the name of the tensor (or vector) accompanied by tensorial indices. All tensorial indices are in lower position, as there is no need of a distinction between covariant and contravariant components. Upper indices have other meanings. The summation convention with respect to repeated tensorial indices is used.

2. Virtual power principle written in the current configuration

2.1. *Virtual power equation*

If the position (configuration) of a body is known and if this body is in equilibrium then the classical strong form of the balance equation holds. Thus it is possible in this equilibrated configuration to write down the weak form (or the virtual power form) of these equations without any further assumption. This is done in this first section.

Let us define Ω^t the current configuration of a body at a time t. The boundary of Ω^t is denoted Γ^t. If the body is in equilibrium then for every kinematically admissible virtual velocity field $\dot{u}^*$:

$$\int_{\Omega^t} \sigma_{ij}^t \frac{\partial \dot{u}_i^*}{\partial x_j^t} \, d\Omega^t - \int_{\Gamma_\sigma^t} t_i^t \dot{u}_i^* \, d\Gamma^t - \int_{\Omega^t} \rho^t f_i^t \dot{u}_i^* \, d\Omega^t = 0, \qquad [1]$$

where :

– Γ_σ^t is the part of the boundary Γ^t where surface tractions are known at time t. The other part is denoted Γ_u^t. Moreover Γ^t is such that : $\Gamma_u^t \cup \Gamma_\sigma^t = \Gamma^t$ and $\Gamma_u^t \cap \Gamma_\sigma^t = \emptyset$.

– σ_{ij}^t are the components of the current Cauchy stress tensor (which in the following means at time t).

– $\dot{u}_i^*$ are the components of the kinematically admissible virtual velocity field at time t. The virtual fields have to be sufficiently smooth and such that $\dot{u}^* = 0$ on Γ_u^t.

– t_i^t are the components of the traction per unit current surface. It is assumed that they are known at time t.

– ρ^t is the mass per unit current volume.

– f_i^t are the body forces per unit mass at time t.

$-\frac{\partial \dot{u}_i^*}{\partial x_j^t}$ is the virtual rate of deformation. (if instead of using a virtual velocity field we use a real velocity field then this term yields the classical rate of deformation).

2.2. *The problem to be solved and the constitutive equation*

The previous section gave the balance equation. Let us recall a continuum mechanics problem (often called initial boundary value problem). The initial configuration of a body Ω^0 is known (which means that all the quantities: stresses, tractions, body forces ...) which meets equation 1. An history of the conditions (traction on Γ_σ^t, position of Γ_u^t, body forces ...) is known (see, however, the discussion in section 3.4). We are looking for the history of configurations such that equation 1 holds for every time t. In order to solve the problem it is necessary to have a constitutive equation which gives the stress (here the Cauchy stress) as a function of the local kinematic history [TRU 65].

$$\sigma_{ij}^t = \mathcal{F}_{ij}[\frac{\partial x_m^\tau}{\partial x_p^0}(\tau = 0 \to t)]. \qquad [2]$$

Here the superscript 0 refers to the initial position of the body as τ refers to any intermediate position between the initial and the final one denoted here as t. Let us notice that for simplicity, we intentionally don't use in the following a more rigorous notation which should be $x_m^\tau = \Phi_m^\tau(x_p^0)$ distinguishing the one to one mapping Φ_m^τ from its value x_m^τ.

2.3. *Comments*

Defining the virtual strain rate as $\dot{\varepsilon}_{ij}^* = \frac{1}{2}(\frac{\partial \dot{u}_i^*}{\partial x_j^t} + \frac{\partial \dot{u}_j^*}{\partial x_i^t})$ equation 1 can be rewritten as

$$\int_{\Omega^t} \sigma_{ij}^t \dot{\varepsilon}_{ij}^* d\Omega^t - \int_{\Gamma_\sigma^t} t_i^t \dot{u}_i^* d\Gamma^t - \int_{\Omega^t} \rho^t f_i^t \dot{u}_i^* d\Omega^t = 0. \qquad [3]$$

From the theoretical point of view these equations are completely satisfactory. A finite element step by step computation can be based on them. For every time step, it is possible to build up an iterative procedure provided the coordinates are updated from one iteration to the following one. Many computer codes are based on these ideas. However, in order to achieve a quadratic convergence of this iterative procedure which means a quick and efficient way of obtaining solutions, it is necessary to use a full Newton Raphson iterative procedure. In this case it is necessary to be careful in deducing the auxiliary linear system. To deduce this system it is useful to write equations 1 or 3 in another configuration, but such a derivation is out of the scope of this paper. The reader willing do it can use as preliminary results the ones of section 3.

3. Virtual power principle written in the initial configuration

Let us assume that we know a mapping from Ω^0 to Ω^t denoted $x_i^t(x_j^0)$. Let us define now the displacement field which is useful in the following.

$$u_i^{0t}(x_j^0) = x_i^t - x_i^0 \qquad [4]$$

We can consider now all the quantities defined in equation 1 as functions of x_l^0.

$$\sigma_{ij}^t(x_k^t) = \sigma_{ij}^t(x_k^t(x_l^0)) = \sigma_{ij}^t(x_l^0),$$

$$t_i^t(x_k^t) = t_i^t(x_k^t(x_l^0)) = t_i^t(x_l^0),$$

$$f_i^t(x_k^t) = f_i^t(x_k^t(x_l^0)) = f_i^t(x_l^0),$$

$$\dot{u}_i^*(x_k^t) = \dot{u}_i^*(x_k^t(x_l^0)) = \dot{u}_i^*(x_l^0). \qquad [5]$$

Let us notice that a kinematically admissible field in configuration Ω^t is also kinematically admissible in Ω^0 provided the mapping $x_i^t(x_j^0)$ is smooth enough. It follows that the set of all kinematically admissible fields in Ω^0 is the set of all kinematically admissible fields in Ω^t.

It is now only a matter of mathematics to transform equation 1. This is done for the three terms of the equation in the following.

3.1. New form of the internal virtual power

The first term in equation 1 is called the internal virtual work. We have:

$$\int_{\Omega^t} \sigma_{ij}^t \frac{\partial \dot{u}_i^*}{\partial x_j^t} d\Omega^t = \int_{\Omega^0} \sigma_{ij}^t \frac{\partial \dot{u}_i^*}{\partial x_k^0} \frac{\partial x_k^0}{\partial x_j^t} det(J^{0t}) d\Omega^0, \qquad [6]$$

where $det(J^{0t})$ is the Jacobian of the mapping $x_i^t(x_j^0)$.

As $\delta_{il} = \frac{\partial x_m^0}{\partial x_l^t} \frac{\partial x_i^t}{\partial x_m^0}$,

$$\sigma_{ij}^t \frac{\partial \dot{u}_i^*}{\partial x_k^0} \frac{\partial x_k^0}{\partial x_j^t} det(J^{0t}) = \frac{\partial x_m^0}{\partial x_l^t} \frac{\partial x_i^t}{\partial x_m^0} \sigma_{lj}^t \frac{\partial \dot{u}_i^*}{\partial x_k^0} \frac{\partial x_k^0}{\partial x_j^t} det(J^{0t}). \qquad [7]$$

3.2. The first Piola-Kirchhoff stress tensor

In equation 6, $\frac{\partial \dot{u}_i^*}{\partial x_k^0}$ is the material time derivative of the virtual deformation gradient. The real (not virtual) corresponding quantity is $\frac{\partial \dot{u}_i^{0t}}{\partial x_k^0}$, which is the material time derivative of the deformation gradient $\frac{\partial x_i^t}{\partial x_k^0}$.

Then let us define:

$$s_{ij}^{0t} = \sigma_{ij}^t \frac{\partial x_k^0}{\partial x_j^t} \, det(J^{0t}),$$ [8]

the so called first Piola-Kirchhoff stress tensor (which in matrical form is the transposed of the Hill nominal stress [HIL 78]). Let us notice that it is necessary to know two different configurations to define s_{ij}^{0t} as it appears in the two superscripts. More over, we have to be careful as this tensor is not symmetric. With this new notation the first term of the virtual power equation yields:

$$\int_{\Omega^t} \sigma_{ij}^t \frac{\partial \dot{u}_i^*}{\partial x_j^t} \, d\Omega^t = \int_{\Omega^0} s_{ik}^{0t} \frac{\partial \dot{u}_i^*}{\partial x_k^0} \, d\Omega^0.$$ [9]

3.3. The second Piola-Kirchhoff stress tensor

Let us first recall the definition of the Green strain tensor:

$$\gamma_{mk}^{0t} = \frac{1}{2} \left(\frac{\partial x_l^t}{\partial x_k^0} \frac{\partial x_l^t}{\partial x_m^0} - \delta_{mk} \right).$$ [10]

It has to be recalled that this tensor is the one which gives all the change of the metrics characterizing in an intrinsic manner the *deformation* of a body as:

$$dx_i^t dy_i^t - dx_i^0 dy_i^0 = 2\gamma_{mk}^{0t} dx_m^0 dy_k^0.$$ [11]

The material time derivative of this tensor reads:

$$\dot{\gamma}_{mk}^{0t} = \frac{1}{2} \left(\frac{\partial \dot{x}_l^t}{\partial x_k^0} \frac{\partial x_l^t}{\partial x_m^0} + \frac{\partial x_l^t}{\partial x_k^0} \frac{\partial \dot{x}_l^t}{\partial x_m^0} \right).$$ [12]

The virtual material time derivative of this tensor reads:

$$\dot{\gamma}_{mk}^{0t*} = \frac{1}{2} \left(\frac{\partial \dot{x}_l^{t*}}{\partial x_k^0} \frac{\partial x_l^t}{\partial x_m^0} + \frac{\partial x_l^t}{\partial x_k^0} \frac{\partial \dot{x}_l^{t*}}{\partial x_m^0} \right) = \frac{1}{2} \left(\frac{\partial \dot{u}_l^*}{\partial x_k^0} \frac{\partial x_l^t}{\partial x_m^0} + \frac{\partial x_l^t}{\partial x_k^0} \frac{\partial \dot{u}_l^*}{\partial x_m^0} \right).$$ [13]

Let us define now:

$$\Pi_{mk}^{0t} = \frac{\partial x_m^0}{\partial x_l^t} \sigma_{lj}^t \frac{\partial x_k^0}{\partial x_j^t} \, det(J^{0t}),$$ [14]

the so called second Piola-Kirchhoff stress tensor. As for the first one it is necessary to know two different configurations to define it. Contrary to s_{ij}^{0t}, Π_{mk}^{0t} is symmetric as it is clear from equation 14 and the symmetry property of the Classical Cauchy stress tensor. Then

$$\int_{\Omega^t} \sigma_{ij}^t \frac{\partial \dot{u}_i^*}{\partial x_j^t} \, d\Omega^t = \int_{\Omega^0} \Pi_{mk}^{0t} \frac{\partial \dot{u}_l^*}{\partial x_k^0} \frac{\partial x_l^t}{\partial x_m^0} d\Omega^0.$$ [15]

As the second Piola-Kirchhoff stress tensor is symmetric finally:

$$\int\limits_{\Omega^t} \sigma^t_{ij}\dot{\varepsilon}^*_{ij}d\Omega^t = \int\limits_{\Omega^0} s^{0t}_{ik}\frac{\partial \dot{u}^*_i}{\partial x^0_k}d\Omega^0 = \int\limits_{\Omega^0} \Pi^{0t}_{mk}\dot{\gamma}^{0t*}_{mk}d\Omega^0. \qquad [16]$$

In the previous equation, the second equality is the main reason for introducing the second Piola-Kirchhoff stress tensor.

3.4. *New form of the external virtual power*

The last two terms of equation 1 are called the external virtual power. They are related to external (known) forces.

Concerning the body forces, we have:

$$\int\limits_{\Omega^t} \rho^t f^t_i \dot{u}^*_i\, d\Omega^t = \int\limits_{\Omega^0} \rho^t f^t_i \dot{u}^*_i det(J^{0t})\, d\Omega^0. \qquad [17]$$

And concerning the surface tractions we have:

$$\int\limits_{\Gamma^t_\sigma} t^t_i \dot{u}^*_i\, d\Gamma^t = \int\limits_{\Gamma^0_\sigma} t^t_i \dot{u}^*_i \frac{dA^t}{dA^0}\, d\Gamma^0, \qquad [18]$$

where $\frac{dA^t}{dA^0}$ is the ratio between the area in the current configuration Ω^t and the reference configuration Ω^0.

In equations 17 and 18, it is assumed that external forces are known in the current configuration. However other assumptions can be made. For instance, volume forces can be known as a function of the position and then are dependent on the configuration itself. In a centrifuge for instance the body forces depend on the position and so f^t_i is not given. About the surface traction it is often assumed that the external loading is a dead load which means that $t^t_i \frac{dA^t}{dA^0}$ is known. This can be realistic for forces but a little bit theoretical for traction per unit area.

We do not want to study all the cases here but let us have an insight into the difficulties coming from different definitions of the surface traction by studying the realistic case of imposed pressure on the boundary:

$$\int\limits_{\Gamma^t_\sigma} t^t_i \dot{u}^*_i\, d\Gamma^t = \int\limits_{\Gamma^t_\sigma} p^t N^t_i \dot{u}^*_i\, d\Gamma^t, \qquad [19]$$

where N^t_i is the internal unit normal in the current configuration. Let V^0_i and W^0_i unit vectors in the reference configuration such that

$$N^0_i d\Gamma^0 = \mathcal{E}_{ijk} V^0_j W^0_k, \qquad [20]$$

where $\mathcal{E}$ is the alternator tensor such that $\mathcal{E}_{ijk} = 1$ for an even permutation of ijk, $\mathcal{E}_{ijk} = -1$ for an odd permutation of ijk and $\mathcal{E}_{ijk} = 0$ if any index is repeated.

$$p^t N_i^t d\Gamma^t = p^t \mathcal{E}_{ijk} \frac{\partial x_j^t}{\partial x_l^0} V_l^0 \frac{\partial x_k^t}{\partial x_m^0} W_m^0, \qquad [21]$$

using the Nanson formulae (see [MAL 69]) yields:

$$N_i^t d\Gamma^t = det(J^{0t}) \frac{\partial x_n^0}{\partial x_i^t} \mathcal{E}_{nlm} V_l^0 W_m^0, \qquad [22]$$

and finally:

$$\int_{\Gamma_\sigma^t} p^t N_i^t \dot{u}_i^* \, d\Gamma^t = \int_{\Gamma_\sigma^0} p^t \dot{u}_i^* det(J^{0t}) \frac{\partial x_n^0}{\partial x_i^t} N_n^0 d\Gamma^0. \qquad [23]$$

4. Summary

Finally we have three equations completely equivalent:

$$\int_{\Omega^t} \sigma_{ij}^t \frac{\partial \dot{u}_i^*}{\partial x_j^t} \, d\Omega^t - \int_{\Gamma_\sigma^t} t_i^t \dot{u}_i^* \, d\Gamma^t - \int_{\Omega^t} \rho^t f_i^t \dot{u}_i^* \, d\Omega^t = 0,$$

$$\int_{\Omega^0} s_{ik}^{0t} \frac{\partial \dot{u}_i^*}{\partial x_k^0} \, d\Omega^0 - \int_{\Gamma_\sigma^0} t_i^t \dot{u}_i^* \frac{dA^t}{dA^0} \, d\Gamma^0 - \int_{\Omega^0} \rho^t f_i^t \dot{u}_i^* det(J^{0t}) \, d\Omega^0 = 0,$$

$$\int_{\Omega^0} \Pi_{mk}^{0t} \dot{\gamma}_{mk}^{0t*} d\Omega^0 - \int_{\Gamma_\sigma^0} t_i^t \dot{u}_i^* \frac{dA^t}{dA^0} \, d\Gamma^0 - \int_{\Omega^0} \rho^t f_i^t \dot{u}_i^* det(J^{0t}) \, d\Omega^0 = 0. \qquad [24]$$

Clearly the constitutive equation 2 can be expressed by using the first and second Piola-Kirchhoff stress tensors:

$$s_{ik}^{0t} = \mathcal{G}_{ik}[\frac{\partial x_n^\tau}{\partial x_p^0}(\tau = 0 \to t)], \qquad [25]$$

and

$$\Pi_{mk}^{0t} = \mathcal{H}_{mk}[\frac{\partial x_n^\tau}{\partial x_p^0}(\tau = 0 \to t)], \qquad [26]$$

which show that these three points of view are mathematically equivalent.

Discretizing equation 24-1 yields the updated Lagrangian finite element formulation. In a step by step computation this means that for every step the coordinates have to be updated. This means also that in order to solve the one step problem with a full Newton-Raphson method it is necessary to be careful in deducing the auxiliary linear system, as already mentioned in section 2.3.

Discretizing equations 24-2 or 24-3 yields the so called total Lagrangian formulation. The choice of one formulation instead of the other one is only a matter of convenience. Both are, as far as the continuum problem is considered, exact formulation. No assumptions, no simplifications have been made. This point is important as in the following we will study the small strain assumption by comparing it with the two Lagrangian formulations.

5. Comparison between large strain formulation and small strain one

5.1. *The small strain assumption*

Basically the so called small strain assumption has nothing to do with the magnitude of strain. In a small strain formulation, it is assumed that instead of doing computations in deformed configuration, the computations are done in the initial configuration. This means that the virtual power principle which should be equation 1 is written like:

$$\int_{\Omega^0} \sigma_{ij}^t \frac{\partial \dot{u}_i^*}{\partial x_j^0} \, d\Omega^0 - \int_{\Gamma_\sigma^0} t_i^t \dot{u}_i^* \, d\Gamma^0 - \int_{\Omega^0} \rho^0 f_i^0 \dot{u}_i^* \, d\Omega^0 = 0 \qquad [27]$$

In order to clarify the comparison between a small strain and a large strain (i.e. without any approximation) computation, let us consider that the reference configuration Ω^0 which is the beginning of a loading step is well balanced. The corresponding Cauchy stress field is denoted σ^0. The balance equation written at the end of the loading step are supposed to correspond to configuration Ω^t which in the case of small strain is assumed to be Ω^0.

5.2. *Comparison between large and small strain for the internal virtual power*

For the complete computation (large strain) the initial internal virtual power is $\int_{\Omega^0} \sigma_{ij}^0 \frac{\partial \dot{u}_i^*}{\partial x_j^0} \, d\Omega^0$ and the final one $\int_{\Omega^t} \sigma_{ij}^t \frac{\partial \dot{u}_i^*}{\partial x_j^t} \, d\Omega^t$. Using equation 6 yields for the difference :

$$\int_{\Omega^0} (\sigma_{ij}^t \frac{\partial \dot{u}_i^*}{\partial x_k^0} \frac{\partial x_k^0}{\partial x_j^t} det(J^{0t}) - \sigma_{ij}^0 \frac{\partial \dot{u}_i^*}{\partial x_j^0}) d\Omega^0 = \int_{\Omega^0} (\sigma_{ik}^t \frac{\partial x_j^0}{\partial x_k^t} det(J^{0t}) - \sigma_{ij}^0) \frac{\partial \dot{u}_i^*}{\partial x_j^0} d\Omega^0.$$

$$[28]$$

Now for the computation done within the small strain assumption, the similar difference reads:

$$\int_{\Omega^0} (\sigma_{ij}^t - \sigma_{ij}^0) \frac{\partial \dot{u}_i^*}{\partial x_j^0} d\Omega^0 = \int_{\Omega^0} \Delta\sigma_{ij} \frac{\partial \dot{u}_i^*}{\partial x_j^0} d\Omega^0, \qquad [29]$$

denoting $\Delta\sigma_{ij} = \sigma_{ij}^t - \sigma_{ij}^0$.

We have in equation $\sigma_{ik}^t \frac{\partial x_j^0}{\partial x_k^t} det(J^{0t}) - \sigma_{ij}^0$ 28 instead of $\Delta\sigma_{ij}$ in equation 29. Our aim is to compare these two terms. Let us write:

$$\sigma_{ik}^t \frac{\partial x_j^0}{\partial x_k^t} det(J^{0t}) - \sigma_{ij}^0 = \Delta\sigma_{ij} + \sigma_{ik}^t \frac{\partial x_j^0}{\partial x_k^t} det(J^{0t}) - \sigma_{ij}^t. \qquad [30]$$

Now:

$$\sigma_{ik}^t \frac{\partial x_j^0}{\partial x_k^t} det(J^{0t}) - \sigma_{ij}^t = \sigma_{ik}^t (\frac{\partial x_j^0}{\partial x_k^t} - \delta_{jk}) + \sigma_{ik}^t \frac{\partial x_j^0}{\partial x_k^t}(det(J^{0t}) - 1). \qquad [31]$$

Finally using the so called small strain assumption, we keep the term $\Delta\sigma_{ij}$ and $\sigma_{ik}^t(\frac{\partial x_j^0}{\partial x_k^t} - \delta_{jk}) + \sigma_{ik}^t \frac{\partial x_j^0}{\partial x_k^t}(det(J^{0t}) - 1)$ is discarded. Now, we do approximations to evaluate the relative magnitudes of both terms, assuming that the loading step is small. $\Delta\sigma_{ij}$ can be seen as the product of some moduli coming from the constitutive equation and the incremental displacement gradient (term like strain and/or rotations between the beginning and the end of the step). The terms discarded are products of the stress and terms of the order of magnitude of the incremental strain. Finally a small strain assumption is sound only if these "moduli" are negligible with respect to stress. Let us notice that these moduli are not the ones entering the so called consistent stiffness matrix [SIM 00] used in a full Newton-Raphson method (2.3), they are related to the incremental form of the constitutive equation. The small strain assumption has nothing to do with the magnitude of strains (this is a very old result which can be seen (for the restricted case of elasticity) in the books of Brillouin [BRI 38] or more recently of Biot [BIO 65]). In industrial elastic computations, metal moduli are much larger than the usual stresses and this assumption is founded. Buckling (even elastic) is however a well known case which needs a large strain assumption and which cannot be modelled in a small strain one. When a large area of a domain is in a plastic state, the corresponding "moduli" are very small, possibly less than the stress values, so for such increments, the small strain assumption is not justified.

5.3. Comparison between large strain and small strain for the external virtual work

In this section, we follow a similar line as in the previous section. However as seen in section 3.4 it is difficult to deal with every possible external force. We consider only the cases of pressure forces acting onto the boundary of the studied domain.

For the complete computation this part of the initial external virtual power is $\int_{\Gamma_\sigma^0} p^0 N_i^0 \dot{u}_i^* d\Gamma^0$ whereas the final one is $\int_{\Gamma_\sigma^t} p^t N_i^t \dot{u}_i^* d\Gamma^t$. Using equation 23 yields for the difference:

$$\int_{\Gamma_\sigma^0} (p^t det(J^{0t}) \frac{\partial x_n^0}{\partial x_i^t} N_n^0 - p^0 N_i^0)\dot{u}_i^* d\Gamma^0. \qquad [32]$$

For a computation carried out within the small strain assumption, the corresponding difference reads:

$$\int_{\Gamma_\sigma^0} (p^t N_i^0 - p^0 N_i^0)\dot{u}_i^* d\Gamma^0 = \int_{\Gamma_\sigma^0} \Delta p N_i^0 \dot{u}_i^* d\Gamma^0, \qquad [33]$$

denoting $\Delta p = p^t - p^0$.

Within the small strain assumption $\Delta p N_i^0$ is used instead of $p^t det(J^{0t})\frac{\partial x_n^0}{\partial x_i^t} N_n^0 - p^0 N_i^0$. Let us compare these two terms:

$$p^t det(J^{0t})\frac{\partial x_n^0}{\partial x_i^t} N_n^0 - p^0 N_i^0 = \Delta p N_i^0 + p^t[(det(J^{0t})-1)\frac{\partial x_n^0}{\partial x_i^t} N_n^0 + (\frac{\partial x_n^0}{\partial x_i^t} - \delta_{in})N_n^0]$$
$$[34]$$

We are now facing a problem similar to the one discussed in section 5.2. The small stress assumption is justified if the products $p^t(det(J^{0t}) - 1)$ and $p^t(\frac{\partial x_n^0}{\partial x_i^t} - \delta_{in})$ are negligible with respect to Δp. Once more if the body under computation is prestressed it can be necessary to deal with a large strain formulation even if the strains are small.

Let us notice that forces related with pressure are called following forces. They are not conservative, which means that they can supply energy in the studied system. Consequences can be catastrophic, one of them is the triggering of an instability called flutter instability [BIO 48].

5.4. Some remarks about constitutive equations

As already mentioned, a study of constitutive equations within the large strain framework is out of the scope of this paper. However it is important to have some insight in two problems related with large strain and which have been discussed intensively in the literature, the most often in a wrong manner.

A first point which deserves discussion is the problem of extension of elastoplastic constitutive equations for large strain. Some people in the past took an elastoplastic model written in a rate form and changed the stress derivative into an objective stress rate. Such a way is dangerous because it does not preserve the thermodynamical meaning of the free energy or the dissipation function. Then the only way to preserve these characteristics of the model is to rewrite these functions in a Lagrangian description and to to use a multiplicative decomposition of the displacement gradient. This is now currently done in metal modelling [SIM 00]. Such developments are very uncommon in geomaterial modelling. To our knowledge, the first rigorous work on large strain elastoplastic constitutive equation is the one of Borja and Tamagnini [BOR 98].

Another point discussed in many papers (more or less related to the previous one) is the question of the suitable stress derivative to be used in constitutive equations.

In our opinion this point is meaningless: let us quote Truesdell [TRU 65], also reported in Simo [SIM 00]. "... Despite the claim and whole papers to the contrary, any advantage claimed for such rate over another is pure illusion ". This question took a new popularity in the 70's when people wanted to generalize elastoplastic constitutive equation in the large strain framework. As previously mentioned, constitutive equations are usually in a rate form where the stress rate is given as a function of the kinematics. As a stress rate can be defined with respect to many frameworks, when we deal with large strain, it is necessary to define clearly the stress rate used and this stress rate has to be objective. Some people concluded at that time, and this "result" is unfortunately often quoted that there is good objective stress derivatives and bad ones. In fact every constitutive equation written using a given stress derivative can be rewritten with another one. Particularly, as mentioned previously, we know that the good way to generalize elastoplastic constitutive equation is to go back to a Lagrangian expression for the free energy and the dissipation functions. However, there can be good constitutive equations and bad ones, depending on the problem solved [CHA 00] but this is another story.

6. Conclusion

The main result of this study is that it is always necessary to compare the magnitude of terms discarded in an approximate study with respect to the one considered as the principal ones. A small strain assumption is justified if the "moduli" are significantly higher than the stresses. Otherwise many observed phenomena such as buckling for instance cannot be modelled. It is interesting to note that the same study has to be done with enhanced models, such as second gradient ones, which now are known to be necessary for studying plastic localization [CHA 01], [MAT 02].

7. References

[BEL 00] BELYTCHKO T., LIU W. K., MORAN B., *Non linear finite elements for continua and structures*, Wiley, Chichester, 2000.

[BIO 48] BIOT M. A., ARNOLD L., "Low speed flutter and its physical interpretation", *Journal of aeronautical science*, vol. 15, num. 4, 1948, p. 232-236.

[BIO 65] BIOT M. A., *Mechanics of incremental deformations*, Wiley, Chichester, 1965.

[BON 97] BONET J., WOOD R. D., *Non linear continuum mechanics for finite element analysis*, Cambridge university press, Cambridge, 1997.

[BOR 98] BORJA R., TAMAGNINI C., "Cam clay plasticity part III Extension of the infinitesimal model to include finite strains", *Computer Methods in Appl. Mechanics and Engineering*, vol. 155, 1998, p. 73-95.

[BRI 38] BRILLOUIN L., *Les tenseurs en mécanique et en élasticité*, Masson, Paris, 1938.

[CHA 00] CHAMBON R., "General presentation of constitutive modelling of geomaterials", *Revue Française de Génie Civil*, vol. 4, 2000, p. 9-31.

[CHA 01] CHAMBON R., CAILLERIE D., TAMAGNINI C., "A finite deformation second gradient theory of plasticity", *C.R.A.S. IIb*, vol. 329, 2001, p. 797-802.

[CRI 97] CRISFIELD M., *Non linear finite element analysis of solids and structures. Volume 2 Advanced topics*, Wiley, Chichester, 1997.

[GER 73a] GERMAIN P., "La méthode des puissances virtuelles en mécanique des milieux continus, Première partie: théorie du second gradient", *Journal de mécanique*, vol. 12, 1973, p. 235-274.

[GER 73b] GERMAIN P., "The method of virtual power in continuum mechanics. Part 2. Microstructure", *S.I.A.M. J. Appl Math.*, vol. 25, 1973, p. 556-575.

[HIL 78] HILL R., "Aspect of invariance in solids", *Advanced in applied mechanics*, vol. 18, 1978, p. 1-75.

[LAG 88] DE LA GRANGE J., *Mechanique Analitique*, veuve Dessaint, Paris, 1788.

[MAL 69] MALVERN L., *Introduction to a mechanicas of a continuum medium*, Prentice-Hall, Englewood Cliffs, N.J., 1969.

[MAT 02] MATSUSHIMA T., CHAMBON R., CAILLERIE D., "Large strain finite element analysis of a local second gradient model: application to localization", *International Journal for Numerical Methods in Engineering*, vol. 54, 2002, p. 499-521.

[SIM 00] SIMO J., HUGHES T., *Computational inelasticity*, Springer, New York, 2000.

[TRU 65] TRUEDELL C., NOLL W., *The non linear field theory of mechanics. in Encyclopaedia of Physics*, Springer-Verlag, berlin, 1965.

[ZIE 00] ZIENKIEWICZ O., TAYLOR R., *The finite element method. Volume 2 Solid mechanics*, Butterworth Heineman, Woburn, 2000.

Chapter 8

Implicit Integration of Constitutive Equations in Computational Plasticity

Claudio Tamagnini
Dipartimento di Ingegneria Civile e Ambientale, Università di Perugia, Italy

Riccardo Castellanza and Roberto Nova
Politecnico di Milano, Italy

1. Introduction

In recent years the parallel development of: i) advanced constitutive theories for the mechanical behavior of geomaterials, ii) robust and accurate numerical methods for the solution of partial differential equations, and iii) powerful computer architectures has led to a radical change in the analysis of geotechnical problems, notably in some areas such as the design of deep excavations or the analysis of complex soil-structure interaction problems where traditional design methods – based on the classical distinction between "failure" and "deformation" problems – are not able to capture the most relevant aspects of the soil–structure system behavior.

A rather detailed outline of the most widely used theoretical approaches to the constitutive modelling of geomaterials is provided by the papers of Tamagnini [TAM 02c] and Pijaudier–Cabot [PIJ 02] in this volume. From an overview of the aforementioned works, it is immediately apparent that a common and almost universal feature of the constitutive models proposed for geomaterials – from those which have now became a standard design tool in geotechnical practice to the ones which were mainly developed for research purposes – is the fact that they are cast in a *incremental* form. Rather than providing the state of stress associated to a specific state of strain, they define the *evolution laws* for the state variables. Therefore, the quantitative evaluation of the mechanical effects of a given "load", be it an imposed stress increment, strain increment or a combination of both, requires the solution of an initial value problem, consisting of the *integration* of the constitutive equation along the assigned loading path, with prescribed *initial conditions*. As this task cannot be performed analytically, except in very special cases, the development of a numerical algorithm is a crucial part of any computational procedure for the solution of non–linear problems in geomechanics.

More specifically, in the application of numerical methods – such as the finite element method – to the solution of a non–linear initial/boundary value problem, the following general strategy is usually adopted, see [SIM 97]:

1) from the original system of governing PDE's, a non–linear system of algebraic *balance equations* is obtained by the introduction of appropriate space and time (if required) discretizations. Such a system is typically solved by adopting an incremental–iterative approach;

2) for any given *global* iteration, the discretized equilibrium equations generate incremental motions, which, in turn, are used to determine the incremental strain history by purely kinematic relationships;

3) for a given strain increment, updated values of the state variables are obtained by integrating numerically the constitutive equations at the *local* level, with given initial conditions;

4) the discrete balance equations are then checked for convergence, and if the convergence criterion is not met, the iteration process is continued by returning to step (2).

As first pointed out by Hughes [HUG 84], the integration of the constitutive equation at the local level – i.e., step (3) – represents the central problem of computational plasticity, since it corresponds to the main role played by the constitutive equation in actual computations. There are of course many other important computational ingredients in the overall procedure, but they are particular to the type of solution strategy employed, and involve the constitutive theory only in a limited way, if at all. Moreover, the precision with which the constitutive equations are integrated has a direct impact on the overall accuracy of the analysis.

Since the early works on metal plasticity, summarized in [HUG 84], a number of fundamental studies have been published on this subject. Implicit algorithms based on the concepts of operator split and closest point projection return mapping, as discussed for example in [SIM 87, SIM 91], have been applied to computational geomechanics in a number of recent papers, of which refs. [BOR 90, BOR 91, BOR 98, ALA 92, MAC 97, JER 97] represent by no means a complete account. Explicit algorithms, either of the return mapping type or based on higher order Runge–Kutta methods with error control have been suggested, e.g., in [ORT 86, SLO 87]. Comparative studies concerning the performance of these different strategies as applied to classical plasticity have been presented in [GEN 88, POT 97].

The objective of this paper is to discuss in some detail a generalization of the classical Generalized Backward Euler (GBE) method for a specific constitutive framework, namely the theory of elastoplasticity with extended hardening rules (see [TAM 02c], sect. 6.4). This particular class of constitutive equations has been choosen for this purpose because not only it embeds the classical theory of plasticity as a special case, but, in addition, it includes a number of special constitutive theories proposed in the geomechanics context to describe a number of practically relevant aspects of the mechanical behavior of geomaterials, such as the brittle/ductile transition of weak rocks subject to thermal loading, the swelling/collapse phenomena in partially saturated soils subject to wetting, and the chemical degradation of weak rocks or bonded soils.

In the following, the discussion will be restricted to deformation processes characterized by linear kinematics ("small" transformations). The general scheme of notation adopted is the same as in [TAM 02c], in this volume. The symbol $\nabla^s(v)$ denotes the symmetric part of the gradient of the vector field v.

2. Evolution equations

For completeness, the evolution equations of the theory of plasticity with extended hardening rules are briefly summarized below (see [TAM 02b, TAM 02c] for details). Let ϵ and θ be the strain tensor and the additional (scalar) variable which affects the mechanical response of the material, i.e., temperature, suction or chemical degradation. Also, let q be the "vector" (of dimension n_{int}) of the internal state variables accounting for the effects of the previous loading history, and:

$$\mathbb{E}_\sigma := \left\{ (\sigma, q) \mid f(\sigma, q) \leq 0 \right\} \tag{1}$$

be the elastic domain, defined through a suitable yield function $f(\sigma, q) = 0$. Taking into account the usual additive decomposition of the strain rate tensor, $\dot{\epsilon}$, into an elastic ($\dot{\epsilon}^e$) and a plastic ($\dot{\epsilon}^p$) part, we have:

$$\dot{\sigma} = D^e(\sigma, \theta) [\dot{\epsilon} - \dot{\epsilon}^p] + m(\sigma, \theta) \dot{\theta} \tag{2}$$

$$\dot{\epsilon}^p = \dot{\gamma} \frac{\partial g}{\partial \sigma}(\sigma, q) = \dot{\gamma} Q(\sigma, q) \tag{3}$$

$$\dot{q} = \dot{\gamma} h(\sigma, q, \theta) + \dot{\theta} \eta(\sigma, q, \theta) \tag{4}$$

subject to the following Kuhn–Tucker complementarity conditions:

$$\dot{\gamma} \geq 0, \quad f(\sigma, q) \leq 0, \quad \dot{\gamma} f(\sigma, q) = 0 \tag{5}$$

which state that plastic processes ($\dot{\gamma} > 0$) can occur only for states on the yield surface, and to the consistency condition:

$$\dot{\gamma} \dot{f} = 0 \tag{6}$$

requiring that the state of the material remains on the yield surface ($f = 0$) whenever plastic loading occurs. Eq. (2) is the elastic constitutive equation of the material in incremental form. The fourth–order tensor $D^e(\sigma, \theta)$ represents the elastic tangent stiffness, while $m(\sigma, \theta)$ is a coupling coefficient (when θ is the temperature, this term defines the thermal stress coefficient). Eq. (3) provides the flow rule for the plastic strain rate, defined in terms of the plastic potential function $g = \hat{g}(\sigma, q)$. The non–negative scalar $\dot{\gamma}$ is the plastic multiplier, while the symbol Q will be used in the following to denote the plastic flow direction (gradient of the plastic potential). The evolution of the internal variables q is provided by the hardening law (4), in which h and η are prescribed hardening functions. The first term on the RHS of eq. (4) quantifies the changes in the internal variables associated with plastic strains, the second term accounts for all non–mechanical hardening/softening processes induced by a change of θ.

3. Numerical integration: basic principles

Let $\mathbb{I} = \bigcup_{n=0}^{N} [t_n, t_{n+1}]$ a partition of the time interval of interest into time steps. It is assumed that at time $t_n \in \mathbb{I}$ the state of the material (σ_n, q_n) is known at any quadrature point in the adopted finite element discretization. Also, let:

$$\{\epsilon_i : i = 0, 1, \ldots, n+1\} \qquad \text{and} \qquad \{\theta_i : i = 0, 1, \ldots, n+1\}$$

be the prescribed histories of ϵ and θ, respectively. The computational problem to be addressed is the update of the state variables:

$$\sigma_{n+1}^{(k)} \rightarrow \hat{\sigma}\left(\epsilon_{n+1}^{(k)}, \theta_{n+1}^{(k)}; \sigma_n, q_n\right) \tag{7}$$

$$q_{n+1}^{(k)} \rightarrow \hat{q}\left(\epsilon_{n+1}^{(k)}, \theta_{n+1}^{(k)}; \sigma_n, q_n\right) \tag{8}$$

for *given* increments $\Delta\epsilon_{n+1}^{(k)} := \epsilon_{n+1}^{(k)} - \epsilon_n$ and $\Delta\theta_{n+1}^{(k)} := \theta_{n+1}^{(k)} - \theta_n$, relative to the global iteration (k), through the integration of the system of ODE's (2)–(4), subject to the algebraic constraints (5). Due to the presence of these constraints, the evolution problem defined by the eqs. (7), (8) belongs to the category of the so–called *stiff differential–algebraic systems* – see [HAI 91] for details – for which *implicit* methods are ideally suited.

Whenever the existence of a free energy function $\psi = \hat{\psi}\left(\epsilon^e, \theta\right)$, such that:

$$\sigma(\epsilon^e, \theta) = \frac{\partial\psi}{\partial\epsilon^e}\left(\epsilon^e, \theta\right) \tag{9}$$

can be postulated, the stress tensor can be considered a *dependent* quantity, and can be replaced in the set of state variables by the elastic strain tensor ϵ^e. The evolution equations (7), (8) can then be replaced by the following ones:

$$\epsilon_{n+1}^{e(k)} \rightarrow \hat{\epsilon}^e\left(\epsilon_{n+1}^{(k)}, \theta_{n+1}^{(k)}; \epsilon_n^e, q_n\right) \tag{10}$$

$$q_{n+1}^{(k)} \rightarrow \hat{q}\left(\epsilon_{n+1}^{(k)}, \theta_{n+1}^{(k)}; \epsilon_n^e, q_n\right) \tag{11}$$

Without loss of generality, we will assume in the following that such a free energy function exists – i.e., the elastic constitutive equation is of the form (9) – and consider the numerical solution of the evolution problem cast by eqs. (10) and (11).

4. Implicit Generalized Backward Euler method

4.1. *Operator split and product formula algorithm*

The additive structure of the elastic constitutive equation in rate form, eq. (3), and of the hardening law, eq. (4), suggest the *elastic–plastic operator split* of the original problem of evolution into an *elastic predictor* problem and a *plastic corrector* problem, as shown in Table 1. Starting from the above operator split, a product formula algorithm is constructed in a standard fashion as follows. First, the elastic predictor problem is solved and a so–called *trial elastic state* is obtained. Then, the constraints (5) are checked for the trial state, and if they are violated, the trial state is taken as the initial condition for the plastic corrector problem. As compared to classical elastoplasticity (see, e.g., [SIM 87, SIM 97]), in the present formulation the plastic corrector problem remains unchanged, while the elastic predictor stage is modified to account for the effects of non–mechanical hardening.

Total	$\equiv$	Elastic predictor	+	Plastic corrector
$\dot{\epsilon} = \nabla^s(\dot{u})$		$\dot{\epsilon} = \nabla^s(\dot{u})$		$\dot{\epsilon} = 0$
$\dot{\epsilon}^e = \dot{\epsilon} - \dot{\gamma}Q$		$\dot{\epsilon}^e = \dot{\epsilon}$		$\dot{\epsilon}^e = -\dot{\gamma}Q$
$\dot{q} = \dot{\gamma}h + \dot{\theta}\eta$		$\dot{q} = \dot{\theta}\eta$		$\dot{q} = \dot{\gamma}h$
$\epsilon^e(t_n) = \epsilon_n^e$		$\epsilon^e(t_n) = \epsilon_n^e$		$\epsilon^e\big|_{(\dot{\gamma}=0)} = \epsilon_{n+1}^{e,tr}$
$q(t_n) = q_n$		$q(t_n) = q_n$		$q\big|_{(\dot{\gamma}=0)} = q_{n+1}^{tr}$

Table 1. *Operator split of the evolution problem.*

4.2. Problem 1: elastic predictor

The elastic predictor problem is the only stage of the numerical procedure in which the effect of non–mechanical hardening is taken into account. From the physical point of view, it can be derived from the original problem of evolution by *freezing* the plastic flow (i.e., setting $\dot{\gamma} = 0$), and taking an incremental *elastic* step which ignores the constraints placed on the stress state by the yield function. The solution of the predictor stage (*trial state*) in terms of elastic strains is trivially given by $\epsilon_{n+1}^{e,tr} = \epsilon_n^e + \epsilon_{n+1} - \epsilon_n$. As for the internal variables, in most of the constitutive models proposed in the framework of plasticity with extended hardening laws, the function η is sufficiently simple that the value of q_{n+1}^{tr} can also be evaluated in closed form. The trial state of stress, σ_{n+1}^{tr}, is obtained by mere function evaluation from $(\epsilon_{n+1}^{e,tr}, \theta_{n+1})$ as:

$$\sigma_{n+1}^{tr} := \frac{\partial \psi}{\partial \epsilon^e}\left(\epsilon_{n+1}^{e,tr}, \theta_{n+1}\right) \tag{12}$$

At the end of the elastic predictor stage, the trial state is checked for consistency with the yield locus. If $f_{n+1}^{tr} := f\left(\sigma_{n+1}^{tr}, q_{n+1}^{tr}\right) \leq 0$, the trial state satisfies the constraints imposed by the Kuhn–Tucker conditions. The process is then declared *elastic* and the trial state represents the actual final state of the material. If, on the contrary, $f_{n+1}^{tr} > 0$, the process is declared plastic, and consistency is restored by solving the plastic corrector problem. Differently from classical elastoplasticity, where the internal variables remain constant during the predictor stage, in the present case q_{n+1}^{tr} can be different from q_n due to non–mechanical effects, see eq. (4). This change might give rise to a violation of the constraint (5), and thus lead to the development of plastic strains and to changes in the stress state, even for processes in which no total strain changes occur, and $\sigma_{n+1}^{tr} = \sigma_n$.

4.3. Problem 2: plastic corrector

If $f^{tr}_{n+1} > 0$, the convexity of the yield surface implies that the trial state lies outside the yield locus, and thus cannot be accepted. Consistency is restored by solving the plastic corrector problem, which takes place at *fixed total strain* ($\dot{\epsilon} = 0$). Since the objective of the plastic corrector stage is to map the trial state back to the yield surface, the algorithms performing such task are commonly referred to as *return mapping algorithms*. In the present case, the plastic corrector problem is solved numerically by integrating the corresponding system of ordinary differential equations by an implicit *backward Euler* scheme, taking the trial state as the new initial condition:

$$\epsilon^e_{n+1} = \epsilon^{e,tr}_{n+1} - \Delta\gamma_{n+1}Q_{n+1} \tag{13}$$

$$q_{n+1} = q^{tr}_{n+1} + \Delta\gamma_{n+1}h_{n+1} \tag{14}$$

As $\Delta\gamma_{n+1} > 0$, the constraints (5) now reduce to:

$$f_{n+1} = f(\sigma_{n+1}, q_{n+1}) = 0 \tag{15}$$

Equations (13)–(15) represent a system of $7 + n_{int}$ non–linear algebraic equations in the $7 + n_{int}$ unknowns ϵ^e_{n+1}, $\Delta\gamma_{n+1}$, and q_{n+1}, which can be solved iteratively by Newton's method, at the Gauss point level.

Let:

$$x_{n+1} := \left\{\epsilon^{eT}_{n+1} \quad q^T_{n+1} \quad \Delta\gamma_{n+1}\right\}^T \in \mathbb{R}^{7+n_{int}} \tag{16}$$

be a vector containing the the unknowns of the problem. The return mapping equations (13)–(15) require the vanishing of the following *residual vector*:

$$R_{n+1}(x_{n+1}) := \begin{Bmatrix} r^\epsilon_{n+1} \\ r^q_{n+1} \\ f_{n+1} \end{Bmatrix} := \begin{Bmatrix} -\epsilon^e_{n+1} + \epsilon^{e,tr}_{n+1} - \Delta\gamma_{n+1}Q_{n+1} \\ -q_{n+1} + q^{tr}_{n+1} + \Delta\gamma_{n+1}h_{n+1} \\ f_{n+1} \end{Bmatrix} = 0 \tag{17}$$

The steps required for the iterative solution of eq. (17) via Newton's method are outlined in Table 2. A first difficulty in applying the procedure outlined in Table 2 is that Step 3 requires the inversion of a $(7 + n_{int}) \times (7 + n_{int})$ square matrix. By observing that the last component of the residual vector $R^{(j)}_{n+1}$ does not depend on $\Delta\gamma_{n+1}$, the resulting linearized system of equation can be reduced in size by one via static condensation. However, the inversion of the resulting tangent operator in closed form can still be very difficult, especially in presence of a large number of internal variables (i.e., in the case of anisotropic hardening models). In the most difficult cases, this problem can be solved by resorting to symbolic computation tools (as, e.g., MATHEMATICA or MAPLE) or by numerical methods, as in [TAM 02b]. Another classical problem in the application of GBE algorithms to complex, three–invariants plasticity models lies in the need for computing the gradients with respect to σ of the plastic flow direction Q (i.e., the second gradient of the plastic potential function) and of the hardening function h. In the most complex situations, this can be solved by the numerical evaluation of the relevant derivatives, as suggested, e.g., in [PER 00].

1. Initialize:
$$\epsilon_{n+1}^e = \epsilon_{n+1}^{e,\mathrm{tr}} \qquad q_{n+1} = q_{n+1}^{\mathrm{tr}} \qquad \Delta\gamma_{n+1} = 0$$

2. Check for convergence:

$$\text{IF: } \begin{cases} \left\| r_{n+1}^{\epsilon(j)} \right\| < TOL_\epsilon \cdot \left\| \epsilon_{n+1}^{e,\mathrm{tr}} \right\| \\ \left\| r_{n+1}^{q(j)} \right\| < TOL_q \cdot \left\| q_{n+1}^{\mathrm{tr}} \right\| \\ f_{n+1}^{(j)} < TOL_f \cdot p_s^{\mathrm{tr}} \end{cases} \text{ THEN exit, ELSE:}$$

3. Find update at local iteration (j):

$$\delta x_{n+1}^{(j)} = - \left[\left(\frac{\partial R}{\partial x} \right)_{n+1}^{(j)} \right]^{-1} R_{n+1}^{(j)}$$

4. Update state variables and plastic multiplier:

$$x_{n+1}^{(j+1)} = x_{n+1}^{(j)} + \delta x_{n+1}^{(j)}$$

5. Set: $j \leftarrow j + 1$, GO TO 2.

Table 2. *Iterative solution of the plastic corrector problem.*

5. Return mapping algorithm in principal elastic strain space

A considerable simplification in the application of the GBE algorithm to complex plasticity models can be obtained in the case of isotropic–hardening plasticity, by formulating the return mapping stage in principal elastic strain space. By exploiting the spectral decomposition of the tensors Q_{n+1}, ϵ_{n+1}^e and $\epsilon_{n+1}^{e,\mathrm{tr}}$, eq. (13) transforms into:

$$\sum_{A=1}^{3} (\epsilon_A^e)_{n+1} \, n_{n+1}^{(A)} \otimes n_{n+1}^{(A)} = \sum_{A=1}^{3} (\epsilon_A^{e,\mathrm{tr}})_{n+1} \, n_{n+1}^{(A),\mathrm{tr}} \otimes n_{n+1}^{(A),\mathrm{tr}} -$$

$$\Delta\gamma_{n+1} \sum_{A=1}^{3} (Q_A)_{n+1} \, n_{n+1}^{(A)} \otimes n_{n+1}^{(A)} \quad (18)$$

in which $n_{n+1}^{(A)}$ and $n_{n+1}^{(A),\mathrm{tr}}$ are the A–th unit eigenvectors of ϵ_{n+1}^e and $\epsilon_{n+1}^{e,\mathrm{tr}}$. Then, it follows at once that:

$$n_{n+1}^{(A)} = n_{n+1}^{(A),\mathrm{tr}} \quad (19)$$

and:

$$(\epsilon_A^e)_{n+1} = (\epsilon_A^{e,\mathrm{tr}})_{n+1} - \Delta\gamma_{n+1} (Q_A)_{n+1} \tag{20}$$

for $A = 1$, 2 or 3. Note that, as the trial elastic strain is known, so are its principal directions. Therefore, the only unknown quantities to be determined remain the three principal elastic strains $(\epsilon_A^e)_{n+1}$, the n_{int} internal variables q_{n+1} and the plastic multiplier $\Delta\gamma_{n+1}$. Introducing for convenience the following vector notation:

$$\hat{\epsilon}^e := \begin{Bmatrix} \epsilon_1^e \\ \epsilon_2^e \\ \epsilon_3^e \end{Bmatrix} \qquad \hat{\epsilon}^{e,\mathrm{tr}} := \begin{Bmatrix} \epsilon_1^{e,\mathrm{tr}} \\ \epsilon_2^{e,\mathrm{tr}} \\ \epsilon_3^{e,\mathrm{tr}} \end{Bmatrix} \qquad \hat{\sigma} := \begin{Bmatrix} \sigma_1 \\ \sigma_2 \\ \sigma_3 \end{Bmatrix} \qquad \hat{Q} := \begin{Bmatrix} \partial g/\partial\sigma_1 \\ \partial g/\partial\sigma_2 \\ \partial g/\partial\sigma_3 \end{Bmatrix} \tag{21}$$

the return mapping problem in principal elastic strain space can be recast as follows:

$$\hat{\epsilon}_{n+1}^e = \hat{\epsilon}_{n+1}^{e,\mathrm{tr}} - \Delta\gamma_{n+1}\hat{Q}_{n+1} \tag{22}$$

$$q_{n+1} = q_{n+1}^{\mathrm{tr}} + \Delta\gamma_{n+1}h_{n+1} \tag{23}$$

$$f_{n+1} := f(\hat{\sigma}_{n+1}, q_{n+1}) = 0 \tag{24}$$

The iterative solution of the return mapping equations (24) follows a scheme similar to that in Table 2. The number of equations to be solved is now reduced by 3. Moreover, only the evaluation of the (3×3) matrix:

$$\nabla\nabla g = \frac{\partial^2 g}{\partial\hat{\sigma} \otimes \partial\hat{\sigma}} \tag{25}$$

is now required to compute the tangent operator $\partial R/\partial x$.

6. Consistent tangent matrix

In a standard finite element context, the starting point for the solution of a static equilibrium problem is the weak form of the balance of momentum equation, which, for the problem at hand, is stated as follows. To find the unknown functions $u(x)$ and $\theta(x)$ such that for any test function $w(x)$ satisfy homogeneous boundary conditions on the appropriate part of the boundary, the following nonlinear functional equation is satisfied:

$$G(u, \theta, w) = \int_\Omega \nabla^s w \cdot \sigma(u, \theta) \, dV - \int_\Omega \rho\, w \cdot b \, dV - \int_{\Gamma_t} w \cdot t \, dA = 0 \tag{26}$$

In the above equation, non–linearity stems from the non–linear dependence of the stress tensor on u and θ induced by the constitutive equation. The iterative solution via Newton's method of the nonlinear algebraic problem resulting after the introduction of a standard finite element discretization, requires the linearization of the non–linear functional G with respect to the independent fields u and θ:

$$D_u G \left(u_{n+1}^{(k)}, \theta_{n+1}^{(k)}; w \right) \cdot \delta u_{n+1}^{(k)} = \int_\Omega \left\{ \nabla^s w \cdot \left(\widetilde{D} \right)_{n+1}^{(k)} \nabla^s (\delta u)_{n+1}^{(k)} \right\} dV \quad (27)$$

$$D_\theta G \left(u_{n+1}^{(k)}, \theta_{n+1}^{(k)}; w \right) \delta \theta_{n+1}^{(k)} = \int_\Omega \left\{ \nabla^s w \cdot \left(\widetilde{D}_\theta \right)_{n+1}^{(k)} \delta \theta_{n+1}^{(k)} \right\} dV \quad (28)$$

in which:

$$\left(\widetilde{D} \right)_{n+1}^{(k)} := \frac{\partial \sigma_{n+1}^{(k)}}{\partial \epsilon_{n+1}^{(k)}} \qquad\qquad \left(\widetilde{D}_\theta \right)_{n+1}^{(k)} := \frac{\partial \sigma_{n+1}^{(k)}}{\partial \theta_{n+1}^{(k)}} \quad (29)$$

In eq. (29), $\widetilde{D}_{n+1}^{(k)}$ is the so–called *consistent tangent stiffness matrix* for the proposed elastoplastic model, see [SIM 85], while $(\widetilde{D}_\theta)_{n+1}^{(k)}$ represents the tangent map of the non–linear function $\sigma_{n+1}(\theta_{n+1})$ as defined by the stress–point algorithm. As discussed by Simo & Taylor [SIM 85], both quantities in eq. (15) depend crucially on the adopted integration procedure. One of the advantages of the proposed algorithm is the possibility of evaluating the consistent tangent operators *in closed form*. In the following, the derivation of the tangent stiffness $\widetilde{D}_{n+1}^{(k)}$ will be discussed in detail for the case of a plastic loading process[1]. The calculation of $(\widetilde{D}_\theta)_{n+1}^{(k)}$ can be easily performed along the same lines, see [TAM 02b].

In the *global* iteration process, any (infinitesimal) variation in the total strain increment induces, by definition, an equal variation in the trial elastic strain:

$$d\epsilon = d\epsilon^{e,\mathrm{tr}} \quad (30)$$

where the subscript $n+1$ and the superscript (k) have been omitted to ease the notation. Moreover, the return mapping equations associate to each trial elastic strain a well defined elastic strain tensor, obtained as a result of the local iteration process. Therefore, for an infinitesimal variation of $\epsilon_{n+1}^{e,\mathrm{tr}(k)}$ one has:

$$d\epsilon^e = L \, d\epsilon^{e,\mathrm{tr}} \quad (31)$$

On the other hand, from the hyperelastic constitutive equation, we have:

$$d\sigma = D^e \, d\epsilon^e = D^e L \, d\epsilon^{e,\mathrm{tr}} = \Xi \, d\epsilon^{e,\mathrm{tr}} \quad (32)$$

By virtue of the definition (29)$_1$ and of the identity (30), the tensor Ξ is the required consistent tangent stiffness tensor. Differentiation of the return mapping equations (13) and (14) yields:

$$A \, d\widetilde{x} = T \, d\epsilon^{e,\mathrm{tr}} - d(\Delta\gamma) U \quad (33)$$

---

1. The case of an elastic process is trivial.

where:

$$d\widetilde{x} := \left\{ (D^e\, d\epsilon^e)^T \quad dq^T \right\}^T \qquad A := \begin{bmatrix} I + \Delta\gamma\, A_\sigma & +\Delta\gamma\, A_q \\ -\Delta\gamma\, B_\sigma & I_q - \Delta\gamma\, B_q \end{bmatrix} \qquad (34)$$

$$T := \begin{bmatrix} I & 0_q \end{bmatrix}^T \qquad U := \left\{ Q^T, -h^T \right\}^T \qquad (35)$$

$$A_\sigma := \frac{\partial Q}{\partial \sigma} = \frac{\partial^2 g}{\partial \sigma \otimes \partial \sigma} \qquad A_q := \frac{\partial Q}{\partial q} = \frac{\partial^2 g}{\partial \sigma \otimes \partial q} \qquad (36)$$

$$B_\sigma := \frac{\partial h}{\partial \sigma} \qquad B_q := \frac{\partial h}{\partial q} \qquad (37)$$

and I_q and 0_q are the identity tensor and the zero tensor in $\mathbb{R}^{n_{int}}$. The variation in the plastic multiplier $d(\Delta\gamma)$ can be evaluated by enforcing the consistency condition $df_{n+1}^{(k)} = 0$. Defining:

$$P := \frac{\partial f}{\partial \sigma} \qquad W := \frac{\partial f}{\partial q} \qquad V := \{P, W\} \qquad (38)$$

the consistency condition reads:

$$V \cdot d\widetilde{x} = 0 \qquad (39)$$

Substituting $d\widetilde{x}$ from eq. (33) and solving for $d(\Delta\gamma)$, the plastic multiplier increment is obtained as:

$$d(\Delta\gamma) = \frac{1}{V \cdot [A^{-1}]\, U}\, V \cdot [A^{-1}]\, T\, d\epsilon^{e,tr} \qquad (40)$$

and then, from eqs. (33) and (40) we have:

$$d\sigma = D^e\, d\epsilon^e = T^T\, d\widetilde{x} = \left\{ T^T \left[A^{-1} - \frac{(A^{-1}U) \otimes (V A^{-1})}{(V \cdot A^{-1}U)} \right] T \right\} d\epsilon^{e,tr} \qquad (41)$$

By comparing eq. (41) with (32) the expression for the consistent tangent stiffness tensor follows:

$$\widetilde{D}_{n+1}^{(k)} = \Xi_{n+1}^{(k)} = T^T \left[A^{-1} - \frac{(A^{-1}U) \otimes (V A^{-1})}{(V \cdot A^{-1}U)} \right]_{n+1}^{(k)} T \qquad (42)$$

Note that the consistent tangent stiffness is, in general, *non symmetric*, even in the case of associative flow rule ($f \equiv g$). The evaluation of the consistent tangent stiffness tensor for the special case of isotropic hardening is discussed in detail in [TAM 02b] and is not reported here for brevity.

7. Examples of application

As an example of application of the proposed numerical procedure to a specific case, we consider the isotropic hardening model proposed by [TAM 02b] to describe

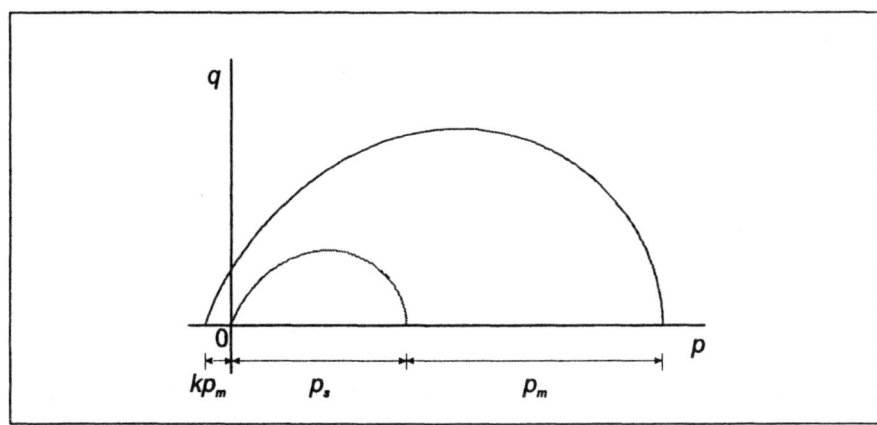

Figure 1. *Yield surface in the triaxial plane.*

the effects of chemical weathering of hard, cemented soils or soft rocks. The key features of the model are: i) the introduction of a (non–increasing) bonding–related internal variable p_m, in addition to the classical preconsolidation pressure p_s, as in, e.g., [LAG 95], in order to account for the effects of bonding on the current size of the yield surface (see Figure 1); and, ii) a generalized hardening law of the type (4), which relates $\dot{p}_m(< 0)$ to a suitable invariant scalar measure of the plastic strain rate (mechanical degradation) and to the time rate of a normalized degradation parameter X_d ($X_d = 0$ denoting the intact material, $X_d = 1$ a state of complete degradation). The specific details of the constitutive model, including the formulation of the elastic constitutive equation and the expressions adopted for the yield surface, f, the plastic potential, g, and the hardening functions h and η, as well as the numerical values of the relevant material constants adopted can be found in [TAM 02b] and in the paper by Castellanza et al. [CAS 02], in this volume.

In the following, the results of a series of numerical tests are briefly discussed to demonstrate the algorithm performance at the single–element level. The performance of the algorithm at the BVP level is examined in [TAM 02b, CAS 02], with reference to the analysis of a circular footing on a soft rock layer undergoing chemical degradation. For each test, the accuracy of the proposed integration algorithm has been evaluated by comparing the different numerical solutions with a reference,"exact" solution of the evolution problem, (σ^*, q^*), obtained by integrating the constitutive equations with an adaptive Runge–Kutta–Fehlberg method of the third order (RKF–23) in connection with an extremely low prescribed relative error tolerance ($TOL_{RKF} = 1.0 \cdot 10^{-10}$), see [TAM 02b] for details.

Figure 2 shows the results of three isochoric axisymmetric compression (TX–CU) tests starting from an initial isotropic state with $p_0 = 200$ kPa, $p_{s0} = 200$ kPa and $p_{m0} = 100$ kPa, in which a total deviatoric strain increment $\Delta\epsilon_s = 1.5 \cdot 10^{-2}$ is applied in 5, 20 and 100 steps, respectively. In all cases, the response is elastic until the

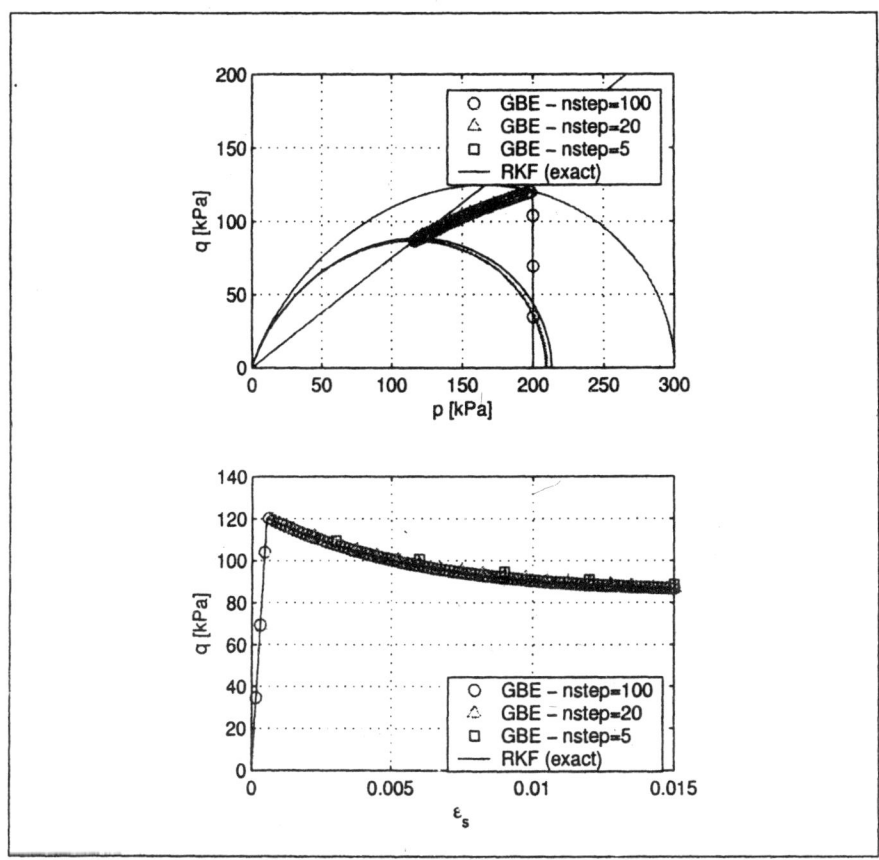

Figure 2. *TX-CU test results.*

stress–path touches the initial yield surface. After yield, a brittle response is predicted, due to the softening induced by mechanical debonding.

The stress–paths computed in a number of *chemical relaxation* tests are plotted in Figure 3, in the $q:p$ plane. In these tests, the geometry of the material is kept unchanged (i.e., no strain increments are applied), while a full degradation of interparticle bonding from $X_d = 0$ to $X_d = 1$ is imposed in 5, 20 and 100 steps, respectively, starting from a (anisotropic) initial state with $p_0 = 200$ kPa, $q_0 = 161$ kPa, $p_{s0} = 200$ kPa and $p_{m0} = 300$ kPa. These last results deserve some comment. In the first part of the tests, up to $X_d \simeq 0.42$, the only effect of chemical degradation is a reduction in size of the yield locus. During this stage, the stress point remains inside the yield surface, and neither plastic nor elastic strains can occur. However, as soon as the yield surface touches the current stress state, the stress point is forced to move downwards, in order to remain on the shrinking yield locus. The consistency condition then requires the development of plastic strains, which – due to the imposed kinematic restrictions – are

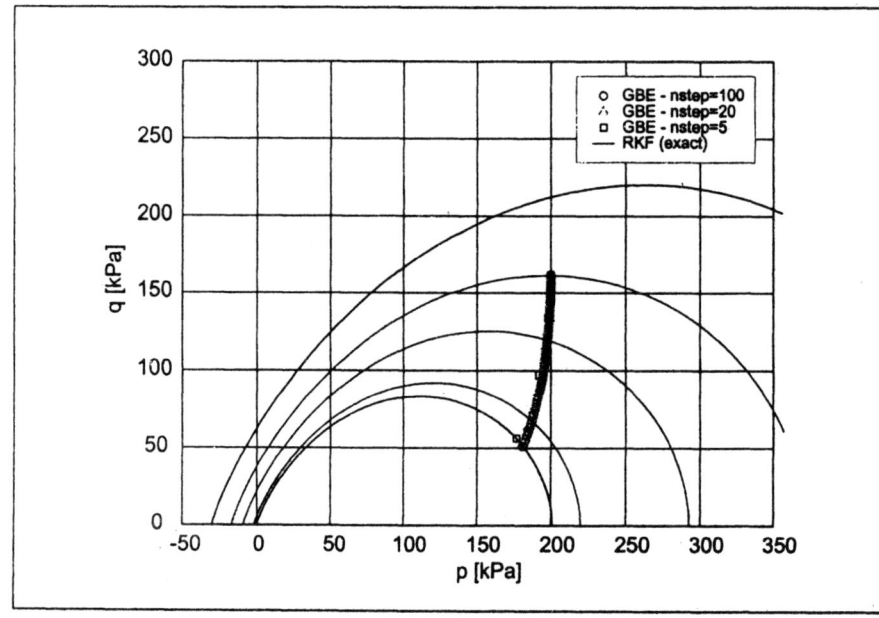

Figure 3. *Chemical relaxation test results.*

associated with elastic strains of equal magnitude and opposite sign. At the beginning of the plastic loading process, the plastic flow direction is almost parallel to the q–axis, and thus the stress changes occur at an almost constant mean stress; however, as the yield locus shrinks, the current state moves progressively to the right side of the yield surface, and the volumetric component of the plastic strain increases. As any plastic volumetric compression corresponds to an equal elastic dilation, the mean stress p is reduced accordingly, and the stress path bends to the left until the final yield locus, at a completely debonded state, is reached.

In all the simulations performed, the numerical solutions are convergent, in the sense that the numerical predictions get closer and closer to the reference "exact" solution as the step size is decreased. As a matter of fact, in all the cases examined, even the results obtained with the larger step size appear quite accurate. The above results can be generalized to an entire class of possible strain and/or chemical degradation paths by means of the so–called *isoerror maps* [SIM 97]. In our case, in order to construct two–dimensional isoerror maps, we have restricted our investigation to plane strain loading paths ($\Delta\epsilon_3 = 0$) in which $\Delta X_d = $ const. The isoerror map corresponding to a initial state with $p_0 = 268.8$ kPa, $q_0 = 63.4$ kPa, $p_{s0} = 189.3$ kPa and $p_{m0} = 100.0$ kPa, and to a degradation increment $\Delta X_d = 11\%$ is shown in Figure 4. In the figure, the coordinate axes represent the imposed principal strain increments $\Delta\epsilon_1$ and $\Delta\epsilon_2$, normalized with respect to a reference volumetric elastic strain $\hat{\kappa} = 0.008$.

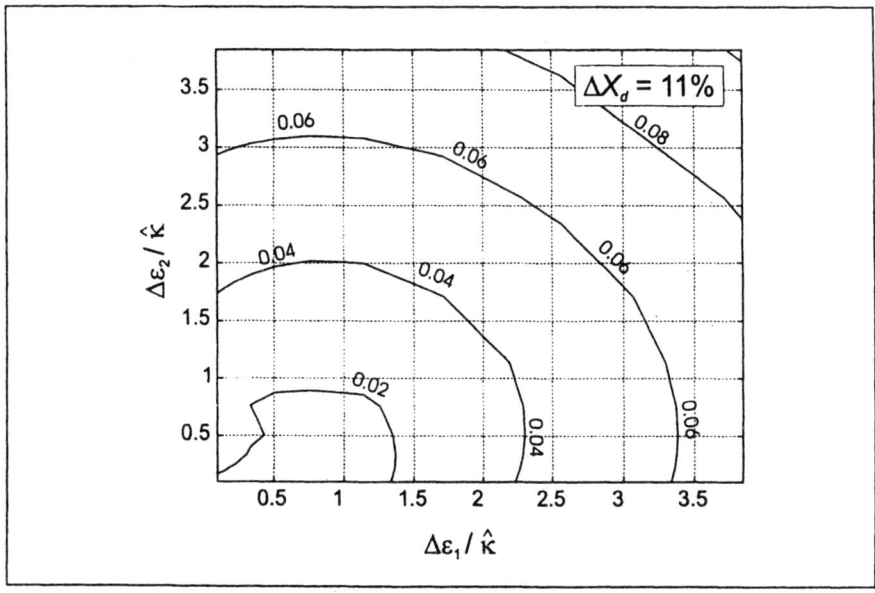

Figure 4. *Isoerror map for plane strain.*

At each point in the diagram, the isolines provide the value of the relative integration error ERR, defined as:

$$ERR := \frac{\sqrt{\|\sigma - \sigma^*\|^2 + \|q - q^*\|^2}}{\sqrt{\|\sigma^*\|^2 + \|q\|^2}} \tag{43}$$

where the pair (σ, q) represents the numerical solution given by the algorithm when the prescribed strain and degradation increments are applied *in a single step*.

Results in Figure 3 show that the integration error remains smaller than 5% for a relatively large region of the loading space, thus generalizing the previous observations to a much larger class of loading paths. Similar tests performed with $\Delta X_d = 0$ seem to indicate that the presence of chemical degradation has only a minor influence on the accuracy of the algorithm, in the particular range of values considered for the principal strain and chemical degradation increments.

8. References

[ALA 92] ALAWAJI H., RUNESSON K., STURE S., "Implicit integration in soil plasticity under mixed control for drained and undrained response", *Int. J. Num. Anal. Meth. Geomech.*, vol. 16, 1992, p. 737–756.

[BOR 90] BORJA R. I., LEE S. R., "Cam–Clay plasticity, Part I. Implicit integration of elasto-plastic constitutive relations", *Comp. Meth. Appl. Mech. Engng.*, vol. 78, 1990, p. 49–72.

[BOR 91]   BORJA R. I., "Cam–Clay plasticity, Part II. Implicit integration of constitutive equation based on a non–linear elastic stress predictor", *Comp. Meth. Appl. Mech. Engng.*, vol. 88, 1991, p. 225–240.

[BOR 98]   BORJA R. I., TAMAGNINI C., "Cam–Clay plasticity, Part III: Extension of the infinitesimal model to include finite strains", *Comp. Meth. Appl. Mech. Engng.*, vol. 155, 1998, p. 73–95.

[CAS 02]   CASTELLANZA R., NOVA R., TAMAGNINI C., "Mechanical effects of chemical degradation of bonded geomaterials in boundary value problems", *Revue Francaise de Génie Civil*, , 2002.

[GEN 88]   GENS A., POTTS D. M., "Critical state models in computational geomechanics", *Engineering Computations*, vol. 5, 1988, p. 178–197.

[HAI 91]   HAIRER E., WANNER G., *Solving Ordinary Differential Equations II. Stiff and Differential–Algebraic Problems, 2nd. Ed.*, Springer Verlag, New York, 1991.

[HUG 84]   HUGHES T. J. R., "Numerical implementation of constitutive models: rate–independent deviatoric plasticity", NEMAT-NASSER S., ASARO R., HEGEMIER G., Eds., *Theoretical Foundations for Large Scale computations of Non Linear Material Behavior*, Martinus Nijhoff Publisher, Dordrecht, 1984, p. 29–57.

[JER 97]   JEREMIĆ B., STURE S., "Implicit integration in elastoplastic geotechnics", *Mech. Cohesive–Frictional Materials*, vol. 2, 1997, p. 165–183.

[LAG 95]   LAGIOIA R., NOVA R., "An experimental and theoretical study of the behavior of a calcarenite in triaxial compression", *Géotechnique*, vol. 45, num. 4, 1995, p. 633–648.

[MAC 97]   MACARI E. J., WEIHE S., ARDUINO P., "Implicit integration of elastoplastic constitutive models for frictional materials with highly non–linear hardening functions", *Mech. Cohesive–Frictional Materials*, vol. 2, 1997, p. 1–29.

[ORT 86]   ORTIZ M., SIMO J. C., "Analysis of a new class of integration algorithms for elastoplastic constitutive relations", *Int. J. Num. Meth. Engng.*, vol. 23, 1986, p. 356–366.

[PER 00]   PEREZ-FOGUET A., RODRIGUEZ-FERRAN A., HUERTA A., "Numerical differentiation for non-trivial consistent tangent matrices: an application to the MRS-Lade model", *Int. J. Num. Meth. Engng.*, vol. 48, 2000, p. 159–184.

[PIJ 02]   PIJAUDIER-CABOT J., "Constitutive models for brittle materials", *Revue Francaise de Génie Civil*, , 2002.

[POT 97]   POTTS D. M., GANENDRA D., "Evaluation of substepping and implicit stress point algorithms", *Comp. Meth. Appl. Mech. Engng.*, vol. 119, 1997, p. 341–354.

[SIM 85]   SIMO J. C., TAYLOR R. L., "Consistent tangent operators for rate independent elasto–plasticity", *Comp. Meth. Appl. Mech. Engng.*, vol. 48, 1985, p. 101–118.

[SIM 87]   SIMO J. C., HUGHES T. J. R., "General return mapping algorithms for rate–independent plasticity", DESAI C. et al., Eds., *Constitutive Laws for Engineering Materials*, Horton, Greece, 1987, Elsevier Science Publishing.

[SIM 91]   SIMO J. C., GOVINDJEE S., "Non–linear B–stability and symmetry preserving return mapping algorithms for plasticity and viscoplasticity", *Int. J. Num. Meth. Engng.*, vol. 31, 1991, p. 151–176.

[SIM 97]   SIMO J. C., HUGHES T. J. R., *Computational Inelasticity*, Springer Verlag, New York, 1997.

[SLO 87] SLOAN S. W., "Substepping schemes for the numerical integration of elastoplastic stress–strain relations", *Int. J. Num. Meth. Engng.*, vol. 24, 1987, p. 893–911.

[TAM 02a] TAMAGNINI C., CASTELLANZA R., NOVA R., "Numerical integration of elasto-plastic constitutive equations for geomaterials with extended hardening rules", PANDE G., PIETRUSCZCZAK S., Eds., *NUMOG VIII*, Rome, Italy, 2002, Balkema, Rotterdam, p. 213–218.

[TAM 02b] TAMAGNINI C., CASTELLANZA R., NOVA R., "A Generalized Backward Euler algorithm for the numerical integration of an isotropic hardening elastoplastic model for mechanical and chemical degradation of bonded geomaterials", *Int. J. Num. Anal. Meth. Geomech.*, , 2002, Page In print.

[TAM 02c] TAMAGNINI C., VIGGIANI G., "Constitutive modelling for rate–independent soils: a review", *Revue Francaise de Génie Civil*, , 2002.

Chapter 9

Non Linear Problems: Advanced Techniques

Pablo Mira and Manuel Pastor

Centro de Estudios y Experimentación de Obras Públicas, CEDEX, Ministerio de Fomento, Madrid, Spain, and ETS Ingenieros de Caminos, Universidad Politécnica de Madrid, Spain

1. Introduction

As mentioned at the beginning of the chapter dealing with the introductory aspects of non linear problems, "standard" non linear analysis techniques as those presented in that chapter are often not sufficient to solve a specific problem. An example of this are limit load problems with a strongly descending post peak behavior. In such cases, special non linear analysis techniques are necessary. The present chapter will deal with two of these special techniques, namely : arc-length control and line search methods.

2. The Arc-length method

The simplest version of the Newton-Raphson method applied to the finite element method in the context of deformable solid mechanics is usually known as load control. These terms refer to the fact that at the beginning of each step the load level is fixed. Having fixed the value of the load vector an iterative process is carried out to compute the displacements at the nodes. This iterative process ends when equilibrium between internal and external forces is satisfied. This version of the Newton-Raphson method was presented in the previous chapter.

However, in problems that exhibit a force-displacement diagram such as the one of figure 1, this type of algorithm is not adequate since the limit load or upper bound of this curve is not known a priori. If the load control method were used to solve such problems, since the value of the limit load is not known a priori, and the load level at the beginning of each step is increased up to a certain level, a step would ultimately be reached where the value of the load vector would be higher than the limit load. At this point it would be impossible to make the iterative process converge because the norm of the residual forces would always have a finite value. In graphical terms, assuming that the solution at each step is obtained by intersection of the curve and a horizontal line drawn at the level of the fixed load value, if this value were higher than the limit load there would be no such intersection.

Following one step further this graphical type of reasoning, one would expect that fixing the displacement level instead of the force level would be a satisfactory solution for the Figure 1 type of problem. However, this is only so for some limit load prob-lems. In an elastoplastic material type context only, in problems exhibiting a positive, zero or very slightly negative material hardening coefficient, would a displacement control approach be adequate. This is so because this type of graphical reasoning is only a 2D approximation of a problem which is really multidimensional and is there-fore difficult to explain on a piece of paper.

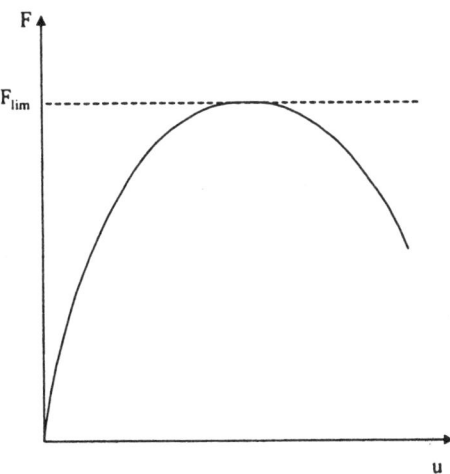

Figure 1. *Limit load problem with descending post peak behaviour*

The arc-length control is an interesting alternative method to overcome limit points in problems such as the one presented in Figure 1. This method is based on a mixed type of control including forces and displacements.

Until the arc-length control appeared, standard techniques based on load or displacement control were able at best to overcome limit points with some difficulty and very frequently simply failed to do so and the process stalled slightly before the limit point. With arc-length control it is possible to overcome limit points in an automatic fashion both for snap-through (Figure 2.a) and snap-back (Figure 2.b) situations.

Very often, limit points constitute the beginning of structural collapse. One could, therefore, pose the following question: What is the use of obtaining the force-displacement diagram after the collapse of the structure has taken place? Crisfield [CRI 91] produces several answers to this question:

1) It is possible that the point identified as a limit point is not such, and is only a local maximum. The only way to confirm that a point is really a limit point is to overcome it. Frequently, when using standard techniques such as load and displacement control, as the process approaches the limit point serious convergence problems appear. Given the impossibility to overcome such points with standard techniques, the assumption is frequently made that convergence problems are caused by the existence of a limit point.

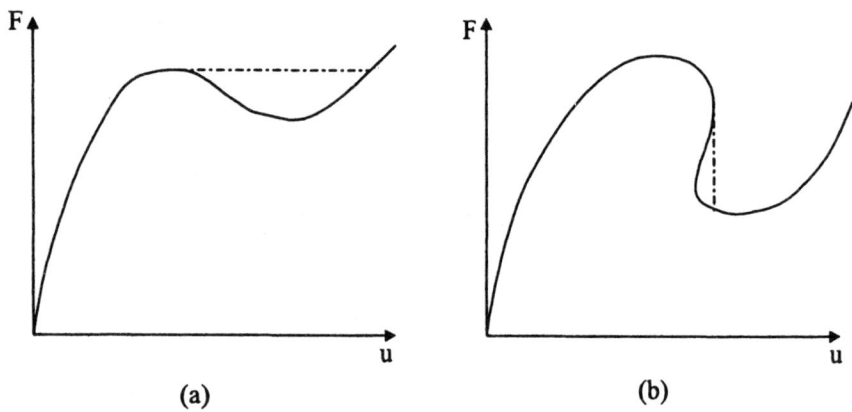

Figure 2. *(a) Snap through problem (b) Snap back problem*

2) It is possible that the point produces only the collapse of a substructure and not global structural collapse. Sometimes it is also important to know whether the collapse is brittle or ductile or to investigate the stress state of different structural components.

The arc-length control consists basically of the following steps :

1) The starting point is the standard set of N equilibrium equations which may be expresssed as:

1)

$$f_{int}(\mathbf{u}) - \lambda f_{ext} = 0$$

Where :

N = Number of degrees of freedom of the system

$\mathbf{u}$ = N component vector representing displacements of the system.

λ =Scalar representing the system load level.

f_{int} = N component vector containing internal forces of the system

$\qquad = A_{e=1,nelem} \left[\int_e \mathbf{B}^T \sigma(\epsilon(\mathbf{u}))d_e \right]$

f_{ext}= N component vector containing external forces of the system .

2) A length restriction is enforced on a norm based on a combination of the displacement and external force incremental vectors.

$$\Delta u^T \Delta u + \Delta \lambda^2 \varphi^2 f_{ext}^T f_{ext} = \Delta l^2 \qquad (1)$$

where:

Δu = Present incremental displacement vector.

$\Delta \lambda$ = Present incremental load level factor

φ = A constant with a value set at the beginning of the computations. This factor controls the relative importance of the displacement and external force vectors in the restriction. Further comments will be made on this value in the following sections

There are many versions of the arc-length method. Although the method was originally introduced by Riks[RIK 79],[RIK 72] and Wempner [WEM 71], in the following sections the version due to Crisfield [CRI 91] will be presented. One of the main differences between the original versions and Crisfield's is that in the former the restriction is enforced at the same time as the equations yielding a N+1 dimension unsymmetric equation system while in the latter the restriction is enforced after solving a symmetric equation system with N unknowns. This, of course, is assuming that the N dimension original equilibrium equation system is symmetric. Symmetry is evidently a very important issue from the point of view of computational cost. The preservation of symmetry of the global equation system reduces significantly the number of operations and the CPU time.

2.1. Analytical description of the method

Before introducing the arc-length method we shall define the residual force vector G as :

$$\Psi(u,\lambda) = f_{int}(u) - \lambda f_{ext} = A_{e=1,nelem} \left[\int_{\Omega e} B^T \sigma(\epsilon(u)) d\Omega_e \right] - \lambda f_{ext} \qquad (2)$$

This vector represents a measure of the degree of convergence of the iterative process. Given the scalar λ representing the external force factor along with the displacements u, and residual force $\Psi(u,\lambda)$ vectors, the problem consists in finding the displacement and load factor corrections δu and $\delta \lambda$ such that the new residual force vector will be zero.

$$\Psi_n = \Psi_o + \frac{\partial \Psi}{\partial u} \delta u + \frac{\partial \Psi}{\partial \lambda} \delta \lambda = \Psi_o + K_t \delta u - f_{ext} \delta \lambda = 0 \qquad (3)$$

Subindices "o" and "n" refer to the old and new situation respectively, or in other words before and after the present iteration. It is easy to prove from equation (2) that

$$\frac{\partial \Psi}{\partial u} = \frac{\partial(A_{e=1,nelem} \left[\int_{\Omega e} \mathbf{B}^T \boldsymbol{\sigma}(\epsilon)d\Omega_e \right])}{\partial u} = A_{e=1,nelem} \left[\int_{\Omega e} \mathbf{B}^T \frac{\partial \boldsymbol{\sigma}(\epsilon)}{\partial u} d\Omega e \right]$$

$$= A_{e=1,nelem} \left[\int_{\Omega e} \mathbf{B}^T \frac{\partial \boldsymbol{\sigma}(\epsilon)}{\partial \epsilon} \frac{\partial \epsilon}{\partial u} d\Omega e \right] = A_{e=1,nelem} \left[\int_{\Omega e} \mathbf{B}^T \mathbf{D}^{ep} \mathbf{B} d\Omega e \right] = \mathbf{K}_t$$

$$\frac{\partial \mathbf{G}}{\partial \lambda} = \mathbf{f}_{ext}$$

One must keep in mind that δu and $\delta \lambda$ are iterative corrections while Δu and $\Delta \lambda$ that will appear in further developments in this section are incremental corrections that come from the accumulation of iterative corrections starting at the beginning of the increment.

Solving for δu in equation (3):

$$\delta u = -\mathbf{K}_t^{-1}(\Psi_o - \delta \lambda \mathbf{f}_{ext})$$

$$\delta u = -\mathbf{K}_t^{-1} \Psi_o + \delta \lambda \mathbf{K}_t^{-1} \mathbf{f}_{ext} = \delta \overline{u} + \delta \lambda \delta u_t \qquad (4)$$

As one may see in equation (4) $\delta \overline{u}$ coincides with the classical displacement iterative correction while the term $\delta \lambda \delta u_t$ allows the procedure to adapt to limit points through a small change in $\delta \lambda$.

It is now possible to update incremental displacements and load factors:

$$\Delta u_n = \Delta u_o + \delta u = \Delta u_o + \delta \overline{u} + \delta \lambda \delta u_t$$

$$\Delta \lambda_n = \Delta \lambda_o + \delta \lambda \qquad (5)$$

Enforcing the restriction from equation (1) :

$$(\Delta u_n^T \Delta u_n + \Delta \lambda_n^2 \varphi^2 \mathbf{f}_{ext}^T \mathbf{f}_{ext}) = \Delta l^2 \qquad (6)$$

and substituting equation (5) into (6) :

$$a_1 \delta \lambda^2 + a_2 \delta \lambda + a_3 = 0 \qquad (7)$$

where:

$$a_1 = \delta u_t^T \delta u_t + [\varphi^2 \mathbf{f}_{ext}^T \mathbf{f}_{ext}] \qquad (8)$$

$$a_2 = 2\delta u_t^T (\Delta u_o + \delta \overline{u}) + [2\Delta \lambda_o \varphi^2 \mathbf{f}_{ext}^T \mathbf{f}_{ext}] \qquad (9)$$

$$a_3 = (\Delta\mathbf{u}_o + \delta\overline{\mathbf{u}})^T(\Delta\mathbf{u}_o + \delta\overline{\mathbf{u}}) - \Delta l^2 + [\Delta\lambda_o^2 \varphi \mathbf{f}_{ext}^T \mathbf{f}_{ext}] \qquad (10)$$

The next step now is to solve (7) for $\delta\lambda$ and choose one of the two possible solutions. The most usual criterion to do so is to choose the $\delta\lambda$ such that the resulting $\Delta\mathbf{u}_n$ is closest to $\Delta\mathbf{u}_o$, or in other words the angle between them is the smallest. It is therefore necessary to compute the cosine of this angle based on the scalar product. The procedure to make the choice can therefore be expressed as :

$$\Delta\mathbf{u}_n^1 = \Delta\mathbf{u}_o + \delta\mathbf{u}_1 = \Delta\mathbf{u}_o + \delta\overline{\mathbf{u}} + \delta\lambda_1\delta\mathbf{u}_t$$

$$\Delta\mathbf{u}_n^2 = \Delta\mathbf{u}_o + \delta\mathbf{u}_2 = \Delta\mathbf{u}_o + \delta\overline{\mathbf{u}} + \delta\lambda_2\delta\mathbf{u}_t$$

$$\cos\theta_1 = \frac{\Delta\mathbf{u}_o \cdot \Delta\mathbf{u}_n^1}{\|\Delta\mathbf{u}_o\| \, \|\Delta\mathbf{u}_n^1\|}$$

$$\cos\theta_2 = \frac{\Delta\mathbf{u}_o \cdot \Delta\mathbf{u}_n^2}{\|\Delta\mathbf{u}_o\| \, \|\Delta\mathbf{u}_n^2\|}$$

choosing $\delta\lambda_i$ such that $\cos\theta_i = \max(\cos\theta_1, \cos\theta_2)$.

The scheme of the procedure is presented in Figure 3.

An aspect of the procedure which has not been covered is the computation of $\Delta\lambda$ y Δl at the beginning of the increment. Assuming $\varphi = 0$, the usual way to procede is to fix the value of $\Delta\lambda$ at the beginning of the first increment and compute the equivalent value of Δl based on the restriction:

$$\left. \begin{array}{c} \Delta\mathbf{u}_1^T\Delta\mathbf{u}_1 = \Delta l^2 \\ \Delta\mathbf{u}_1 = \mathbf{K}_t^{-1}\Delta\lambda_1\mathbf{f}_{ext} = \Delta\lambda_1\delta\mathbf{u}_t \end{array} \right\} \Longrightarrow \Delta l = \Delta\lambda_1\sqrt{\delta\mathbf{u}_t^T\delta\mathbf{u}_t}$$

The subindex of $\Delta\lambda_1$ and $\Delta\mathbf{u}_1$ refers to the first iteration of the increment. For the first iteration of increments different from the first one, the order of the procedure would be the contrary, that is obtaining $\Delta\lambda_1$ from Δl:

$$\Delta\lambda_1 = sign\frac{\Delta l}{\sqrt{\delta\mathbf{u}_t^T\delta\mathbf{u}_t}} \qquad (11)$$

The value of variable sign in equation 11 will be decided based on the following criterion:

a) If $\delta\mathbf{u}_t^T\mathbf{K}_t\delta\mathbf{u}_t > 0 \Rightarrow$ before limit point $\Rightarrow sign = +1$

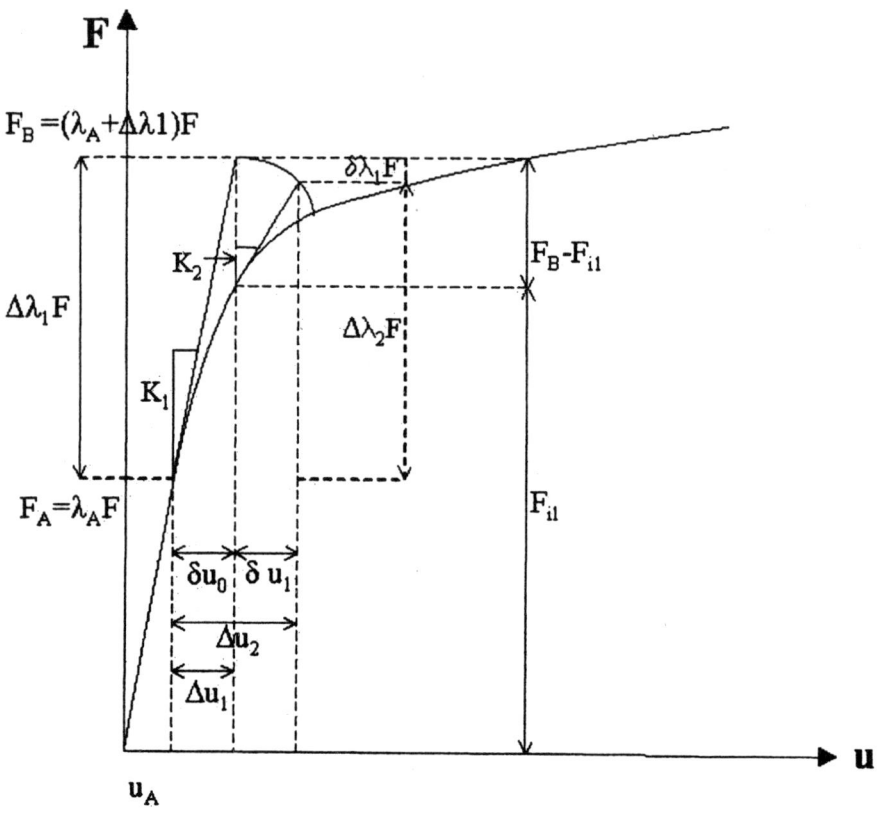

Figure 3. *The arc-length method*

b) If $\delta u_t^T K_t \delta u_t < 0 \Rightarrow$ after limit point $\Rightarrow sign = -1$

Fixing the value of Δl in equation (11) is the same type of decision as fixing the value of load steps in a classical load control analysis. Although it is a decision which should be based on the experience of the analyst it is clear that the size of the step should diminish as the loading process advances, since the degree of non linearity usually increases with the loading factor. Although it requires additional programming, usually the best solution is to establish some sort of automatic time stepping algorithm, based on error estimation or some other criterion. Crisfield [CRI 91] presents in his text a very simple time stepping algorithm which will in many cases solve the problem in a satisfactory manner. The algorithm is based on the concept of "desirable number of iterations". If in the previous increment the number of itereations required to converge has been "too high" the length of the present increment should be smaller.If in the previous increment the number of itereations required to converge has been "too small" the length of the present increment should be bigger. This, of course, requires the analyst to establish what is a "desirable number of iterations" based on his/her computational experience. Following these ideas, the expression controlling the step length would be:

$$\Delta l_j = \Delta l_{j-1} \frac{I_d}{I_{j-1}}$$

where:

Δl_j = length of present increment

Δl_{j-1} = length of previous increment

I_d = Desirable number of iterations, based on experience, usually around 5.

I_{j-1} = Number of iterations of previous increment.

2.2. *Alternative versions*

2.2.1. *Spherical versus cylindrical arc-length*

The arc-length method as presented in the previous section is usually known as spherical arc-length. The term spherical refers to the fact that restriction (1) includes both forces and displacements. It is a restriction applied in a multidimensional space of forces and displacements, and therefore a spherical restriction. When $\varphi = 0$, the restriction includes only displacements, and is therefore referred to as cylindrical. In fact, the latter version is much more frequently used because it is simpler and the spherical arc-length does not have significant advantages. In equations 8, 9 and 10, coefficients for the cylindrical versions may by obtained by omitting the terms in brackets.

Although they will not be covered in detail, one should mention that there are "linear" versions for both cylindrical and spherical arc-length. These versions are linear in the sense that the arclength restriction is only enforced in an approximate fashion. Using a 2D graphical explanation, a linear approximation would be to obtain the intersection between the tangent to a circumference arc and the force-displacement while the full arc-length restriction would be to obtain the intersection between the said circumference arc and the force-displacement curve.

2.2.2. Modified versus full Newton-Raphson

The arc-length method may be implemented following either a full Newton-Raphson algorithm where stiffness matrix is updated at every iteration or a modified Newton-Raphson where the stiffnes matrix is only updated at the first iteration of each increment. The first alternative is obviously more powerful and converges faster but each iteration is computationally more expensive. In spite of the latter disvantage it is usually more reliable than the modified alternative. The updating of vector δu_t is done at the same time as K_t is updated, at each iteration for the full Newton-Raphson and at the beginning of the increment for the modified version.

2.2.3. A displacement control type version of the arc-length method

One the possible versions of the arc-length method, due to Batoz [BAT 79], is similar although not equivalent to a displacement control. The method consists in applying the the length restriction to a specific degree of freedom, that is:

$$\Delta l^2 = \Delta u_i^2$$

where subindex i refers to the degree of freedom which is being controlled.

This method has the limitation of requiring the certainty that the specific degree of freedom chosen for control varies monotonically during the loading process. This means that it will always have to increase or decrease during the the loading process. The possibility exists that no such degree of freedom exists in the model, that is, all the degrees of freedom in the model switch from an increasing to a decreasing pattern or viceversa. That would be the case in a "snap-back" problem.

2.2.4. Bathe-Dvorkin version

An interesting version of the arc-length method was formulated by Bathe and Dvorkin that consists of using a restriction based on a constant external work along the increment. This means that the external work is computed in the first iteration of the increment according to the following expression:

$$\Delta l = \Delta W = (\lambda + \frac{1}{2}\Delta\lambda_p)\mathbf{f}_{ext}^T\Delta\mathbf{u}_p = \Delta\lambda_p(\lambda + \frac{1}{2}\Delta\lambda_p)\mathbf{f}_{ext}^T\delta\mathbf{u}_t$$

and is forced to remain constant during the subsequent iterations:

$$(\lambda_o + \frac{1}{2}\delta\lambda)\mathbf{f}_{ext}^T\delta\mathbf{u} = (\lambda_o + \frac{1}{2}\delta\lambda)(\mathbf{f}_{ext}^T\delta\overline{\mathbf{u}} + \delta\lambda\mathbf{f}_{ext}^T\delta\mathbf{u}_t) = 0$$

where λ_o is the load factor in the previous iteration. This restriction gives rise to second degree equation in $\delta\lambda$, as in the standard arc-length. The difference between this version and the standard arc-length lies in the fact that in this case the roots of the second degree equation are guaranteed to be real. This is why some authors recommend to switch to the Bathe-Dvorkin version when the Crisfield version yields complex solutions.

3. Line search methods

As seen in previous sections the Newton-Raphson procedure is based on an iterative process in which the displacement vector correction is:

$$\delta\overline{u} = -K_t^{-1}\Psi$$

where Ψ is the residual force vector. The line-search method is based on the idea that the optimum prediction for the new displacement vector u_n is not obtained by simply adding the displacement correction vector $\delta\overline{u}$ to the old displacement vector u_o but according to the following expression:

$$u_n = u_o + \eta\delta\overline{u} \tag{12}$$

where η is a new parameter obtained by minimization of a certain potential function ϕ :

$$\phi_n(\eta + \delta\eta) = \phi_o(\eta) + \frac{\partial\phi}{\partial\eta}\delta\eta + ... = \phi_o + \frac{\partial\phi}{\partial u}\frac{\partial u}{\partial\eta}\delta\eta + ...$$

The minimization condition for ϕ in terms of η is:

$$\frac{\partial\phi}{\partial\eta} = 0$$

In the case of non linear elastic problems ϕ is taken as the total potential energy which may be precisely define from solid mechanics point of view. This choice is consistent with the classical variational approach to a non linear elastic problem as a total potential energy minimization problem. In this context it is possible to prove that the previous equation may be expressed as:

$$\frac{\partial\phi}{\partial u}\frac{\partial u}{\partial\eta} = \Psi(\eta)^T\delta\overline{u} = 0$$

or similarly as:

$$\delta(\eta) = \frac{\partial\phi}{\partial\eta} = \delta\overline{u}^T\Psi(\eta) = 0 \tag{13}$$

Strictly speaking this approach cannot be used with problems such as elastoplasticity in a materially non linear context. However, for this type of problems it is possible to use an algorithm with an energy approach such as line search, with a potential function of which it is known that the derivative with respect to the displacement vector $\frac{\partial \phi}{\partial u}$ is the residual force vector Ψ. Following this approach and according to equation (13) it is possible to define δ_0 as $\delta(\eta = 0)$ according to the following expression:

$$\delta_0 = \delta(\eta = 0) = \delta \overline{u}^T \Psi(\eta = 0) = \delta \overline{u}^T \Psi_0 = -\Psi_0^T K_t^{-1} \Psi_0 = -\delta \overline{u}^T K_t \delta \overline{u}$$

Although the strict minimization condition is the one expressed in (13) from a numerical point of view it is more interesting to enforce a lax or non strict minimization condition such as :

$$|r(\eta)| = \left| \frac{\delta(\eta)}{\delta_0} \right| < \beta \tag{14}$$

where β is a constant fixed by the analyst usually between 0.1 and 0.5. To solve equation (14) the following steps are necessary:

1°) Obtain $\delta_0 = \delta \overline{u}^T \Psi_0$. where Ψ_0 is available from the previous iteration.

2°) Compute $\delta_1 = \delta(\eta = 1)$.

3°) Obtain:

$$\eta_2 = \frac{-\eta_1 \delta_0}{\delta_1 - \delta_0} \tag{15}$$

Equation (15) comes from performing a linear interpolation of the value η_2 between η_0 and η_1 by enforcing $\delta(\eta_2) = 0$.

4°) $\delta_2 = \delta(\eta_2)$ is obtained

5°) Check if δ_2 satisfies or not the non strict minimization condition $\left| \frac{\delta(\eta)}{\delta_0} \right| < \beta$

6°) If the answer is :

a) **Yes** : the line search algorithm has ended and displacements are updated according to equation (12)

b) **No** : Return to step 3 obtaining $\eta_3 = \frac{-\eta_2 \delta_0}{\delta_2 - \delta_0}$ and continuing with the iterative process until convergence is reached.

As one may see from the described procedure, each iteration in a line search algorithm requires an additional update of the internal force vector, with computational cost that this implies. This is why a non strict line search with a few iterations is usually performed instead of strict minimation line search. The computaional cost of the latter would be prohibitive.

Combination of arc-length and line-search

The application of the line search method in combination with arc-length has also been treated by Crisfield [CRI 83]. The combination of these two algorithms implies the application of equation minimization condition (13) to the total iterative displacement in equation (4), that is:

$$\delta = (\delta\overline{u} + \delta\lambda\delta u_t)^T \Psi(\lambda, \eta) = 0 \tag{16}$$

The problem is that every time equation (15) is applied in search of a new value for η that approximately satisfies equation (16) the arc-length condition ceases to be satisfied. This means that a new value for $\delta\lambda$ has to be obtained such that:

$$(\Delta u_0 + \eta(\delta\overline{u} + \delta\lambda\delta u_t))^T(\Delta u_0 + \eta(\delta\overline{u} + \delta\lambda\delta u_t)) = \Delta l^2 \tag{17}$$

This gives rise to an additional iterative procedure nested in the algorithm presented in the previous section. As expected equation (17) ends up in a second degree equation in $\delta\lambda$ with similar coefficients to (8),(9) and (10) but including η. The additional iterative procedure consists in:

1) Application of equation (15).η is obtained.

2) Solution of proble (17). A new value for $\delta\lambda_n$ is obtained.

If $\left(\frac{|\delta\lambda_n - \delta\lambda_o|}{|\delta\lambda_o|} < tolerancia\right) \Rightarrow$ End of the iterative process.

Else $\Rightarrow$ return to step 1

4. References

[BAT 79] J. L. BATOZ AND G. DHATT, " Incremental displacement algorithms for non linear problems, *International Journal for Numerical and Analytical Methods in Engineering*, vol. 14, 1979, p. 1262-1266.

[CRI 81] M. A. CRISFIELD, "A fast incremental/iterative solution procedure that handles 'snap-through' , *Computers and Structures*, vol. 13, 1981, p. 55-62.

[CRI 83] M. A. CRISFIELD, " An Arc-length method including line-searches and accelerations, *International Journal of Solids for Numerical Methods in Engineering*, vol. 19, 1983, p. 1269-1289.

[CRI 91] M. A. CRISFIELD, *Non-linear Finite Element Analysis of Solids and Structures. Volume 1: Essentials*, John Wiley and Sons,1991.

[RIK 79] E. RIKS, " An incremental approach to the solution of snapping and buckling problems, *International Journal for Solids Structures*, vol. 15, 1979, p. 529-551.

[RIK 72] E. RIKS, " The application of Newton's method to the problem of elastic stability, *Journal of Applied Mechanics*, vol. 39,1972, p. 1060-1069.

[WEM 71] G. A. WEMPNER, " Discrete approximations related to non linear theories of solids, *International Journal for Solids Structures*, vol. 7, 1971, p. 1581-1599.

Chapter 10

A Finite Element Model for Water Saturated and Partially Saturated Geomaterials

Space and Time Discretisation for a Multiphase Porous Material Model

Lorenzo Sanavia and Bernhard A. Schrefler
Department of Structural and Transportation Engineering, University of Padua, Italy

1. Introduction

This paper presents the finite element model for a saturated and partially saturated porous material capable to sustain large elastic or elasto-plastic strains, extending the previous work of Sanavia et al. [SAN 02a], [SAN 02b].

The porous medium is treated as an isothermal multiphase continuum with the pores of the solid skeleton filled by water and air, the last at constant pressure (passive air phase assumption). This pressure may either be the atmospheric pressure or the cavitation pressure (isothermal monospecies approach). Quasi-static loading conditions are considered. The governing equations at macroscopic level are described in Section 2 in a spatial setting. This model follows from the general thermo-hydro-mechanical model developed in [LEW 98], which is described in the contribution *Coupling equations for water saturated and partially saturated geomaterials* by the authors in this publication. Solid displacements and water pressures are the primary variables. The solid grains and water are assumed to be incompressible at microscopic level. The elasto-plastic behaviour of the solid skeleton is described by the multiplicative decomposition of the deformation gradient into an elastic and a plastic part, as shown in Section 3. The modified effective stress in partially saturated conditions (*Bishop* like stress) in the form of *Kirchhoff* measure of the stress tensor and the logarithmic principal strains are used in conjunction with an hyperelastic free energy function. The effective stress state is limited by the *von Mises* or the *Drucker-Prager* yield surface for simplicity. Water is assumed to obey *Darcy*'s law. In the partially saturated state, the water degree of saturation and the relative permeability are dependent on the capillary pressure by experimental functions. The spatial weak form of the governing equations, the temporal integration of the mixture mass balance equation, which is time dependent because of the seepage process of water, and the consistent linearization are described in Sections 4, 5, and 6, respectively. In particular, the generalized trapezoidal method is used for the time integration. Finally, the finite element discretization in space is obtained by applying a *Galerkin* procedure in the spatial setting, using different shape functions for solid and water (see Section 7).

2. Balance equations for an isothermal saturated/partially saturated medium

In this section the macroscopic balance equations of the simplified model that we shall use in the sequel are obtained. In this model, the main features are that water pressure and solid displacements are chosen as the primary variables and that the elastoplastic behaviour of the solid skeleton is developed in the framework of the hyperelastoplasticity. Moreover, it permits one to outline the main guidelines used in modern computational mechanics.

The following assumptions are now introduced in the general model previously presented:

– All the processes are isothermal. This means that the energy balance equation is no more necessary and the phase changed are neglected.

– Gas phase is assumed to remain at constant pressure and flows without resistance in the partially saturated zone ([LEW 98], [OCZ 99]). This means that the mass balance equations for dry air and vapour are neglected. The gas pressure may either be the atmospheric pressure or the cavitation pressure at a certain temperature (e. g. the ambient temperature). The first case is a common assumption in soil mechanics because in many cases occurring in practice the air pressure is close to the atmospheric pressure as the pores are interconnected [OCZ 99]. The second case can be derived from the experimental observations [MOK 98] and the obtained model is also called *Isothermal Monospecies Approach*, which can be used to simulate cavitation at localization in initially water saturated dense sands under globally undrained conditions, as first developed in [SCH 96] for the geometrically linear case. In fact, in this situation, neglecting air dissolved in water, only two fluid phases are present after cavitation: liquid water and water vapour at cavitation pressure, which is then considered constant and is neglected because of its small value.

– At the micro level, the porous medium is assumed to consist of incompressible solid and water constituents. The averaged intrinsic density $\rho^\pi(\mathbf{x}, t)$ $(\pi = s, w)$ is hence constant, while the averaged density $\rho_\pi(\mathbf{x}, t)$ can vary due to the volume fraction $\eta^\pi(\mathbf{x}, t)$. Consequently, the density of the mixture $\rho(\mathbf{x}, t)$ (12) and the porosity $n(\mathbf{x}, t)$ can change during the deformation of the porous medium.

– The process is considered as quasi-static, so the solid and fluids accelerations are neglected.

The formulation in terms of spatial co-ordinates is now presented.

2.1. *Mass balance equation*

Taking into account the incompressibility constraint of the solid and water constituents in eq. (16) and eq. (18) of [SCH 02], the mass balance equation for the solid and water phases becomes

$$\frac{\partial}{\partial t}(1-n) + \operatorname{div}\left[(1-n)\,\mathbf{v}^s\right] = 0 \qquad [1]$$

$$\frac{\partial}{\partial t}(n\,S_w) + \operatorname{div}(n\,S_w\,\mathbf{v}^w) = 0 \qquad [2]$$

where the definition of the phase average density $\rho_\pi(\mathbf{x}, t) = \eta^\pi(\mathbf{x}, t)\,\rho^\pi(\mathbf{x}, t)$, $(\pi = s, w)$ has been introduced, thus eliminating the intrinsic (constant) average density $\rho^\pi(\mathbf{x}, t)$. Using the concept of the material time derivative, (1) is rewritten as

$$\frac{\mathrm{D}^s}{\mathrm{D}t}(1-n) + (1-n)\,\operatorname{div}\mathbf{v}^s = 0 \qquad [3]$$

where the classical relationship

$$\text{div } \mathbf{v}^s = \frac{D^s J^s}{Dt} (J^s)^{-1}$$ [4]

can be introduced for the solid deformation [MAR 83]. The time integration of (3) gives the evolution law for the porosity $n(x, t)$ related to the determinant $J^s(\mathbf{X}^s, t)$ of the deformation gradient $\mathbf{F}^s(\mathbf{X}^s, t)$:

$$n = 1 - (1 - n_0) (J^s)^{-1}$$ [5]

where $n_0(\mathbf{X}^s)$ is the porosity in the reference configuration at $t = t_0$ (or initial porosity). Because of the relation $\eta^s(\mathbf{x}, t) = 1 - n(\mathbf{x}, t)$, (5) can be rewritten as

$$\eta^s = \eta_0^s (J^s)^{-1}$$ [6]

where $\eta_0^s(\mathbf{X}^s)$ is the solid volume fraction in the reference configuration at $t = t_0$.

The sum of the mass balance equation of the two constituents (1) and (2) produces the following mass balance equation for the mixture under consideration:

$$\frac{\partial}{\partial t} (1 - n + n S_w) + \text{div } [(1 - n) \mathbf{v}^s + n S_w \mathbf{v}^w] = 0$$ [7]

Introducing the water velocity relative to the solid, i. e. $\mathbf{v}^{ws} = \mathbf{v}^w - \mathbf{v}^s$ and the definition of material time derivative with respect to the solid, the mixture mass balance equation (7) becomes

$$n \frac{D^s}{Dt} (S_w) + S_w \text{ div } \mathbf{v}^s + \text{div } (n S_w \mathbf{v}^{ws}) = 0$$ [8]

The term $n S_w \mathbf{v}^{ws}(\mathbf{x}, t)$ represent the filtration water velocity. The water velocity relative to the solid is related to the water pressure by the linear momentum balance equation for water phase, which gives *Darcy*'s law (15), as will be demonstrated in the sequel.

In the case of fully saturated conditions, $S_w = 1$ and hence the previous equation is reduced to the one of the saturated model.

2.2. Linear momentum balance equation

Neglecting the inertial term in eq. (23) and eq. (24) of [SCH 02], the linear momentum balance equation for the solid and water constituents are respectively

$$\operatorname{div} t^s + (1 - n)\, \rho^s\, g + (1 - n)\, \rho^s\, \hat{t}^s = 0 \qquad [9]$$

$$\operatorname{div} t^w + n\, S_w\, \rho^w\, g + n\, S_w\, \rho^w\, (\hat{t}^w + e^w) = 0 \qquad [10]$$

The linear momentum balance equation for the mixture

$$\operatorname{div}(t^s + t^w) + \rho g = 0 \qquad [11]$$

is obtained by summation of (9) and (10), taking into account the constraint (25) of [SCH 02].

In (11), $\rho(x, t)$ is the density of the mixture:

$$\rho = (1 - n)\, \rho^s + n\, S_w\, \rho^w \qquad [12]$$

and $t^s + t^w = \sigma$ is the total *Cauchy* stress, which can be decomposed into the effective and pressure (equilibrium) parts following the principle of effective stress

$$\sigma = \sigma' - S_w\, p^w\, 1 \qquad [13]$$

where $\sigma'(x, t)$ is the modified effective *Cauchy* stress tensor, also called *Bishop*'s stress tensor in soil mechanics. The linear momentum balance equation of the mixture in terms of total *Cauchy* stress assumes the form

$$\operatorname{div} \sigma + \rho g = 0 \qquad [14]$$

Using the constitutive equation (29) of [SCH 02] for $n\, S_w\, \rho^w\, \hat{t}^w$ and the definition of t^w (eq.(30) of [SCH 02]), the linear momentum balance equation for water (10) gives *Darcy*'s law

$$n\, S_w\, v^w = -\frac{k\, k^{rw}}{\mu^w}\, (\operatorname{grad} p^w - \rho^w\, g) \qquad [15]$$

for the water, where $k^{rw} = k^{rw}(\mathbf{x}, t)$ is the relative permeability which is an experimentally determined function of the capillary pressure. This law is valid for the transport of the fluid in slow phenomena when the thermal effects are negligible. Moreover, the equilibrium equation for the fluid pressures ($p^c = p^g - p^w$) is simplified as follows because of the assumption on the gas phase

$$p^c \cong -p^w \qquad [16]$$

which states that capillary pressures can be approximated as pore water tractions. Hence, the water pressure can change sign, which means that a partially saturated zone is developing in the porous medium. The effect of the capillary pressure on the stiffness of the medium is taken into account by the constitutive laws for $S_w(p^c)$ and $k^{rw}(p^c)$.

As a consequence of the above assumptions, the independent fields of the model are the solid displacements $\mathbf{u}(\mathbf{x}, t)$ and the water pressure $p^w(\mathbf{x}, t)$.

In case of fully saturated conditions, $S_w = 1$ and $k^{rw} = 1$ and hence (12), (13), and (15) are reduced to those of the saturated model.

3. Constitutive equations

The constitutive equations necessary for the model are those related to the solid skeleton and the water in the partially saturated zones. In particular, the structure of the developed model can describe the elasto-plastic behaviour of the solid skeleton at finite strain based on the multiplicative decomposition of the deformation gradient $\mathbf{F}^s(\mathbf{X}^s, t)$ into an elastic and plastic part

$$\mathbf{F}^s = \mathbf{F}^{se}\,\mathbf{F}^{sp} \qquad [17]$$

In the following, the treatment of the isotropic elasto-plastic behaviour for the solid skeleton based on the product formula algorithm proposed for the single phase material by Simo [SIM 98] will be briefly summarized for the interested readers. Experienced readers or non interested readers may wish to turn directly to Section 4. The geometrically linear case can be found e. g. in [LEW 98] and [OCZ 99]. The spatial formulation is used in this section.

In this section, the superscript ' s ' will be neglected and the symbol ' $\cdot$ ' will be used for the material time derivative with respect to the solid skeleton instead of D^s/Dt (as well as in the remaining part of the paper).

The effective *Kirchhoff* stress tensor $\tau'(\mathbf{x}, t) = J\sigma'(\mathbf{x}, t)$ and the logarithmic principal values of the elastic left *Cauchy-Green* strain tensor $\epsilon_A(\mathbf{x}, t)$ are used. In the present sub-section also the prime ' ' ' for the effective stress tensor will be neglected. The yield function restricting the stress state is developed in the form of *von Mises* and *Drucker-Prager* for simplicity, to take into account the behaviour of clays under undrained conditions and the dilatant/contractant behaviour of dense or loose sands, respectively. The return mapping and the consistent tangent operator can be developed, solving the singular behaviour of the *Drucker-Prager* yield surface in the zone of the apex using the concept of multisurface plasticity (see [SAN 02b]).

The elastic behaviour of the solid skeleton is assumed to be governed by an hyperelastic free energy $\psi(\mathbf{x}, t)$ function in the form

$$\psi = \psi(\mathbf{b}^e, \xi) \tag{18}$$

dependent on the elastic left *Cauchy-Green* strain tensor $\mathbf{b}^e(\mathbf{x}, t) = \mathbf{F}^e(\mathbf{F}^e)^{-1}$ and the internal strain like variable $\xi(\mathbf{x}, t)$, the equivalent plastic strain. The second law of thermodynamics yields, under the restriction of isotropy, the constitutive relations

$$\tau = 2 \frac{\partial \psi}{\partial \mathbf{b}^e} \mathbf{b}^e, \quad q = -\frac{\partial \psi}{\partial \xi} \tag{19}$$

and the remaining dissipation inequality

$$-\frac{1}{2}\tau : \left[(L_v \mathbf{b}^e)\,(\mathbf{b}^e)^{-1} \right] + q\dot{\xi} \geq 0, \, \psi \tag{20}$$

where $L_v \mathbf{b}^e = \dot{\mathbf{b}}^e - \mathbf{l}\,\mathbf{b}^e - \mathbf{b}^e \mathbf{l}^T$ is the *Lie* derivative of the elastic left *Cauchy-Green* strain tensor $\mathbf{b}^e(\mathbf{x}, t)$.

The evolution equations for the rate terms of the dissipation inequality (20) can be derived from the postulate of the maximum plastic dissipation in the case of associative flow rules

$$-\frac{1}{2} L_v \mathbf{b}^e = \dot{\gamma} \frac{\partial F}{\partial \tau} \mathbf{b}^e \tag{21}$$

$$\dot{\xi} = \dot{\gamma} \frac{\partial F}{\partial q}, \psi \tag{22}$$

subjected to the classical loading-unloading conditions in *Kuhn-Tucker* form:

$$\dot{\gamma} \geq 0, \quad F = F(\tau, q) \leq 0, \quad \dot{\gamma} F = 0 \qquad [23]$$

where $\dot{\gamma}$ is the plastic multiplier and $F = F(\tau, q)$ the isotropic yield function.

Simple examples for the yield functions are those of *Drucker-Prager* and *von Mises* with linear isotropic hardening, in the form, respectively, of

$$F(p, \mathbf{s}, \xi) = 3\,\alpha_F\,p + \|\mathbf{s}\| - \beta_F \sqrt{\frac{2}{3}} \,(c_0 + h\,\xi) \qquad [24]$$

and

$$F(\mathbf{s}, \xi) = \|\mathbf{s}\| - \sqrt{\frac{2}{3}} \,(\sigma_0 + h\,\xi) \qquad [25]$$

in which $p = \frac{1}{3}(\tau : 1)$ is the mean effective *Kirchhoff* pressure, $\|\mathbf{s}\|$ is the L_2-norm of the deviator effective *Kirchhoff* stress tensor τ, c_0 is the initial apparent cohesion of the *Drucker-Prager* model, α_F and β_F are two parameters related to the friction angle ϕ of the soil,

$$\alpha_F = 2\,\frac{\sqrt{\frac{2}{3}}\,\sin\phi}{3 - \sin\phi}, \quad \beta_F = \frac{6\,\cos\phi}{3 - \sin\phi}, \psi \qquad [26]$$

h the hardening/softening modulus, and σ_0 is the yield stress in the *von Mises* law.

Note: In the present contribution, the effect of the capillary pressure p^c on the evolution of the yield surface is not taken into account. The interested reader can refer to [LEW 98] for a constitutive relationship function of the effective stress and the capillary pressure.

4. Weak form: Variational approach

The weak form of the spatial governing equations presented in the previous section is now derived obtaining the variational equations formally equivalent to the initial-boundary-value problem given by the governing equation and the boundary conditions. This means that the governing equations (7) and (11) are multiplied by independent weighting functions that vanish on the boundary in which *Dirichlet* boundary conditions are applied and are then integrated over the spatial domain B with boundary ∂B. The linear momentum balance equation of the binary porous media (11) is hence weighted on the domain by the test function $\delta\mathbf{u}_s$ corresponding to the solid displacement (or virtual displacement) in the form

$$\int_B (\operatorname{div}\sigma + \rho\,\mathbf{g}) \cdot \delta\mathbf{u}_s\,dv = 0 \quad \forall\ \delta\mathbf{u}_s \qquad [27]$$

Applying partial integration and *Green*'s theorem in the form (e. g. [MAR 83])

$$\int_B \operatorname{div}\sigma \cdot \delta\mathbf{u}_s\,dv = -\int_B \sigma : \operatorname{grad}\delta\mathbf{u}_s\,dv + \int_{\partial B} \bar{\mathbf{t}} \cdot \delta\mathbf{u}_s\,ds\psi \qquad [28]$$

to the divergence part of (27) and taking into account the boundary conditions, this equation is transformed into the weak form

$$-\int_B (\sigma' - S_w\,p^w 1) : \operatorname{grad}\delta\mathbf{u}_s\,dv + \int_B \rho\,\mathbf{g} \cdot \delta\mathbf{u}_s\,dv +$$
$$\int_{\partial B} \bar{\mathbf{t}} \cdot \delta\mathbf{u}_s\,ds = 0 \quad \forall\ \delta\mathbf{u}_s \qquad [29]$$

where the effective stress principle (13) has been introduced. Using the relation $\operatorname{div}\delta\mathbf{u}_s = \operatorname{grad}\delta\mathbf{u}_s : 1$, the previous weak form is transformed into

$$-\int_B \sigma' : \operatorname{grad}\delta\mathbf{u}_s\,dv + \int_B S_w\,p^w \operatorname{div}\delta\mathbf{u}_s\,dv + \int_B \rho\,\mathbf{g} \cdot \delta\mathbf{u}_s\,dv +$$
$$\int_{\partial B} \bar{\mathbf{t}} \cdot \delta\mathbf{u}_s\,ds = 0 \quad \forall\ \delta\mathbf{u}_s \qquad [30]$$

The weak form of the mixture mass balance equation (7) is obtained in a similar way, introducing *Darcy*'s law (15) and using the test function δp^w corresponding to p^w (or virtual water pressure):

$$\int_B n\frac{\mathrm{D}^s}{\mathrm{D}t}(S_w)\,\delta p^w\,dv + \int_B S_w \operatorname{div}\mathbf{v}^s\,\delta p^w\,dv +$$
$$\int_B \operatorname{div}\left[\frac{\mathbf{k}\,k^{rw}}{\mu^w}(-\operatorname{grad}p^w + \rho^w\,\mathbf{g})\right]\delta p^w\,dv = 0 \quad \forall \delta p^w \qquad [31]$$

Applying *Green*'s theorem to the last integral term of the previous equation, the following is obtained

$$
\int_B n \frac{\mathrm{D}^s}{\mathrm{D}t} (S_w)\, \delta p^w \, dv + \int_B S_w \, \mathrm{div}\, \mathbf{v}^s \, \delta p^w \, dv +
$$

$$
\int_B \left[\frac{\mathbf{k}\, k^{rw}}{\mu^w} (\mathrm{grad}\, p^w - \rho^w \, \mathbf{g}) \right] \cdot \mathrm{grad}\, \delta p^w \, dv +
$$
[32]

$$
\int_{\partial B} q^w \, \delta p^w \, ds = 0 \quad \forall \ \delta p^w
$$

where $q^w(\mathbf{x}, t)$ is the water flow draining through the surface ∂B.

Note: It can be observed that the weak forms (30) and (32) are very similar to those of the geometrically linear theory, e. g. [LEW 98], by substituting the deformed integration domain B with the undeformed one B_0. Moreover, in the small strain theory $\mathrm{div}\, \mathbf{v}^s = \dot{\varepsilon} : 1$, where ε is the small strain tensor of the solid skeleton, while in finite strain $\mathrm{div}\, \mathbf{v}^s = \dot{J}^s / J^s$. In the small strain theory the additive decomposition of the strain tensor ε in elastic ε^e and plastic ε^p parts is also possible, thus rendering the computation of the constitutive tangent operator in the linearization of the weak form particularly easy.

5. Time discretization

Time integration of the weak form of the mass balance equation (32) over a finite time step $\Delta t = t_{n+1} - t_n$ is necessary because of the time dependent terms $\mathrm{div}\, \mathbf{v}^s$ and $\mathrm{D}^s S_w / \mathrm{D}t$.

The generalized trapezoidal method is used here, as shown for instance in [LEW 98]. Because of the dependence of the integration domain on time, we rewrite the weak forms (30) and (32) with respect to the undeformed domain as follows:

$$
\int_{B_0} (\tau' - J^s \, S_w \, p^w \, 1) : \mathrm{grad}\, \delta \mathbf{u}_s \, dV - \int_{B_0} \rho_0 \, \mathbf{g} \cdot \delta \mathbf{u}_s \, dV -
$$

$$
\int_{\partial B_0} \bar{\mathbf{T}} \cdot \delta \mathbf{u}_s \, dA = 0 \quad \forall \ \delta \mathbf{u}_s
$$
[33]

$$\int_{B_0} J^s \, S_W \, \text{div} \, \mathbf{v}^s \, \delta p^w \, dV + \int_{B_0} \left[J^s \frac{k \, k^{rw}}{\mu^w} \left(\text{grad} \, p^w - \rho^w \, \mathbf{g} \right) \right] \cdot \text{grad} \, \delta p^w dV +$$

$$\int_{\partial B_0} Q^w \, \delta p^w \, dA + \int_{B_0} J^s \, N \, \frac{D^s}{Dt} \, (S_W) \, \delta p^w \, dV = 0 \quad \forall \; \delta p^w \qquad [34]$$

where τ' is the modified effective *Kirchhoff* stress tensor and $\bar{\mathbf{T}} = \mathbf{P} \, \mathbf{N}$ and $Q^w = N \, S_W \, \bar{\mathbf{V}}^{ws} \cdot \mathbf{N}$ are, respectively, the traction vector and the water flow computed with respect to the undeformed configuration. The form of (33) and (34) is also useful for the subsequent linearization because it will be easily performed with respect to the undeformed (fixed) domain.

Equation (34) is now rewritten at time t_{n+1} using the relationships

$$\dot{J}^s_{n+\beta} \;=\; \frac{J^s_{n+1} - J^s_n}{\Delta t}, \psi \; (\dot{S}_w)_{n+\beta} \;=\; \frac{S_{wn+1} - S_{wn}}{\Delta t} \qquad [35]$$

$$(\cdot)_{n+\beta} = (1 - \beta) \, (\cdot)_n + \beta (\cdot)_{n+1} = (\cdot)_n + \beta \, [(\cdot)_{n+1} - (\cdot)_n] \qquad [36]$$

with $\beta \in [0, 1]$, obtaining

$$\int_{B_0} (S_w)_{n+\beta} \left(J^s_{n+1} - J^s_n \right) \delta p^w \, dV - \Delta t \int_{B_0} \left(J^s \, \mathbf{v}^D \cdot \text{grad} \, \delta p^w \right)_{n+\beta} dV +$$

$$\int_{B_0} (J^s \, n)_{n+\beta} \, [S_{wn+1} - S_{wn}] \, \delta p^w \, dV + \Delta t \int_{\partial B_0} Q^w_{n+\beta} \, \delta p^w \, dA = 0 \quad \forall \; \delta p^w \qquad [37]$$

where $\mathbf{v}^D = -k \, k^{rw} / \mu^w \, (\, \text{grad} \, p^w - \rho^w \, \mathbf{g})$ is *Darcy*'s velocity of the water.

The weak form of the linear momentum balance equation (33) is directly written at time t_{n+1} because it is time independent

$$\int_{B_0} [(\tau' - J^s \, S_w \, p^w \, \mathbf{1}) : \text{grad} \, \delta \mathbf{u}_s]_{n+1} \, dV - \int_{B_0} \rho_{0n+1} \, \mathbf{g} \cdot \delta \mathbf{u}_s \, dV -$$

$$\int_{\partial B_0} \bar{\mathbf{T}}_{n+1} \cdot \delta \mathbf{u}_s \, dA = 0 \quad \forall \; \delta \mathbf{u}_s \qquad [38]$$

Linearized analysis of accuracy and stability suggest the use of $\beta \geq \frac{1}{2}$. In the examples section, implicit one-step time integration has been performed ($\beta = 1$).

The weak forms (37) and (38) represent a non-linear coupled equations system where the non-linearities are introduced by the finite kinematics and the constitutive laws.

6. Consistent linearization

The non-linear equation system (37) and (38) can be written in the following compact form

$$\mathbf{G}(\chi, \eta) = 0, \quad \text{with} \quad \chi = (\chi^s, p^w)^T \quad \text{and} \quad \eta = (\delta \mathbf{u}_s, \delta p^w)^T \qquad [39]$$

where $\chi^s(\mathbf{X}^s, t)$ is the motion function (deformation map) of the solid. For its numerical solution, iterative methods have to be employed and the linearization at $\bar{\chi}$ is hence necessary

$$\mathbf{G}(\bar{\chi}, \eta, \Delta \mathbf{u}) \cong \mathbf{G}(\bar{\chi}, \eta) + D\mathbf{G}(\bar{\chi}, \eta) \cdot \Delta \mathbf{u} \cong 0 \qquad [40]$$

where $\Delta \mathbf{u} = (\Delta \mathbf{u}_s, \Delta p^w)^T$ and $D\mathbf{G} \cdot \Delta \mathbf{u} = \frac{d}{d\alpha} \mathbf{G}(\bar{\chi} + \alpha \Delta \mathbf{u})|_{\alpha=0}$ is the directional derivative or *Gateaux* derivative of G at $\bar{\chi}$ in the direction of $\Delta \mathbf{u}$ (e. g. [MAR 83, WRI 93] for single-phase material). Since the equation system G is composed of the weak form of the linear momentum balance equation (G_{LBE}) and of the mass balance equation (G_{MBE}), then

$$DG \cdot \Delta \mathbf{u} = \begin{bmatrix} DG_{\text{LBE}} \cdot \Delta \mathbf{u}_s + DG_{\text{LBE}} \cdot \Delta p^w \\ DG_{\text{MBE}} \cdot \Delta \mathbf{u}_s + DG_{\text{MBE}} \cdot \Delta p^w \end{bmatrix} \qquad [41]$$

Using the symbol $(\cdot)_{n+1}^{k+1}$ to indicate the current iteration in the current time step, the linearization on the configuration $(\cdot)_{n+1}^{k}$ is written as

$$DG_{n+1}^{k} \cdot \Delta \mathbf{u}_{n+1}^{k+1} = -\mathbf{G}_{n+1}^{k} \qquad [42]$$

and the solution vector $\mathbf{u} = (\mathbf{u}_s, p^w)^T$ is then updated by the incremental relationship

$$\mathbf{u}_{n+1}^{k+1} = \mathbf{u}_{n+1}^{k} + \Delta \mathbf{u}_{n+1}^{k+1} \qquad [43]$$

For an efficient numerical performance of the scheme (42), the consistent linearization is applied [WRI 93] in which the linearization of the integrated constitutive equation plays a central role (this concept was first pointed out in [SIM 85] for the geometrically linear case).

The linearization of (37) and (38), performed in the undeformed configuration B_0 and then pushed forward in the deformed configuration B, gives the following result:

– For the linear momentum balance equation:

$$
\int_B \left(\operatorname{grad} \delta \mathbf{u}_s : \mathbf{c}^{ep} : \operatorname{sym} \left(\operatorname{grad} \Delta \mathbf{u}_s \right) + \boldsymbol{\sigma}' : \operatorname{grad}^T \delta \mathbf{u}_s \operatorname{grad} \Delta \mathbf{u}_s \right) \, dv +
$$

$$
\int_B S_w \, p^w \operatorname{grad} \delta \mathbf{u}_s : \left(\operatorname{grad}^T \Delta \mathbf{u}_s - \operatorname{div} \Delta \mathbf{u}_s \, \mathbf{1} \right) \, dv -
$$

$$
\int_B \rho^w S_w \, \delta \mathbf{u}_s \cdot \mathbf{g} \operatorname{div} \Delta \mathbf{u}_s \, dv - \int_B \left(p^w \frac{\partial S_w}{\partial p^w} + S_w \right) \operatorname{div} \delta \mathbf{u}_s \, \Delta p^w \, dv
$$

[44]

– For the mass balance equation (in case of isotropic permeability):

$$
\int_B \delta p^w \left(1 + S_{w n+\beta} + \beta \Delta S_w \right) \operatorname{div} \Delta \mathbf{u}_s \, dv +
$$

$$
\beta \Delta t \int_B \frac{k \, k^{rw}}{\mu^w} \operatorname{grad} \delta p^w \cdot \operatorname{grad} \Delta p^w \, dv +
$$

$$
\beta \Delta t \int_B \operatorname{grad} \delta p^w \cdot \left[\left(\frac{1-n}{k} \frac{\partial k}{\partial n} + 1 \right) \frac{k \, k^{rw}}{\mu^w} \left(\operatorname{grad} p^w - \right. \right.
$$

$$
\left. \left. - \rho^w \, \mathbf{g} \right) \operatorname{div} \Delta \mathbf{u}_s \right] \, dv -
$$

[45]

$$
\beta \Delta t \int_B \operatorname{grad} \delta p^w \cdot \left[\frac{2 \, k \, k^{rw}}{\mu^w} \operatorname{sym} \left(\operatorname{grad} \Delta \mathbf{u}_s \right) \operatorname{grad} p^w \right] \, dv +
$$

$$
\beta \Delta t \int_B \operatorname{grad} \delta p^w \cdot \left(\frac{k \, k^{rw}}{\mu^w} \rho^w \operatorname{grad} \Delta \mathbf{u}_s \, \mathbf{g} \right) \, dv +
$$

$$
\beta \Delta t \int_B \frac{k}{\mu^w} \frac{\partial k^{rw}}{\partial p^w} \operatorname{grad} p^w \cdot \operatorname{grad} \delta p^w \, \Delta p^w \, dv -
$$

$$
\int_B \delta p^w \frac{\left(\beta \Delta J + J_{n+\beta} \, n_{n+\beta} \right)}{J} \frac{\partial S_w}{\partial p^c} \Delta p^w \, dv
$$

In the directional derivative $DG_{\text{LBE}} \cdot \Delta \mathbf{u}_s$ the term

$$\int_B \left(\text{grad}\,\delta \mathbf{u}_s : \mathbf{c}^{ep} : \text{sym}\,(\text{grad}\,\Delta \mathbf{u}_s) + \boldsymbol{\sigma}' : \text{grad}^T \delta \mathbf{u}_s \, \text{grad}\,\Delta \mathbf{u}_s \right) dv\psi \qquad [46]$$

contains $\mathbf{c}^{ep}$, the spatial constitutive operator following the linearization of the computed effective stress

$$\mathbf{c}^{ep}_{n+1} = \sum_{A=1}^{3} \sum_{B=1}^{3} a^{ep}_{AB_{n+1}} \, (\mathbf{n}^{tr}_A \otimes \mathbf{n}^{tr}_A) \otimes (\mathbf{n}^{tr}_B \otimes \mathbf{n}^{tr}_B) +$$
$$2 \sum_{A=1}^{3} \tau_{A_{n+1}} \, \mathbf{c}^{tr(A)}_{n+1} \qquad [47]$$

It is useful to remark that in (47) only the second order tensor $\mathbf{a}^{ep} = \partial \tau_A / \partial \varepsilon^{tr}_B$ depends on the specific model of plasticity and the structure of the return mapping algorithm in principal stretches, while the tensors $\mathbf{c}^{tr(A)}_{n+1}$ and $\mathbf{n}^{tr}_A \otimes \mathbf{n}^{tr}_A$ are independent of the specific plastic model in use. Moreover, it is easy to prove that the moduli $\mathbf{a}^{ep}$ have a form identical to the algorithmic elasto-plastic tangent moduli of the infinitesimal theory [SIM 98]. The expression for $\mathbf{c}^{tr(A)}_{n+1}$ can be obtained by linearization of the eigenbases dyadic $\mathbf{n}^{tr}_A \otimes \mathbf{n}^{tr}_A$ in the spatial setting:

$$\mathbf{c}^{tr(A)}_{n+1} = \frac{\partial (\mathbf{n}^{tr}_A \otimes \mathbf{n}^{tr}_A)}{\partial \mathbf{g}} \qquad [48]$$

where g is the spatial metric, or by pull-back [MAR 83] of $\mathbf{n}^{tr}_A \otimes \mathbf{n}^{tr}_A$, subsequent to linearization in the material setting and then by push-forward of the linearization in spatial setting.

The expressions for the algorithmic moduli $\mathbf{a}^{ep}$ of the *Drucker-Prager* model are derived in [SAN 02b].

7. Finite element discretization in space

The appropriate spatial finite element formulation is derived by applying the well-known *Galerkin* procedure, in which the weighting functions are approximated by the same shape functions used to approximate the driving variables (isoparametric finite elements). This means that the geometry $\mathbf{X}^s$, the current configuration x, the displacement field $\mathbf{u}_s$, the water pressure p^w, the incremental generalized displacement $\Delta \mathbf{u} = (\Delta \mathbf{u}_s, \Delta p^w)^T$ and the variations $\eta = (\delta \mathbf{u}_s, \delta p^w)^T$, are interpolated within a finite element by the same type of functions. In the present setting, different shape functions are chosen for quantities associated respectively with the solid and the fluid, thus satisfying the LBB condition (*Ladyzhenskaya-Babuška-Brezzi* condition) for the locally undrained case. Standard procedures have been applied, following any text books on FEM. With respect to the small strain case, the discretization of the spatial form of the linearized system of equations is made taking into account that each quantity is referred to the spatial co-ordinates x, instead of the co-ordinates of the undeformed configuration X^s. The solid displacement $u_s(x, t)$ and the water pressure $p^w(x, t)$ are hence expressed in the whole domain by global shape function matrices $N_u(x)$ and $N_w(x)$ and the nodal value vectors $\bar{u}(t)$ and $\bar{p}(t)$:

$$u = N_u \bar{u}, \qquad p^w = N_w \bar{p} \qquad [49]$$

The linearized system of equations (42) in matrix form can be expressed as

$$\begin{bmatrix} K_T + K_{sw}^{geom} & -c_{sw}\, Q_{sw} \\ Q_{ws} - \beta\, \Delta t\, Q_{sw}^{geom} & \beta\, \Delta t\, H \end{bmatrix} \begin{bmatrix} \Delta\bar{u} \\ \Delta\bar{p} \end{bmatrix} = - \begin{bmatrix} G^u \\ G^p \end{bmatrix} \qquad [50]$$

which is non-symmetric (details concerning the implementation as well as the matrices and the residuum vectors of (50) will be described in a future paper). Owing to the strong coupling between the mechanical and the pore fluid problem, a monolithic solution of (50) is preferred using a *Newton* scheme.

8. Conclusions

A finite element formulation for the hydro-mechanical behaviour of a water saturated/partially saturated porous material has been presented. This model is obtained as a result of research in progress on the thermo-hydro-mechanical model for multiphase geomaterials undergoing large inelastic strains. For the interested reader, further finite element models as well as the corresponding numerical codes can be found in [LEW 98] and [OCZ 99].

Acknowledgements

This work has been carried out within the research project *Cofin MM08323597* sponsored by the Italian Ministry of Scientific and Technological Research *MIUR*.

9. References

[LEW 98] LEWIS R. W., SCHREFLER B. A., *The Finite Element Method in the Static and Dynamic Deformation and Consolidation of Porous Media*, John Wiley & Sons, Chichester, 1998.

[MAR 83] MARSDEN J. E., HUGHES T. J. R., *Mathematical Foundations of Elasticity*, Prentice-Hall, Englewood Cliffs, 1983.

[MOK 98] MOKNI M., DESRUES J., "Strain Localization Measurements in Undrained Plane-strain Biaxial Tests on Hostun RF Sand", *Mechanics of Cohesive-frictional materials*, vol. 4, 1998, p. 419-441.

[SAN 02a] SANAVIA L., SCHREFLER B. A., STEINMANN, P., A mathematical and numerical model for finite elastoplastic deformations in fluid saturated porous media. In: *Modeling and Mechanics of Granular and Porous Materials*, Series of Modeling and Simulation in Science, Engineering and Technology, G.Capriz, V.N. Ghionna and P. Giovine eds., Birkhäuser, Boston, (in print), p. 297-346.

[SAN 02b] SANAVIA L., SCHREFLER B. A., STEINMANN P., "A formulation for an unsaturated porous medium undergoing large inelastic strains", *Computational Mechanics*, vol. 28, 2002, p. 25-40.

[SCH 96] SCHREFLER B. A., SANAVIA L., MAJORANA, C. E., "A Multiphase Medium Model for Localization and Postlocalization Simulation in Geomaterials", *Mechanics of Cohesive-frictional materials*, vol. 1, 1996, p. 95-114.

[SCH 02] SCHREFLER B.A., SANAVIA L., *Coupling equations for water saturated and partially saturated geomaterials*, Lecture notes of the ALERT course, RFGC Hermes Science Publications, Paris, 2002.

[SIM 98] SIMO J. C., HUGHES T. J. R., *Computational Inelasticity*, Springer-Verlag, 1998.

[SIM 85] SIMO J. C., TAYLOR R., "Consistent Tangent Operators for Rate-Independent Elastoplasticity", *Comp. Meth. In Applied Mech. Eng.*, vol. 48, 1985, p. 101-118.

[OCZ 99] ZIENKIEWICZ O. C., CHAN A., PASTOR M., SCHREFLER B. A., SHIOMI T., *Computational Geomechanics with special Reference to Earthquake Engineering*, John Wiley & Sons, Chichester, 1999.

[WRI 93] WRIGGERS P., *Continuum Mechanics, Non-linear Finite Element Techniques and Computational Stability*. In Stein, E. (ed.): Progress in computational Analysis of Inelastic Structures, CISM 321, Springer-Verlag, Wien, 1993.

Chapter 11

Alternative Formulations in Soil Dynamics

Manuel Pastor
Centro de Estudios y Experimentación de Obras Publicas, CEDEX, Ministerio de Fomento, Madrid, Spain

Mokhtar Mabssout
Faculty of Sciences, University of Tetouan, Morocco

1. Introduction

Soil dynamics is a wide area covering problems of different complexity ranging from simple linear analysis in the frequency domain to nonlinear coupled problems. Here we can mention problems such as liquefaction of foundations or failure of slopes induced by earthquakes. Much effort has been devoted during the past decades to develop suitable mathematical, numerical and constitutive models for soil dynamics problems, including phenomena such as localization.

Dynamic response of geostructures is usually described by a set of PDEs which are of hyperbolic nature. Fundamental solutions are called waves, and in 1D undamped media the solution consists of two waves which propagate in both directions without changing their shape or their amplitude. This is perhaps the main property of hyperbolic problems.

From a numerical point of view, dynamic problems present some important difficulties: (i) Discretization can produce an artificial or "numerical damping", which depends on how large is the wave with respect to mesh size, (ii) Velocity of propagation can be changed by the numerical scheme. Again, this effect depends on the relative wavelength. This effect is known as "numerical dispersion", (iii) In many real situations, the domain extends to infinity. For practical reasons, it has to be limited by artificial boundaries where radiation boundary conditions should be imposed. Otherwise, spurious reflections will be produced.

A dynamic problem can be cast in general in two alternative forms: (i) as a system of second order equations, or (ii) as a system of first order equations. Most of the models described in the literature have been formulated in terms of displacements (or velocities) and pore pressures as primary field variables (see for instance [ZIE 01]).

It has been found that displacement based finite elements present disadvantages when computing failure loads and mechanisms. The problem is more remarkable when using elements of low order such as linear triangles, tetrahedra, bilinear quadrilaterals or trilinear hexahedra. The main reasons are poor accuracy of the elements to reproduce some strain fields (like in bending), and volumetric locking caused by the restriction imposed by the flow rule in plasticity.

Another disadvantage of classical displacement formulations arises when the problem involves short wavelengths and shock propagation. In the case of linear elastic soils, the different Fourier components (i) are propagated at speeds which do not coincide with the physical velocities, and (ii) they are damped. As a consequence, propagation of a shock results in unrealistic oscillations and propagation of short-waves resulting in unacceptable damping.

Concerning shock propagation, Riemann solvers improve the accuracy of numerical simulations. They were introduced by Godunov [GOD 59] in 1959, and since then they have been successfully applied to both fluid dynamics [TOR 88] and solid dynamics problems [MIL 01], [PIT 93] and [TRA 92]. The only problem is the computational overload. Other methods such as the Taylor Galerkin algorithm proposed by [DON 84] and [LOH 84] have been shown to present a reasonable compromise between cost and accuracy for shock propagation problems in fluid dynamics [SAT 98].

Since the early formulations of Lohner et al. and Donea, the algorithm has been improved and applied to a wide variety of hyperbolic problems both in fluid dynamics by Donea, Giuliani, Laval , Quartapelle [DON 84], Baker and Kim [BAK 87], Bottura and Zienkiewicz [BOT 90], Quecedo and Pastor [QUE 01], Donea, Quartapelle and Selmin [DON 87], Donea and Quartapelle [DON 92],and Donea, Roig and Huerta [DON 99] and in solid dynamics where it is worth mentioning the contributions of Safjan and Oden [SAF 93], [SAF 95], Tamma and Namburu [TAM 88] [TAM 90] and Zhang and Tabarrok [ZHA 99].

2. Dynamic Problems

2.1. *Introduction*

We will consider the dynamic behaviour of a 1D elastic bar. The field equations are the following: (i) The balance of linear momentum, which is written as:

$$\frac{\partial \sigma}{\partial x} = \rho \frac{\partial^2 u}{\partial t^2} = \rho \frac{\partial v}{\partial t} \qquad (1)$$

where v is the velocity u the displacement, σ the axial stress and ρ the density; (ii) the constitutive relation between stress and strain, which can be written in terms of displacements and velocities as

$$\sigma = E . \frac{\partial u}{\partial x} \qquad (2)$$

From here we can eliminate the stress σ and arrive at:

$$\frac{\partial^2 u}{\partial t^2} = c^2 \frac{\partial^2 u}{\partial x^2} \qquad (3)$$

where we have introduced the wave celerity c given in terms of the elastic modulus E and the density as

$$c = \sqrt{\frac{E}{\rho}} \qquad (4)$$

Alternatively, we can write the system

$$\frac{\partial}{\partial t} \begin{pmatrix} \sigma \\ v \end{pmatrix} + \begin{pmatrix} 0 & -E \\ -1/\rho & 0 \end{pmatrix} \frac{\partial}{\partial x} \begin{pmatrix} \sigma \\ v \end{pmatrix} = \begin{pmatrix} 0 \\ 0 \end{pmatrix} \qquad (5)$$

which can be expressed in a more compact manner as

$$\frac{\partial}{\partial t} \Phi + A \frac{\partial}{\partial x} \Phi = 0 \qquad (6)$$

where we have introduced the vector of unknowns

$$\Phi = \begin{pmatrix} \sigma \\ v \end{pmatrix}$$

and the matrix

$$A = \begin{pmatrix} 0 & -E \\ -1/\rho & 0 \end{pmatrix}$$

Another representation, referred to as conservation form, is the following:

$$\frac{\partial}{\partial t}\Phi + \frac{\partial}{\partial x}F = 0 \tag{7}$$

where F is the vector of fluxes.

2.2. First Order Equations. Riemann Invariants and Radiation BCs

We will consider the general case where the vector Φ has n components. The eigenvalues $\{\lambda_1, \lambda_2, \lambda_3, ..\lambda_n\}$ and eigenvectors $\{x^{(1)}, x^{(2)}, ...x^{(n)}\}$ of A are obtained from

$$A.x^{(k)} = \lambda_k.x^{(k)} \tag{8}$$

We will define a matrix P as

$$P = \left\{ x^{(1)}, x^{(2)}, ...x^{(n)} \right\}$$

where the term (i, j) is the i-th component of the j-th eigenvector

$$P_{ij} = x_i^{(j)} \tag{9}$$

and we will introduce a vector Ψ such as:

$$\Phi = P.\Psi$$

We arrive then at:

$$\frac{\partial}{\partial t}(P.\Psi)\Phi + A\frac{\partial}{\partial x}(P.\Psi) = 0$$

Multiplying above equation by P^{-1} we obtain:

$$\frac{\partial}{\partial t}(\Psi)\Phi + \Lambda\frac{\partial}{\partial x}(\Psi) = 0$$

where

$$\Lambda = P^{-1}A.P$$

is a diagonal matrix with

$$\Lambda_{kk} = \lambda_k$$

The system of PDEs is therefore transformed into an uncoupled system

$$\frac{\partial \Psi^{(i)}}{\partial t} + \lambda_i \frac{\partial \Psi^{(i)}}{\partial x} = 0 \tag{10}$$

with solutions of the type:

$$\Psi^{(i)} = F^{(i)}(x - \lambda_i t)$$

Once Ψ is known, Φ is immediately obtained as:

$$\Phi = P.\Psi$$

In the case of 1D compression waves in a bar, the eigenvalues of A are:

$$\lambda_1 = +\sqrt{E/\rho} \text{ and } \lambda_2 = -\sqrt{E/\rho}$$

and the eigenvectors:

$$x^{(1)} = \begin{pmatrix} -\sqrt{E\rho} \\ 1 \end{pmatrix} \text{ and } x^{(2)} = \begin{pmatrix} +\sqrt{E\rho} \\ 1 \end{pmatrix}$$

Matrix P is obtained as:

$$P = \begin{pmatrix} -\sqrt{E\rho} & +\sqrt{E\rho} \\ 1 & 1 \end{pmatrix}$$

and its inverse is:

$$P^{-1} = \frac{1}{2}\begin{pmatrix} -\frac{1}{\sqrt{E\rho}} & 1 \\ \frac{1}{\sqrt{E\rho}} & 1 \end{pmatrix}$$

From here, we obtain Ψ as

$$= \frac{1}{2\sqrt{E\rho}}\begin{pmatrix} -\sigma + v\sqrt{E\rho} \\ \sigma + v\sqrt{E\rho} \end{pmatrix}$$

The uncoupled equations are:

$$\frac{\partial \Psi^{(1)}}{\partial t} + c\frac{\partial \Psi^{(1)}}{\partial x} = 0$$

$$\frac{\partial \Psi^{(1)}}{\partial t} - c\frac{\partial \Psi^{(1)}}{\partial x} = 0$$

They describe the convective transport of two magnitudes $\Psi^{(1)}$ and $\Psi^{(2)}$ with velocities $+c$ and $-c$, which are referred to as Riemann invariants of the problem. One important application of Riemann invariants is the derivation of simple radiation and incident wave boundary conditions, which are:

– **Radiation BCs** The condition to be imposed is that no wave is entering the domain from the exterior. If we want to specify such condition at $x = L$, we make zero the corresponding Riemann invariant:

$$\Psi^{(2)} = \sigma + v\sqrt{E\rho} \,|_{x=L} = 0$$

or

$$\Psi_P^{(2)} = \sigma + \rho c \, v = 0 \tag{11}$$

– **Incoming Wave BCs** The condition consists of specifying the known value of the Riemann invariant:

$$\Psi^{(2)} = \sigma + v\sqrt{E\rho} = \Psi_{inc}^{(2)}$$

Concerning two and three dimensional problems in isotropic elastic materials, the method is based on the property that longitudinal and shear waves are uncoupled. A local reference system with axes along the normal and tangent to the surface is defined on the boundary. The equations are then projected on this reference system, and the conditions to be imposed are:

$$\Psi_P^{(2)} \;=\; t_n + \rho c_P \, v_n = 0 \quad \text{and} \tag{12}$$
$$\Psi_S^{(2)} \;=\; t_t + \rho c_S \, v_t = 0$$

2.3. *Elastodynamics: Second order equation*

As in the case of 1D problems, the field equations for 2D and 3D problems are: (i) The balance of momentum equation

$$\mathbf{S}^T \sigma + \mathbf{b} = \rho \frac{\partial^2 \mathbf{u}}{\partial t^2} = \rho \frac{\partial \mathbf{v}}{\partial t} \tag{13}$$

where $\mathbf{S}^T$ is the discrete divergence operator, (ii) the constitutive equation, which in the case of linear elastic materials is

$$\sigma = \mathbf{D}.\varepsilon \tag{14}$$

and

$$\varepsilon = \mathbf{S}.\mathbf{u} \tag{15}$$

For the problem to be well posed, we have to add the following initial and boundary conditions:

– Initial conditions

$$\mathbf{u}(\mathbf{x}, t_0) = \mathbf{u}_0(\mathbf{x})$$
$$\mathbf{v}(\mathbf{x}, t_0) = \mathbf{v}_0(\mathbf{x})$$

– Boundary conditions:

- Prescribed displacements (or velocities) on $\partial\Omega_u$:

$$\mathbf{u}(\mathbf{x}, t) - \bar{\mathbf{u}}(t) = \mathbf{0}$$

- Prescribed tractions on $\partial\Omega_q$

$$\sigma_{ij}.n_j - \bar{t}_i = 0$$

- Radiation BC's on $\partial\Omega_R$ where we meet waves arriving from the interior to leave the domain without spurious reflections. This condition is given by (12), and it can be seen that relates tractions to velocities. Therefore, it can be interpreted as a "numerical damper". If we introduce the transformation matrix T and the matrix V

$$T = \begin{pmatrix} n_x & n_y \\ -n_y & n_x \end{pmatrix} \quad V = \begin{pmatrix} \rho c_P & 0 \\ 0 & \rho c_S \end{pmatrix} \tag{16}$$

the condition can be expressed as

$$\bar{t} + T^{-1}VT\, v = 0 \tag{17}$$

where $t_i = \sigma_{ij}.n_j$.

3. Discretization of the second order equation

3.1. Weak Formulation

The weak formulation is obtained in a similar way as described in the case of the stationary elastic problem.

(WF) Given the domain Ω with boundary $\partial\Omega = \partial\Omega_u \cup \partial\Omega_q$ such that $\partial\Omega_u \cap \partial\Omega_q = \{\varnothing\}$ and the time interval $I = (0, T)$
Find $u(t) \in U = \left\{u : I \times \bar{\Omega} \to \mathbb{R}^{ndim} \mid u \in H^1(\bar{\Omega}) \text{ and } u - \bar{u} = 0 \text{ on } \partial\Omega_u\right\}$ such that

$$\left(\rho\frac{\partial^2 u}{\partial t^2}, w\right) + a(u, w) = \int_{\partial\Omega_q} w^T \bar{t}\, d\Gamma \tag{18}$$

$\forall w \in V = \left\{w : \bar{\Omega} \to \mathbb{R}^{ndim} \mid w \in H^1(\bar{\Omega}) \text{ and } w = 0 \text{ on } \partial\Omega_u\right\}$

In the above, we have used the bilinear form $a(w, u) = \int_\Omega \nabla w^T \mathbf{D}^e \mathbf{S}\, u\, d\Omega$.

3.2. Discretization

Discretization of the weak form 18 will be carried out using Galerkin method. We will assume that the trial functions $u^h(t)$, with $t \in \bar{I}$ belong to the space U^h

$$U^h = \left\{ u^h \in H^1\left(\bar{\Omega}\right) \;\middle|\; u^h = w^h + \psi^h, \; u^h \in V^h \text{and } \psi^h - \bar{u} = 0 \text{ on } \partial\Omega_u \right\}, \text{where}$$

the space of test functions V^h is

$$V^h = \left\{ w^h \in H^1\left(\bar{\Omega}\right) \;\middle|\; w^h = 0 \text{ on } \partial\Omega_u \right\}$$

If the dimension of V^h is nh, and we choose a basis $\{N_j\}$, the approximate solution u^h can be expressed as

$$u^h\left(x,t\right) = \sum_{j=1}^{nh} \hat{u}_j(t)\, N_j(x) + \psi^h \tag{19}$$

We arrive at:

Find $u^h \in U^h$ such that

$$\left(\rho\frac{\partial^2 u^h}{\partial t^2}, w^h\right) + a(u^h, w^h) = \int_{\partial\Omega_q} w^{hT}\, \bar{t}\, d\Gamma - \left(\rho\frac{\partial^2 \psi^h}{\partial t^2}, w^h\right) - a(\psi^h, w^h) \; \forall w^h \in V$$

with $\left(u^h\left(x,0\right), w^h\right) = (u_0, w^h)$ and $\left(\dot{u}^h\left(x,0\right), w^h\right) = (v_0, w^h)$ for all $v^h \in V^h$

The nh unknowns $\left\{\hat{\phi}_j\right\}$ can be obtained applying the $\cdot nh$ conditions:

$$\left(\rho\sum_{j=1}^{nh} N_j \frac{\partial^2 \hat{u}_j}{\partial t^2} + \ddot{w}^h, N_i\right) + a\left(\sum_{j=1}^{nh} \hat{\phi}_j(t)\, N_j + w^h, \phi_i^h\right) = \int_{\partial\Omega_q} \phi_i^h q\, d\Gamma \tag{20}$$

from which we obtain the following linear system:

$$\mathbf{M}\frac{d^2\hat{\mathbf{u}}}{dt^2} + \mathbf{K}\hat{\mathbf{u}} = \mathbf{f} \tag{21}$$

It is sometimes convenient to introduce an approximation to the material damping which is not present in the linear elastic constitutive model. This can be done by adding the term $\mathbf{C}\frac{d\hat{\mathbf{u}}}{dt}$

$$\mathbf{M}\frac{d^2\hat{\mathbf{u}}}{dt^2} + \mathbf{C}\frac{d\hat{\mathbf{u}}}{dt} + \mathbf{K}\hat{\mathbf{u}} = \mathbf{f} \tag{22}$$

There is a wide variety of time integration schemes which can be applied to discretize equation. The interested reader can find in the texts by Zienkiewicz and Taylor [ZIE 00] and Hughes [HUG 87] excellent descriptions.

We will describe here a particular case (GN22) of a family of schemes which has been used in Soil Dynamics [ZIE 01]. They are often referred to as "Generalized Newmark" and were proposed by Katona and Zienkiewicz [ZIE 00].

We will start with the dynamic equation particularized at time t^{n+1}, and will assume that the second derivatives with respect to time can be approximated by

$$\overset{..}{\hat{\mathbf{u}}}{}^{n+1} = \overset{..}{\hat{\mathbf{u}}}{}^{n} + \Delta\overset{..}{\hat{\mathbf{u}}}{}^{n}$$

and we approximate the displacements and velocities as:

$$\dot{\hat{u}}^{n+1} = \dot{\hat{u}}^n + \Delta t \ddot{\hat{u}}^n + \beta_1 \Delta t \Delta \ddot{\hat{u}}^n$$

$$\hat{u}^{n+1} = \hat{u}^n + \Delta t \dot{\hat{u}}^n + \frac{1}{2}\Delta t^2 \ddot{\hat{u}}^n + \frac{1}{2}\beta_2 \Delta t^2 \Delta \ddot{\hat{u}}^n$$

we finally arrive at

$$\mathbf{K}\left(\hat{u}^n + \Delta t \dot{\hat{u}}^n + \frac{1}{2}\Delta t^2 \ddot{\hat{u}}^n + \frac{1}{2}\beta_2 \Delta t^2 \Delta \ddot{\hat{u}}^n\right) + $$

$$\mathbf{C}\left(\dot{\hat{u}}^n + \Delta t \ddot{\hat{u}}^n + \beta_1 \Delta t \Delta \ddot{\hat{u}}^n\right) + $$

$$+\mathbf{M}\left(\ddot{\hat{u}}^n + \Delta \ddot{\hat{u}}^n\right) = \mathbf{f}^{n+1}$$

which can be written as:

$$\left(\mathbf{M} + \beta_1 \Delta t \mathbf{C} + \frac{1}{2}\beta_2 \Delta t^2 \mathbf{K}\right)\Delta \ddot{\hat{u}}^n = \mathbf{F}^n \tag{23}$$

where

$$\mathbf{F}^n = \mathbf{f}^{n+1} - \left(\mathbf{K}\hat{u}^n_{pred} + \mathbf{C}\dot{\hat{u}}^n_{pred} + \mathbf{M}\ddot{\hat{u}}^n_{pred}\right) \tag{24}$$

and

$$\hat{u}^n_{pred} = \hat{u}^n + \Delta t \dot{\hat{u}}^n + \frac{1}{2}\Delta t^2 \ddot{\hat{u}}^n \tag{25}$$

$$\dot{\hat{u}}^n_{pred} = \dot{\hat{u}}^n + \Delta t \ddot{\hat{u}}^n$$

$$\ddot{\hat{u}}^n_{pred} = \ddot{\hat{u}}^n$$

4. Two-step Taylor-Galerkin algorithm

Here we will present a simple 2 step explicit Taylor Galerkin algorithm formulated in terms of velocities and stresses as primary variables. This two step algorithm has been widely used for advection dominated problems in fluid dynamics [PER 86], [ZIE 00], but to the author's knowledge, this is the first time it is used in solid dynamics. We will present next the algorithm for completeness.

We will start by writing the momentum and constitutive equations as:

$$\frac{\partial}{\partial t}\begin{bmatrix} \sigma_{11} \\ \sigma_{22} \\ \sigma_{33} \\ v_1 \\ v_2 \end{bmatrix} - \frac{\partial}{\partial x}\begin{bmatrix} D_{11}v_1 \\ D_{12}v_1 \\ D_{33}v_2 \\ \frac{\sigma_{11}}{\rho} \\ \frac{\sigma_{12}}{\rho} \end{bmatrix} - \frac{\partial}{\partial y}\begin{bmatrix} D_{12}v_2 \\ D_{22}v_2 \\ D_{33}v_1 \\ \frac{\sigma_{12}}{\rho} \\ \frac{\sigma_{22}}{\rho} \end{bmatrix} = \begin{bmatrix} 0 \\ 0 \\ 0 \\ 0 \\ 0 \end{bmatrix} \tag{26}$$

where D_{ij} are the components of the elastic matrix $\mathbf{D}^e$

The above equation can be written in conservation form as

$$\frac{\partial \bar{\phi}}{\partial t} + \frac{\partial F_x}{\partial x} + \frac{\partial F_y}{\partial y} = \bar{S} \tag{27}$$

where we have introduced the vectors of unknowns $\bar{\phi}$, fluxes F_x and F_y, and source $\bar{S}$.

The Taylor-Galerkin algorithm for solving the conservation equation (27)

$$\frac{\partial \bar{\phi}}{\partial t} + div\mathbf{F} = \bar{S} \tag{28}$$

where $\mathbf{F}$ is the advective flux tensor and $\bar{S}$ is the vector of sources, starts from a second order expansion in time

$$\bar{\phi}^{n+1} = \bar{\phi}^n + \Delta t \left.\frac{\partial \bar{\phi}}{\partial t}\right|^n + \frac{1}{2}\Delta t^2 \left.\frac{\partial^2 \bar{\phi}}{\partial t^2}\right|^n \tag{29}$$

where the first order time derivative of the unknowns can be calculated using equation (27) as

$$\left.\frac{\partial \bar{\phi}}{\partial t}\right|^n = \left(\bar{S} - div\,\mathbf{F}\right)^n \tag{30}$$

To obtain the second order time derivative, the Two-Step Taylor-Galerkin procedure considers an intermediate step between t^n and t^{n+1}. The aim of this first time step is to calculate the solution at a time $t^{n+1/2}$. This step is followed by a second one that brings the solution to t^{n+1}.

In this way, the first step results in

$$\bar{\phi}^{n+1/2} = \bar{\phi}^n + \frac{\Delta t}{2}\left(\bar{S} - div\,\mathbf{F}\right)^n \tag{31}$$

which allows the calculation of $\mathbf{F}^{n+1/2}$ and $\bar{S}^{n+1/2}$.

Considering now a Taylor series expansion of the flux and source terms,

$$\mathbf{F}^{n+1/2} = \mathbf{F}^n + \left(\frac{\partial \mathbf{F}}{\partial t}\right)^n \frac{\Delta t}{2}$$

$$\bar{S}^{n+1/2} = \bar{S}^n + \left(\frac{\partial \bar{S}}{\partial t}\right)^n \frac{\Delta t}{2}$$

where the values of $\mathbf{F}^{n+1/2}$ and $\bar{S}^{n+1/2}$ are calculated using $\bar{\phi}^{n+1/2}$, the flux and sources time derivatives are

$$\left(\frac{\partial \mathbf{F}}{\partial t}\right)^n = \frac{2}{\Delta t}\left(\mathbf{F}^{n+1/2} - \mathbf{F}^n\right)$$

$$\left(\frac{\partial \bar{S}}{\partial t}\right)^n = \frac{2}{\Delta t}\left(\bar{S}^{n+1/2} - \bar{S}^n\right)$$

Incorporating these expressions into the second order time derivative

$$\frac{\partial^2 \bar\phi}{\partial t^2}\bigg|^n = \frac{\partial}{\partial t}\left(\bar S - div\,\mathbf{F}\right)^n$$

results in

$$\frac{\partial^2 \bar\phi}{\partial t^2}\bigg|^n = \frac{2}{\Delta t}\left(\bar S^{n+1/2} - \bar S^n - div\left(\mathbf{F}^{n+1/2} - \mathbf{F}^n\right)\right) \tag{32}$$

Substituting now the expressions obtained for the first (30) and second (32) order time derivatives in the Taylor series expansion (29) results in

$$\bar\phi^{n+1} = \bar\phi^n + \Delta t\left(\bar S^{n+1/2} - div\,\mathbf{F}^{n+1/2}\right)$$

This equation is discretized in space using the conventional Galerkin weighting to finally result in the system of equations to be solved to obtain the unknown increments in the variables at the time step. The resulting system of equations is :

$$\underline M\,\Delta\phi = \Delta t \int_\Omega \mathbf{N}\,\mathbf{S}^{n+1/2} d\Omega - \int_{\Gamma_N} \mathbf{N}\left(\mathbf{F}^{n+1/2}\cdot\bar n\right) d\gamma + \int_\Omega \mathbf{F}^{n+1/2}\,grad\,\underline N\,d\Omega \tag{33}$$

The system of equations to be solved during each timestep is of the type

$$\underline M \mathbf{x} = \mathbf{f}$$

and can be economically solved using a Jacobi iteration scheme [PER 86]

$$\mathbf{x}^{(k+1)} = \mathbf{x}^{(k)} + M_L^{-1}(f - \underline M\mathbf{x}^{(k)})$$

where the superscript k is an iteration counter, if an approximate inverse matrix, M_L^{-1}, is known in advance. As in the case of equation (33) an approximate inverse of the system matrix, $\underline M$, is the lumped mass matrix, the equation system (33) can be solved using this algorithm. Typically, less than six iterations are enough to obtain an accurate solution.

We will consider the simple one dimensional model problem

$$\frac{\partial\phi}{\partial t} + A\frac{\partial\phi}{\partial x} = G\phi$$

where A and G are constant.

The stability limits to be imposed on the timestep are obtained, according to [PER 86], as follows. The timestep should fulfill the limitation

$$\Delta t|_{Global} \leq f * \min\left(\alpha\frac{\Delta x}{A}; \frac{2}{G}\right)$$

where f is a safety factor with a typical value of 0.8-0.9, α is taken as 1 when using a lumped mass matrix and $\alpha = 1/\sqrt{3}$ when using the consistent mass matrix. The first component of the timestep limit is due to advection, and the second to the source term.

5. Examples

5.1. Propagation of a shock wave on a soil layer

In this example, we will solve the problem of propagation of shear wave through an elastic soil stratum using the two-step Taylor-Galerkin scheme. Figure 1 depicts the problem lay out. A constant horizontal velocity of $v_x = 1\ m/s$ is prescribed at the top boundary for $t \leqslant 2.5\ 10^{-3}\ s$ and $v_x = 0$ for $t > 2.5\ 10^{-3}\ s$. In order to ensure a pure shear wave, the vertical movement of all nodes is restrained. The material under consideration has the following parameters: Young modulus $E = 8\ 10^7\ Pa$,the density is $\rho = 2000\ kg/m^3$ the Poisson's ratio is $\nu = 0.3$. Linear triangular elements are used to discretize the stratum. The shear stress wave travels with a velocity of

$$c = \sqrt{\frac{E}{2\,\rho(1+\nu)}} = 124,03\ m/s$$

Figure 2 presents the shear stress evolution at $x = 0.75\ m$ for which there is an available analytical solution $\tau = \rho c v = 2.48\ 10^5\ Pa$ .The time step used for computation $\Delta t = 1.58\ 10^{-5}\ s$ and the element size used is $h = 0.01\ m$. The solution is compared against the results obtained with a Newmark scheme, and a one-step TG algorithm. We observe that Two-Step Taylor-Galerkin introduce minimum diffusion and dispersion and presents a better behaviour than that of classical Newmark algorithm.

5.2. Vibration isolation

One practical problem of importance in the design of underground railroads is the transmission of vibrations to buildings. In this example, we analyze the effect of vibrations caused by an underground train on a neighbouring building. We will use the two step algorithm introduced in this paper. Figure 4 sketches the position of the tunnel and the building. The tunnel of radius 3 m is centered at a depth of 15 m below the surface. Trains induce vibrations which will arrive at the building located at point A. Should the amplitude not be acceptable, the building would have to be isolated. One possibility is the construction of an screening wall such as the one depicted in Figure 4.

The purpose of the analysis we will perform is to compare the amplitude of the vibrations arriving at A in both cases without and with screen. For simplicity, we have

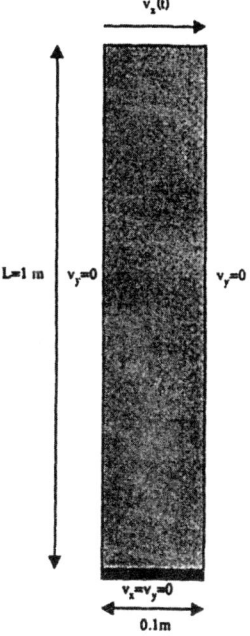

Figure 1. *Shear layer. Problem layout*

used a single mesh where we will change the material properties of the wall elements in the second case.

We have assumed the soil is elastic and isotropic, with a Young modulus $E = 3.0\,10^7 Pa$, density $\rho = 2000\ kgm^{-3}$ and Poisson's ratio $\nu = 0.3$. The wall properties are:

Young modulus	$E = 10^{10}\ Pa$
Density	$\rho = 2500\ kgm^{-3}$
Poisson's ratio	$\nu = 0.3$

Boundary conditions used for the Two-step Taylor-Galerkin algorithm are the following :

(i) Prescribed vertical displacement at point B

$$v_y = v_o \sin \omega' t \sin \omega t$$
$$\omega' = \frac{\omega}{n}$$

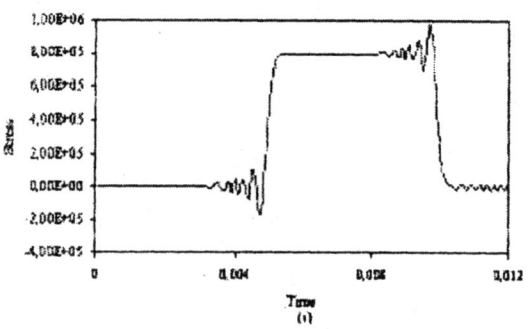

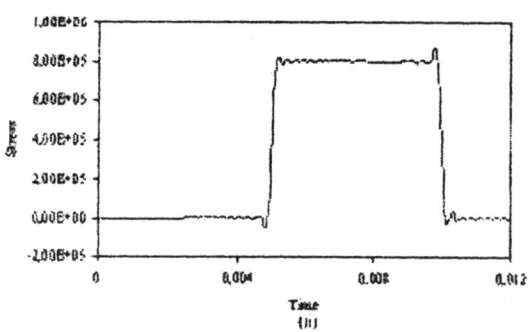

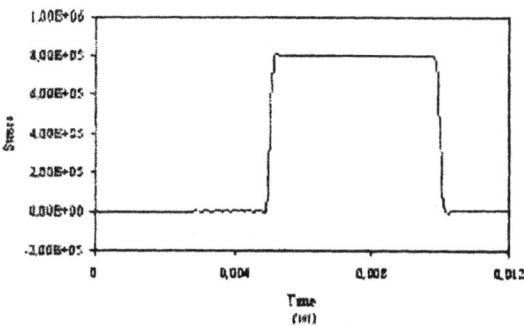

Figure 2. *Stress history at x=0 in the elastic case: (i) Newmark (ii) Taylor-Galerkin One-Step (iii) Taylor-Galerkin Two-Step*

History of prescribed vertical displacement is given in Figure 3.

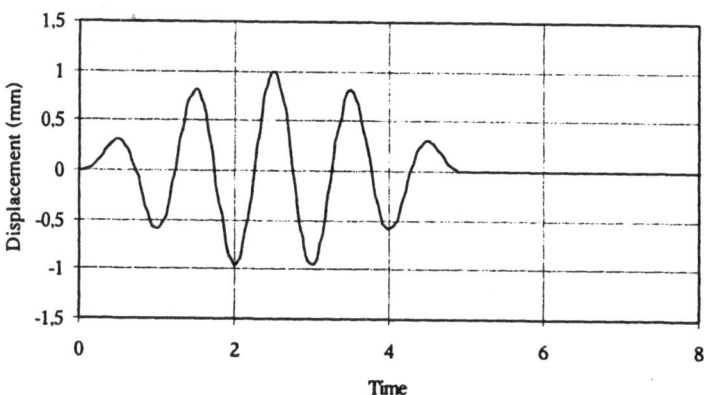

Figure 3. *Displacement at point B*

(ii) Boundaries Γ_1 and Γ_5 have been assumed to be traction free

(iii) Finally, boundaries Γ_2, Γ_3 and Γ_4 arc artificial boundaries from which waves should leave the domain without spurious reflections. The radiation boundary conditions are of first order type, and consist of specifying that Riemann invariants of waves entering the domain are zero. In one dimensional situations, this condition is:

$$\Psi^{(2)} = \sigma + v\sqrt{E\rho} = \Psi^{(2)}_{inc}$$

Concerning two and three dimensional problems in isotropic elastic materials, the method is based on the property that longitudinal and shear waves are uncoupled. A local reference system with axes along the normal and tangent to the surface is defined on the boundary. The equations are then projected on this reference system, and the conditions to be imposed are:

$$\Psi^{(2)}_P = t_n + \rho c_P\, v_n = 0 \quad \text{and} \tag{34}$$

$$\Psi^{(2)}_S = t_t + \rho c_S\, v_t = 0$$

The results of the computations are given in Figures 5(a) (without screen) and 5(b).

On the other hand, in order to protect a building situated on the surface, at point A for example, from the incident wave, a pile of 1 m of diameter and 10 m length was installed from the ground surface in the soil and at 5 m from point A (see Figure 4).

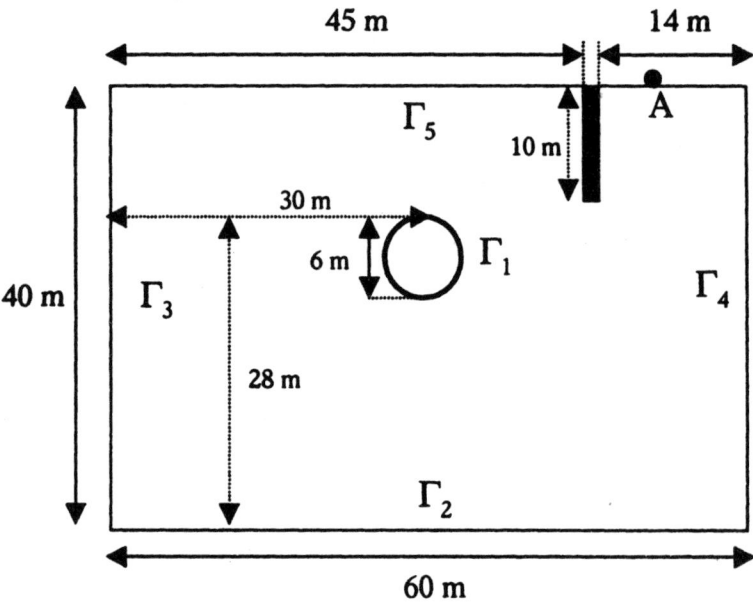

Figure 4. *Tunnel geometry*

Figure 5 (a) depicts the time history of vertical displacement at point A without the screening wall, and Figure 5 (b) gives the vertical displacement at the same point after the wall has been built. We can observe that without protection the amplitude of the displacement is about 0.0045 *mm* and with protection the displacement of the point A decreases and it is about 0.0015 *mm*.

6. Acknowledgments

The authors gratefully acknowledge the support of the spanish Agencia Española de Cooperación Internacional (AECI) and the Ministerio de Ciencia y Tecnología (MCYT) for the economic support granted.

7. References

[ZIE 01] O.C.Zienkiewicz, A.H.C.Chan, M.Pastor, B.A.Schrefler and T.Shiomi, Computational Geomechanics with Special Reference to Earthquake Engineering

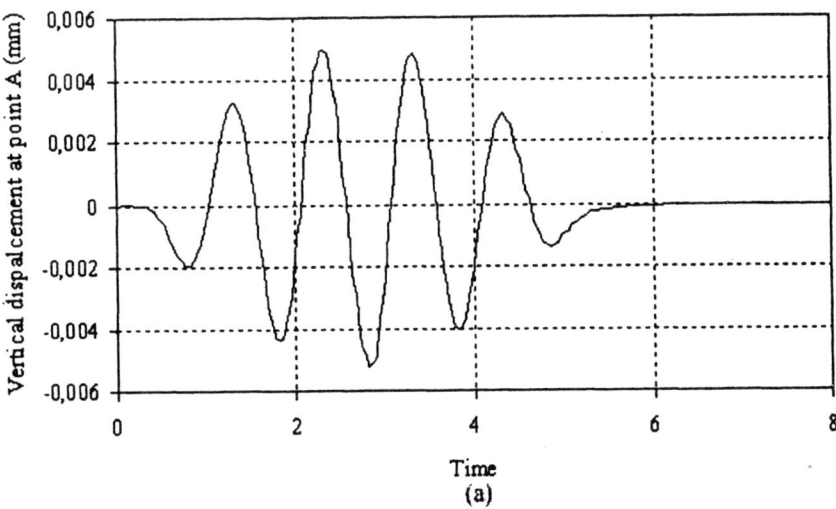

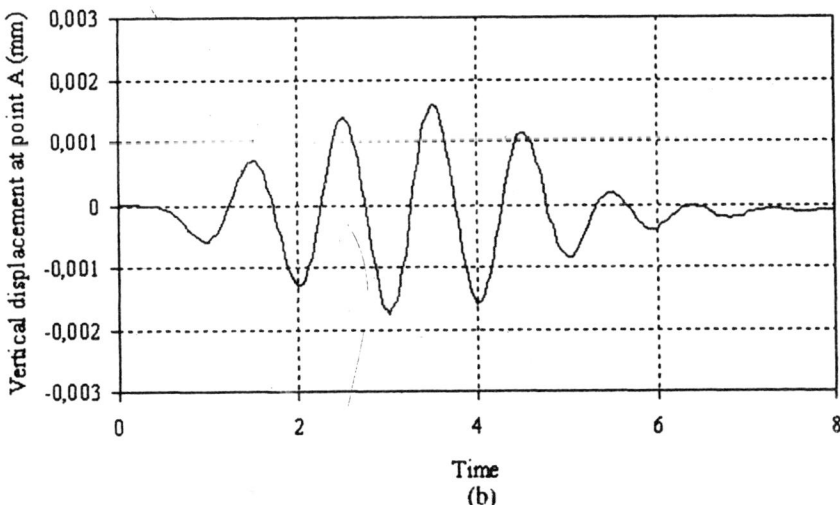

Figure 5. *Vertical displacement at point A. (a) without (b) With screening wall*

[GOD 59] S.K.Godunov, "A difference method for numerical calculation of discontinuous equations of hydrodynamics", **Matematicheskii Sbornik 47**, 271-300 (1959) (in Russian)

[TOR 88] E.F.Toro, **Shock-Capturing Methods for Free-Surface Shallow Flows**, John Wiley & Sons (2001)

[MIL 01] G.H. Miller & P. Colella, " A High-Order Eulerian Godunov Method for Elastic-Plastic flow in solids", **J.Comp.Phys.** 167, 131-176 (2001)

[PIT 93] E.B. Pitman, "A Godunov Method for Localization in Elastoplastic Granular Flow", **Int. J. Num. Anal.Meth.Geomech.** 17, 385-400 (1993)

[TRA 92] J.A. Trangenstein & R.B. Pember, "Numerical Algorithms for Strong Discontinuities in Elastic-Plastic Solids"**J.Comp.Phys.** 103, 63-89 (1992)

[DON 84] J. Donea, "A Taylor-Galerkin method for convection transport phenomena", **Int. J. Num. Meth. Engng.**, 20, 101-119 (1984)

[LOH 84] R.L. Lohner, K. Morgan and O. C. Zienkiewicz, "The solution of non-linear hyperbolic equation systems by the finite element method", **Int. J. Num. Meth. Fluids 4**, 1043-1063 (1984)

[SAT 98] B.V.K.Satya Sai, O.C.Zienkiewicz, M.T.Manzari, P.Lyra and K.Morgan, General purpose versus special algorithms for high speed flows with shocks, **Int. J. Num. Meth. Fluids 27**, 57-80, 1998.

[DON 84] J. Donea, S. Giuliani, H. Laval and L. Quartapelle, "Time-accurate solution of advection-diffusion problems by finite elements", **Comp. Meth. Appl.Mech. Engng.**, 45, 123-145 (1984)

[BAK 87] A. J. Baker and J. W. Kim, "A Taylor weak-statement algorithm for hyperbolic conservation laws" Int .**J. Num. Meth. Fluids, 7**, 489-520 (1987)

[BOT 90] L. Bottura and O. C. Zienkiewicz, "Experiments on iterative solution for the semi-implicit characteristic-Galerkin Algorithm", **Comm. Appl. Num. Meth.,** 6. 387-393 (1990)

[QUE 01] M.Quecedo and M.Pastor, "A reappraisal of Taylor-Galerkin algorithm for dry-wetting areas in shallow water computations", **Int.J.Num.Meth.Fluids.38**, 515-531 (2002)

[DON 87] J.Donea, L.Quartapelle and V.Selmin, "An analysis of Time Discretization in the Finite Element Solution of Hyperbolic Problems", **J.Comp.Phys.** 70, 463-499 (1987)

[DON 92] J.Donea and L.Quartapelle and V.Selmin, "An introduction to finite element methods for transient advection problems", **Comp.Meth.Appl.Mech.Engng.** 95, 169-203 (1992)

[DON 99] J.Donea,B.Roig and A.Huerta , Pade/Least-squares schemes for convective transport problems, **Métodos Numéricos en Ingeniería**, R.Abascal, J.Domínguez and G.Bugeda (Eds.), SEMNI, Barcelona, Spain 1999

[SAF 93] A.Safjan and J.T.Oden, "High-order Taylor-Galerkin and adaptive h-p methods for second order hyperbolic systems: Applications to elastodynamics", **Comp.Meth.Appl.Mech.Engng.** 103, 187-230 (1993)

[SAF 95] A.Safjan and J.T.Oden, "High-Order Taylor-Galerkin Methods for Linear Hyperbolic Systems", **J.Comp.Phys.** 120, 206-230 (1995)

[TAM 88] K.K.Tamma and R.R.Namburu, "A New Finite Element based Lax-Wendroff/Taylor- Galerkin methodology for Computational Dynamics", **Comp.Meth.Appl.Mech.Engng.** 71, 137-150 (1988)

[TAM 90] K.K.Tamma and R.R.Namburu, "A Robust self-starting explicit computational methodology for structural dynamic applications: Architecture and representations", **Int. J. Num. Meth. Engng.** 29, 1441-1454 (1990)

[ZHA 99] Y.Zhang and B.Tabarrok, "Modifications to the Lax-Wendroff scheme for hyperbolic systems with source terms", Int.J.Num.Meth.Engng. 44, 27-40 (1999)

[PER 86] J. Peraire, "A Finite Element Method for Convection Dominated Flows", Ph.D. Thesis, University of Wales, Swansea (1986)

[HUG 87] T.J.R.Hughes, The Finite Element Method. Linear static and Dynamic Finite Element Analysis, Prentice-Hall Int.Ed. London, 1987

[ZIE 00] O. C. Zienkiewicz and R. L. Taylor, The Finite Element Method, 5th Edition , Butterworth-Heinmann (2000).

Chapter 12

Objective Modelling of Strain Localization

Milan Jirásek

Laboratory of Structural and Continuum Mechanics, Swiss Federal Institute of Technology (EPFL), Lausanne, Switzerland

1. Introduction

In many structures subjected to extreme loading conditions, the initially smooth distribution of strain changes into a highly localized one. Typically, the strain increments are concentrated in narrow zones while the major part of the structure experiences unloading. Such *strain localization* can be caused by geometrical effects (e.g., necking of metallic bars) or by material instabilities (e.g., microcracking, frictional slip, or nonassociated plastic flow). Here we concentrate on the latter case. To keep the presentation simple, we consider only the static response in the small-strain range.

After an illustrative example that explains why the conventional theories fail to provide an objective description of strain localization, we will give a general overview of modeling approaches that are appropriate for this purpose. Attention will then focus on models regularized by spatial integrals or by gradient terms. Models that treat highly localized strains as strong discontinuities (jumps in the displacement field) will be discussed in the subsequent article [JIR 02a].

2. Problems with objective description of strain localization

To introduce the basic concepts and to illustrate the behavior of various models, we will analyze a simple one-dimensional localization problem, which can be interpreted either as a bar of a constant cross section A and of total length L under uniaxial tension (Figure 1a), or as a material layer under shear (Figure 1b). In the examples we will discuss the tensile bar, but all the results can be reinterpreted in terms of the shear problem, simply by replacing the normal stress σ by shear stress τ, normal strain ε by shear angle γ, Young's modulus E by shear modulus G, etc.

The material is assumed to obey a simple stress-strain law with linear elastic behavior up to the peak stress, f_t, followed by linear softening; see Figure 2a. The strain at which the transmitted stress completely disappears is denoted by ε_f. The peak stress is attained at strain $\varepsilon_0 = f_t/E$ where E is Young's modulus of elasticity. If the bar is loaded in tension by an applied displacement u at one of the supports, the response remains linear elastic up to $u_0 = L\varepsilon_0$. At this state, the force transmitted by the bar (reaction at the support) reaches its maximum value, $F_0 = Af_t$. After that, the re-

(a)   (b)

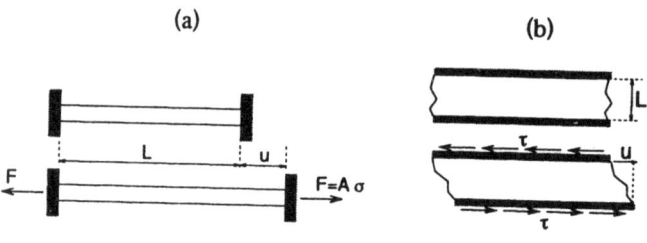

Figure 1. *(a) Bar under uniaxial tension, (b) shear layer*

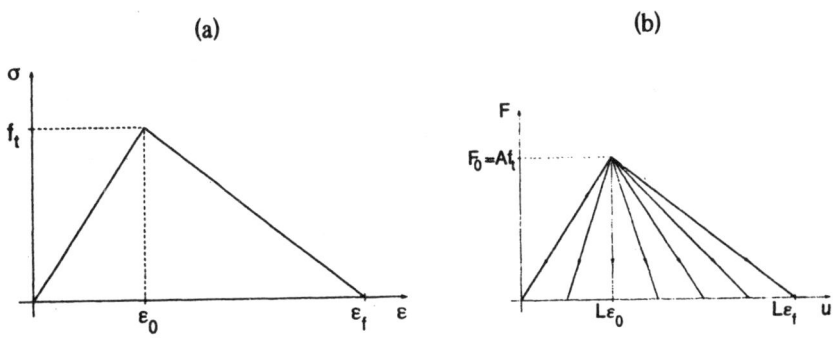

Figure 2. *(a) Stress-strain diagram with linear softening, (b) fan of possible post-peak branches of the load-displacement diagram*

sistance of the bar starts decreasing. At each cross section, stress can decrease either at increasing strain (softening) or at decreasing strain (elastic unloading). The static equation of equilibrium implies that the stress profile must remain uniform along the bar. However, the same stress $\bar{\sigma}$ between 0 and f_t can be generated by strain histories with different final values of strain, and so the strain profile does not need to be uniform. For example, the material can be softening in an interval of length L_s and unloading everywhere else. When the stress is completely relaxed to zero, the strain in the softening region is $\varepsilon_s = \varepsilon_f$ and the strain in the unloading region is $\varepsilon_u = 0$; the total elongation of the bar is therefore $u = L_s\varepsilon_s + L_u\varepsilon_u = L_s\varepsilon_f$. The length L_s remains undetermined, and it can have any value between 0 and L. This means that the problem has infinite solutions, and the corresponding post-peak branches of the load-displacement diagram fill the fan shown in Figure 2b.

The ambiguity is removed if imperfections are taken into account. Real material properties and sectional dimensions cannot be perfectly uniform. Suppose that the strength in a small region is slightly lower than in the remaining portion of the bar. When the applied stress reaches the reduced strength, softening starts and the stress decreases. Consequently, the material outside the weaker region must unload elastically, because its strength has not been exhausted. This leads to the conclusion that the size of the softening region cannot exceed the size of the region with minimum strength. Such a region can be arbitrarily small, and the corresponding softening branch can be arbitrarily close to the elastic branch of the load-displacement diagram. Thus the standard strain-softening continuum formulation leads to a solution that has several pathological features: (i) the softening region is infinitely small; (ii) the load-displacement diagram always exhibits snapback, independently of the structural size and of the material ductility; (iii) the total amount of energy dissipated during the failure process is zero.

From the mathematical point of view, these annoying features are related to the so-called *loss of ellipticity* of the governing differential equation. In the present one-

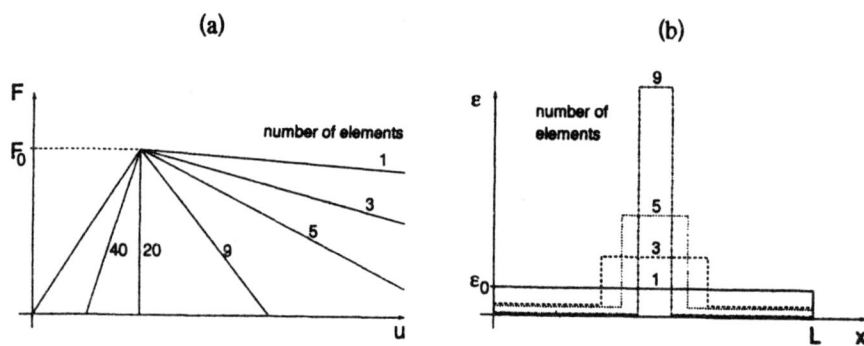

Figure 3. *Effect of mesh refinement on the numerical results: (a) load-displacement diagrams, (b) strain profiles*

dimensional setting, loss of ellipticity occurs when the tangent modulus ceases to be positive. The boundary value problem becomes ill-posed, i.e., it does not have a unique solution with continuous dependence on the given data. From the numerical point of view, ill-posedness is manifested by pathological sensitivity of the results to the size of finite elements. For example, suppose that the bar is discretized by N_e two-node elements with linear displacement interpolation. If the numerical algorithm properly captures the most localized solution, the softening region extends over one element, and we have $L_s = L/N_e$. The slope of the post-peak branch therefore strongly depends on the number of elements, and it approaches the initial elastic slope as the number of elements tends to infinity; see Figure 3a, constructed for a stress-strain law with $\varepsilon_f/\varepsilon_0 = 20$. The strain profiles at $u = 2u_0$ for various mesh refinements are plotted in Figure 3b (under the assumption that the imperfection is located at the center of the bar). In the limit, the profiles tend to $2u_0\, \delta(x - L/2)$ where δ denotes the Dirac distribution. The limit solution represents a displacement jump at the center, with zero strain everywhere else.

3. Classification of models

In real materials, inelastic processes typically localize in narrow bands that initially have a small but finite width. Propagation and coalescence of microdefects in the localization band can eventually lead to the formation of a displacement discontinuity, e.g., of a macroscopic stress-free crack or a sharp slip line. The initial thickness of the localization band depends on the material microstructure and is usually of the same order of magnitude as the characteristic material length, determined by the size or spacing of dominant heterogeneities. Mathematically, narrow zones of highly concentrated evolving microdefects can be represented in many different ways. To establish a systematic classification, we will first look at the kinematic description and then discuss the corresponding material models.

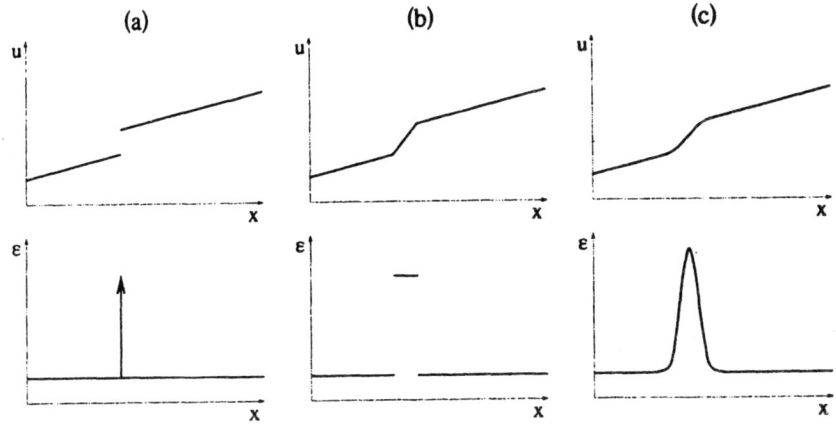

Figure 4. *Kinematic description with (a) one strong discontinuity, (b) two weak discontinuities, (c) no discontinuities*

3.1. *Kinematic description*

Depending on the regularity of the displacement field, $\mathbf{u}(\mathbf{x})$, we can distinguish three types of kinematic descriptions. The first one incorporates *strong discontinuities*, i.e., jumps in displacements across a discontinuity curve (in two dimensions) or discontinuity surface (in three dimensions). The strain field, $\boldsymbol{\varepsilon}(\mathbf{x})$, then consists of a regular part, obtained by standard differentiation of the displacement field, and a singular part, having the character of a multiple of the Dirac delta distribution. This is schematically shown for the one-dimensional case in Figure 4a. In physical terms, the strong discontinuity corresponds to a sharp crack (not necessarily a stress-free one).

Another possible kinematic description represents the region of localized deformation by a band of a small but finite thickness, separated from the remaining part of the body by two *weak discontinuities*, i.e., curves or surfaces across which certain strain components have a jump but the displacement field remains continuous. This is shown in Figure 4b. Since the displacement is continuous, the strain components in the plane tangential to the discontinuity surface must remain continuous as well, and only the out-of-plane components can have a jump. In physical terms, the band between the weak discontinuities corresponds to a damage process zone with an almost constant density of microdefects.

Finally, the most regular description uses a continuously differentiable displacement field, and the strain field remains *continuous*. Strain localization is manifested by high strains in a narrow band, with a continuous transition to much lower strains in the surrounding parts of the body. A typical strain profile of this type is shown in Figure 4c. In physical terms, this corresponds to a damage process zone with a continuously varying concentration of defects.

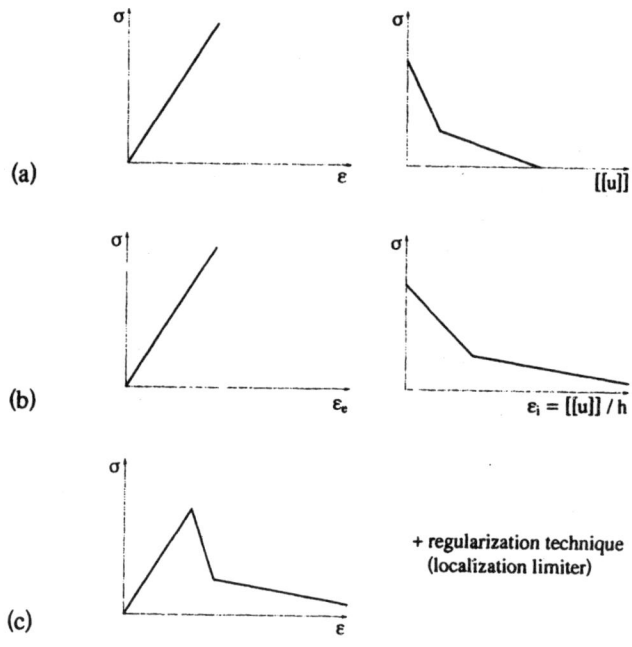

Figure 5. *Schematic representation of the constitutive description: (a) cohesive crack model, (b) smeared crack model, (c) regularized continuum model with softening*

3.2. *Constitutive models*

Each of the kinematic descriptions discussed in the preceding section requires a different type of constitutive law for the cracking material. Of course, one could also consider the material law as the primary component of the model and say that the regularity of the displacement field obtained as the solution of the corresponding boundary value problem depends on the choice of that law.

In the simplest case, the strong discontinuity can be considered as stress-free, which necessarily leads to a stress singularity at the discontinuity front (crack tip). This is the linear elastic fracture mechanics approach, applicable at very large scales, when the process zone is negligible with respect to the characteristic dimensions of the body (such as the crack length and ligament size). Usually it is necessary to take into account the finite size of the process zone at least in the direction tangential to the discontinuity surface. In the direction normal to the discontinuity, all inelastic effects are lumped and replaced by the equivalent displacement jump (crack opening), $[[u]]$. Before the growing microdefects coalesce and form a true physical crack, the process zone still carries some stresses, which are represented in the model as cohesive tractions transmitted by the discontinuity. The usual assumption of *cohesive crack models* or *cohesive zone models* is that the cohesive traction depends only on the displacement jump (separation vector), even though the evolution of the process zone is in general affected also by strains (or stresses) in the tangential plane. This means that the strong discontinuity is governed by its own constitutive law, formulated as a traction-separation law, which complements the stress-strain law valid for the continuous part of the body; see Figure 5a. Advanced numerical techniques for the resolution of strong discontinuities will be presented in the next contribution.

Models with localization bands bounded by weak discontinuities can be considered as simple regularizations of models with strong discontinuities. Instead of lumping all the inelastic effects into a surface, it is possible to distribute them uniformly across the width of a band of a finite thickness h. This naturally leads to the *smeared crack models*, which transform the traction-separation law into a law that links the stress transmitted by the localization band to the average inelastic strain in that band; see Figure 5b. The transformation of the displacement jump $[[u]]$ into the equivalent inelastic strain $\varepsilon_i = [[u]]/h$ is affected by the thickness h of the numerical localization band, which is proportional to the size of the finite elements. Consequently, the softening part of the equivalent stress-strain law is not a material property but depends also on the numerical discretization. This technique is known as the crack band model or the fracture energy approach.

Strain fields that remain continuous even after the onset of localization can be obtained with more sophisticated regularization techniques, called localization limiters. They are usually based on various forms of *enriched continuum theories*, e.g., on nonlocal or higher-order gradient continua. Such enrichments typically introduce a parameter defining a characteristic length of the material. They will be discussed in more detail in the following section.

4. Regularized softening continua

Full regularization of the localization problem can be achieved by a proper generalization of the underlying continuum theory. Generalized continua in the broad sense can be classified according to the following criteria:

1) Generalized kinematic relations (and the dual equilibrium equations).

   a) Continua with microstructure, e.g., Cosserat-type continua or strain-gradient theories.

   b) Continua with nonlocal strain, e.g., nonlocal elasticity.

2) Generalized constitutive equations.

   a) Material models with gradients of internal variables (in some cases also with gradients of thermodynamic forces).

   b) Material models with nonlocal internal variables (in some cases also with nonlocal thermodynamic forces).

Here we focus on the second class of models, with enhancements on the level of the constitutive equations. Their advantage is that the kinematic and equilibrium equations remain standard, and the notions of stress and strain keep their usual meaning.

4.1. Nonlocal models

Even though the idea of a nonlocal continuum has a much longer history, nonlocal material models of the integral type were first exploited as localization limiters in the 1980s. After some preliminary formulations exploiting the concept of an imbricate continuum [BAŽ 84], the nonlocal damage theory emerged [PIJ 87]. Nonlocal formulations were then developed for a number of constitutive theories, including softening plasticity, smeared cracking, microplane models, etc. For a list of references, see e.g. [BAŽ 02] or Chapter 26 in [JIR 01].

Generally speaking, the nonlocal approach consists in replacing a certain variable by its nonlocal counterpart obtained by weighted averaging over a spatial neighborhood of each point under consideration. If $f(x)$ is some "local" field in a domain V, the corresponding nonlocal field is defined as

$$\bar{f}(x) = \int_V \alpha(x, \xi) f(\xi) \, d\xi \qquad [1]$$

where $\alpha(x, \xi)$ is a given *nonlocal weight function*. In an infinite specimen, the weight function depends only on the distance between the "source" point, ξ, and the "receiver" point, x. In the vicinity of a boundary, the weight function is usually rescaled such that the nonlocal operator does not alter a uniform field. This can be achieved by setting

$$\alpha(x, \xi) = \frac{\alpha_0(\|x - \xi\|)}{\int_V \alpha_0(\|x - \zeta\|) \, d\zeta} \qquad [2]$$

where $\alpha_0(r)$ is a monotonically decreasing nonnegative function of the distance $r = \|x - \xi\|$. It is often taken as the Gaussian function

$$\alpha_0(r) = \exp\left(-\frac{r^2}{2\ell^2}\right) \qquad [3]$$

where ℓ is called the *internal length* of the nonlocal continuum. Another possible choice is the bell-shaped function

$$\alpha_0(r) = \begin{cases} \left(1 - \dfrac{r^2}{R^2}\right)^2 & \text{if } 0 \leq r \leq R \\ 0 & \text{if } R \leq r \end{cases} \qquad [4]$$

where R is a parameter related to the internal length. Since R corresponds to the largest distance of point ξ that affects the nonlocal average at point x, it is called the *interaction radius*. The Gauss function [3] has an unbounded support, i.e., its

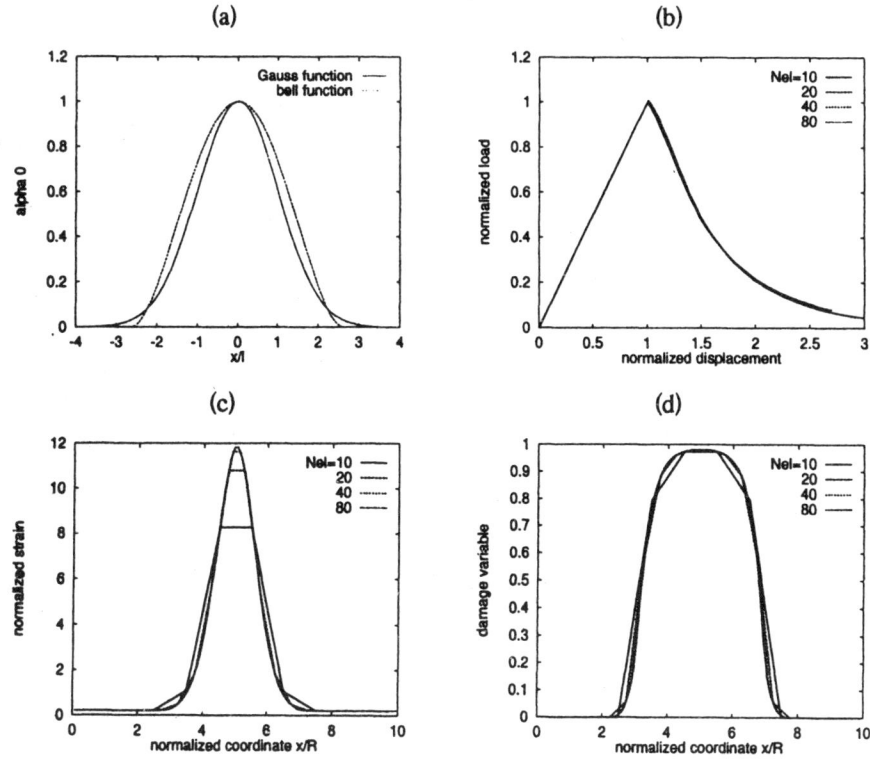

Figure 6. *(a) Nonlocal weight functions α_0, (b) convergence of load-displacement diagram, (c) convergence of strain profile, (d) convergence of damage profile; Nel = number of elements*

interaction radius is $R = \infty$. A normalized plot of both functions in one dimension is shown in Figure 6a.

From a purely phenomenological point of view, the choice of the variable to be averaged remains to some extent arbitrary, provided that a few basic requirements are satisfied. First of all, we want the enriched model to coincide with the standard local elastic continuum as long as the material behavior remains in the elastic range. For this reason, it is not possible to simply replace the local strain by nonlocal strain and apply the usual constitutive law. Except for the case of homogeneous strain, nonlocal strain differs from the local one and the model behavior would be altered already in the elastic range. Second, the model should give a realistic response in elementary loading situations such as uniaxial tension. This aspect has been studied for nonlocal damage models in [JIR 98], and for nonlocal plasticity models in [JIR 02b].

As a simple example of a softening continuum, consider the one-parameter isotropic damage model described by the stress-strain law

$$\boldsymbol{\sigma} = (1 - \omega)\boldsymbol{D_e}\,\boldsymbol{\varepsilon} \tag{5}$$

damage law

$$\omega = f(\kappa) \tag{6}$$

and loading-unloading conditions

$$g(\boldsymbol{\varepsilon}, \kappa) \equiv \tilde{\varepsilon}(\boldsymbol{\varepsilon}) - \kappa \le 0, \qquad \dot{\kappa} \ge 0, \qquad g(\boldsymbol{\varepsilon}, \kappa)\,\dot{\kappa} = 0 \tag{7}$$

in which $\boldsymbol{\sigma}$ is the column matrix of stress components, $\boldsymbol{\varepsilon}$ is the column matrix of engineering strain components, $\boldsymbol{D_e}$ is the elastic material stiffness matrix, ω is the damage variable, $\tilde{\varepsilon}$ is the equivalent strain (to be specified later), and κ is an internal variable that corresponds to the maximum level of equivalent strain ever reached in the previous history of the material. The loading function g controls the shape of the elastic domain in the strain space, and the damage function f controls the shape of the stress-strain curve. It is often assumed that the damage growth begins only when a certain elastic limit is exceeded, i.e., that $f(\kappa) = 0$ for $\kappa \le \varepsilon_0$ where ε_0 is the strain at the elastic limit. For $\kappa > \varepsilon_0$, the damage function f is monotonically increasing and it reaches the limit value $f = 1$ either at a finite value $\kappa = \kappa_f$, or asymptotically as $\kappa \to \infty$.

The equivalent strain $\tilde{\varepsilon}$ is a scalar measure of the strain level, and under uniaxial stress it could be defined as the normal strain in the direction of loading. However, since damage in many materials propagates much more easily under tension than under compression, the definition of $\tilde{\varepsilon}$ is usually modified such that the influence of tension is emphasized. As a first approximation, one may assume that compressive strains do not lead to any damage, and relate $\tilde{\varepsilon}$ to positive strains only. For instance, Mazars [MAZ 84] defined the equivalent strain as

$$\tilde{\varepsilon} = \sqrt{\sum_{I=1}^{3} \langle \varepsilon_I \rangle^2} \tag{8}$$

where ε_I, $I = 1, 2, 3$, are the principal strains, and the brackets $\langle \ldots \rangle$ denote the positive part, i.e., $\langle \varepsilon \rangle = \varepsilon$ for $\varepsilon > 0$ and $\langle \varepsilon \rangle = 0$ for $\varepsilon \le 0$.

Of course, as soon as the damage evolution induces softening, the damage model in the foregoing local form loses objectivity and leads to pathological mesh sensitivity. A suitable nonlocal formulation that regularizes the problem is obtained if damage is computed from the nonlocal equivalent strain. In the loading-unloading conditions, the local value $\tilde{\varepsilon}$ is replaced by its nonlocal average

$$\bar{\varepsilon}(\boldsymbol{x}) = \int_V \alpha(\boldsymbol{x}, \boldsymbol{\xi})\tilde{\varepsilon}(\boldsymbol{\xi})\,\mathrm{d}\boldsymbol{\xi} \tag{9}$$

According to the modified loading-unloading conditions,

$$\bar{\varepsilon} - \kappa \le 0, \qquad \dot{\kappa} \ge 0, \qquad (\bar{\varepsilon} - \kappa)\,\dot{\kappa} = 0 \tag{10}$$

the internal variable κ has the meaning of the largest previously reached value of *nonlocal* equivalent strain. The corresponding damage variable evaluated from [6] is then substituted into the stress-strain equations [5]. It is important to note that the damage variable is evaluated from the nonlocal equivalent strain but the strain ε that appears in [5] explicitly is kept local. In the elastic range, the damage variable remains equal to zero, and the stress-strain relation is local.

Figure 6b shows the load-displacement diagram for strain localization in a bar under uniaxial tension (Figure 1a), calculated using a nonlocal damage model with the weight function [4]. As the number of finite elements increases, the load-displacement curve rapidly converges to the exact solution. Convergence of strain and damage profiles generated by an applied displacement $u = 2u_0$ is documented in Figures 6c, d. In contrast to the local model, the process zone does not shrink to a single point as the mesh is refined. Its size is controlled by the interaction radius, considered as a material parameter.

4.2. Explicit gradient models

Gradient models can be considered as the differential counterpart of integral nonlocal formulations. Instead of dealing with integrals that represent spatial interactions, we can take the microstructure into account by incorporating the influence of gradients (of the first or higher order) of internal variables into the constitutive relations. For example, in the gradient-dependent plasticity theory that evolved from the original ideas of Aifantis [AIF 84], the yield stress is assumed to depend not only on the value of the hardening variable, κ, but also on its Laplacean, $\nabla^2\kappa$. In the simplest case of linear isotropic softening, the gradient-enhanced softening law reads

$$\sigma_Y = \sigma_0 + H(\kappa + \ell^2\nabla^2\kappa) \qquad [11]$$

where σ_0 is the initial yield stress (analogous to the tensile strength f_t), σ_Y is the current yield stress, H is the plastic modulus (negative in the case of softening), and ℓ is a material parameter with the dimension of length.

The hardening-softening variable κ is usually defined as the cumulative plastic strain, i.e., it is obtained by integrating in time the scaled norm of the plastic strain rate tensor. Under uniaxial tension (Figure 1a) κ is equal to the plastic strain, ε_p, and the Laplacean $\nabla^2\kappa$ reduces to the second derivative ε_p''. If the solution remains uniform, the plastic strain is constant in space, its second derivative vanishes, and [11] reduces to the standard "local" law $\sigma_Y = \sigma_0 + H\varepsilon_p$. After the onset of localization, the higher-order term is activated and prevents localization of plastic strain into a set of measure zero.

The yield condition $\sigma - \sigma_Y = 0$ implies that, inside the plastic zone, the plastic strain must satisfy the second-order differential equation

$$\varepsilon_p + \ell^2\varepsilon_p'' = \frac{\sigma - \sigma_0}{H} \qquad [12]$$

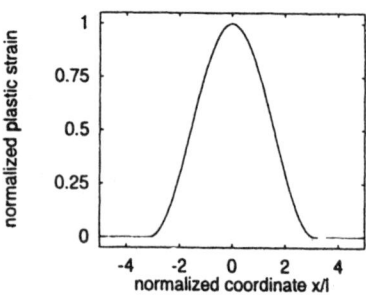

Figure 7. *Distribution of plastic strain according to gradient-enhanced softening plasticity model*

At a given state, the right-hand side is constant (due to static equilibrium), and so the general solution has the form

$$\varepsilon_p(x) = \frac{\sigma - \sigma_0}{H} + C_1 \sin\frac{x}{\ell} + C_2 \cos\frac{x}{\ell} \qquad [13]$$

The integration constants C_1 and C_2 must be determined from certain boundary conditions. The solution [13] is valid only in the plastic zone, and so the boundary conditions are imposed at the elasto-plastic boundary. The distribution of plastic strain must remain continuously differentiable, otherwise the Laplacian in [11] would be unbounded. Consequently, both ε_p and ε_p' must vanish on the boundary of the localized plastic zone. Since the zone has in general two boundary points, we obtain four conditions with four unknowns—two integration constants, C_1 and C_2, and two coordinates of the boundary points (note that they are not given in advance). The problem would normally have a unique solution, but due to the highly idealized and symmetric nature of the structure (prismatic bar with perfectly uniform properties), the exact position of the localized plastic zone remains undetermined and only the size of this zone can be evaluated. If the problem is perturbed by random imperfections, the number of solutions becomes finite.

The solution of the idealized problem reads

$$\varepsilon_p(x) = \begin{cases} \dfrac{\sigma - \sigma_0}{H}\left(1 + \cos\dfrac{x - x_c}{\ell}\right) & \text{if} \quad |x - x_c| \le \pi\ell \\ 0 & \text{if} \quad |x - x_c| \ge \pi\ell \end{cases} \qquad [14]$$

where x_c is the (arbitrary) coordinate of the center of the localization zone. Obviously, parameter ℓ controls the size of the localization zone, $L_s = 2\pi\ell$, and so it plays the role of an internal length. The actual plastic strain profile is plotted in Figure 7 (for $x_c = 0$).

It is instructive to discuss how the gradient term limits localization. Around the point that experiences the largest strain, the curvature of the plastic strain profile is negative, and due to the Laplacean term in [11] the current yield stress is higher than it would be for a standard local model. If the softening zone were too narrow, the negative curvature of the strain profile around its peak would have a large magnitude, and the current yield stress would be higher than at other points. Since the applied stress is constant along the bar, the sections around the center of the localization zone could not yield, and the zone would have to be extended in order to reduce the magnitude of the negative curvature and thus restore a constant value of the current yield stress.

4.3. *Implicit gradient models*

Due to the presence of second derivatives of internal variables, the numerical implementation of explicit gradient models is not easy. It is often necessary to use C^1-continuous finite elements, or mixed elements with an independent interpolation of one or more primary unknown fields in addition to the usual displacement interpolation [PAM 94, COM 96]. There are also problems related to the enforcement of the interface conditions at the evolving elasto-plastic boundary. New developments in this research area indicate that a more robust implementation can be achieved with implicit gradient models. Such formulations, first proposed for gradient damage [PEE 96], have recently been adapted for gradient plasticity [ENG 02].

The implicit gradient damage model has a similar structure to the nonlocal damage model from Section 4.1, but the nonlocal equivalent strain $\bar{\varepsilon}$ is defined as the solution of the *Helmholtz differential equation*

$$\bar{\varepsilon} - \ell^2 \, \nabla^2 \bar{\varepsilon} = \tilde{\varepsilon} \qquad [15]$$

with the homogeneous Neumann boundary condition $\partial \bar{\varepsilon}/\partial n = 0$ imposed on the actual physical boundary of the body.

The solution $\bar{\varepsilon}$ of the above boundary value problem can be expressed in the form of an averaging integral [9] with the weight function $\alpha(x, \xi)$ replaced by the Green function of the boundary value problem. For instance, in one dimension and on an infinite domain, the Green function of the Helmholtz equation [15] is given by

$$G(x, \xi) = \frac{1}{2\ell} \exp\left(-\frac{|x - \xi|}{\ell}\right) \qquad [16]$$

So the implicit gradient models are equivalent to integral-type nonlocal models with special nonlocal weight functions. Despite this formal equivalence, their numerical implementation is quite different [PEE 96].

5. Concluding remarks

This short paper could provide only an elementary introduction into the wide and complex subject of inelastic strain localization and its mathematical modeling. The readers may find more detailed information in the specialized literature, e.g., in [VAR 95, BAŽ 98, BOR 98, JIR 01, BAŽ 02].

6. References

[AIF 84] AIFANTIS E. C., "On the microstructural origin of certain inelastic models.", *Journal of Engineering Materials and Technology, ASME*, vol. 106, 1984, p. 326–330.

[BAŽ 84] BAŽANT Z. P., BELYTSCHKO T. B., CHANG T.-P., "Continuum model for strain softening", *Journal of Engineering Mechanics, ASCE*, vol. 110, 1984, p. 1666–1692.

[BAŽ 98] BAŽANT Z. P., PLANAS J., *Fracture and Size Effect in Concrete and Other Quasibrittle Materials*, CRC Press, Boca Raton, 1998.

[BAŽ 02] BAŽANT Z. P., JIRÁSEK M., "Nonlocal integral formulations of plasticity and damage: Survey of progress", *Journal of Engineering Mechanics, ASCE*, vol. 128, num. 10, 2002.

[BOR 98] DE BORST R., VAN DER GIESSEN E., Eds., *Material Instabilities in Solids*, Wiley, Chichester, 1998.

[COM 96] COMI C., PEREGO U., "A generalized variable formulation for gradient dependent softening plasticity", *International Journal for Numerical Methods in Engineering*, vol. 39, 1996, p. 3731–3755.

[ENG 02] ENGELEN R. A. B., GEERS M. G. D., BAAIJENS F. P. T., "Nonlocal implicit gradient-enhanced elasto-plasticity for the modelling of softening behaviour", *International Journal of Plasticity*, vol. 18, 2002, in press.

[JIR 98] JIRÁSEK M., "Nonlocal models for damage and fracture: Comparison of approaches", *International Journal of Solids and Structures*, vol. 35, 1998, p. 4133–4145.

[JIR 01] JIRÁSEK M., BAŽANT Z. P., *Inelastic Analysis of Structures*, John Wiley and Sons, Chichester, 2001.

[JIR 02a] JIRÁSEK M., "Numerical modeling of strong discontinuities", *Revue française de génie civil*, 2002, this issue.

[JIR 02b] JIRÁSEK M., ROLSHOVEN S., "Comparison of integral-type nonlocal plasticity models for strain-softening materials", *International Journal of Engineering Science*, vol. 40, 2002, in press.

[MAZ 84] MAZARS J., "Application de la mécanique de l'endommagement au comportement non linéaire et à la rupture du béton de structure", Thèse de Doctorat d'Etat, 1984, Université Paris VI., France.

[PAM 94] PAMIN J., "Gradient-dependent plasticity in numerical simulation of localization phenomena", PhD thesis, Delft University of Technology, Delft, The Netherlands, 1994.

[PEE 96] PEERLINGS R. H. J., DE BORST R., BREKELMANS W. A. M., DE VREE J. H. P., "Gradient-enhanced damage for quasi-brittle materials", *International Journal for Numerical Methods in Engineering*, vol. 39, 1996, p. 3391–3403.

[PIJ 87] PIJAUDIER-CABOT G., BAŽANT Z. P., "Nonlocal damage theory", *Journal of Engineering Mechanics, ASCE*, vol. 113, 1987, p. 1512–1533.

[VAR 95] VARDOULAKIS I., SULEM J., *Bifurcation Analysis in Geomechanics*, Blackie Academic & Professional, London, 1995.

Chapter 13

Numerical Modelling of Strong Discontinuities

Milan Jirásek
Laboratory of Structural and Continuum Mechanics, Swiss Federal Institute of Technology (EPFL), Lausanne, Switzerland

1. Introduction

The preceding paper [JIR 02] gave an overview of three main classes of models that provide an objective description of strain localization, and then discussed in more detail one of those classes, namely formulations regularized by spatial integrals or by gradient terms. The present paper is focused on models that treat highly localized strains as strong discontinuities (jumps in the displacement field).

Standard finite element approximations cannot properly capture the discontinuous character of the displacement field corresponding to localized fracture. In the context of smeared-crack models, this deficiency can lead to a spurious stress transfer across a widely open crack [JIR 98]. Discrete-crack models with special interfaces between conventional elements [SAO 81, ČER 94] do not suffer by this pathology, but they require frequent remeshing in order to allow for crack propagation in the correct direction. The recently emerged idea of incorporating strain or displacement discontinuities into standard finite element interpolations triggered the development of powerful techniques that allow efficient modeling of regions with highly localized strains, e.g. of fracture process zones in concrete or shear bands in metals or soils. The discontinuities can have an arbitrary orientation, which makes it much easier to capture a propagating crack or softening band without remeshing. This class of methods, collectively called *elements with embedded discontinuities*, was inspired by the pioneering work of Ortiz et al. [ORT 87] and Belytschko et al. [BEL 88]. The early works used weak (strain) discontinuities, but the idea was later extended to strong (displacement) discontinuities [DVO 90, KLI 91, OLO 94, SIM 94].

A systematic classification and critical evaluation of embedded discontinuity models within a unified framework was presented in [JIR 00a], with the conclusion that there exist three main groups of such formulations, called statically optimal symmetric (SOS), kinematically optimal symmetric (KOS), and statically and kinematically optimal nonsymmetric (SKON). The SOS formulation works with a natural stress continuity condition, but it does not properly reflect the kinematics of a completely open crack. On the other hand, the KOS formulation describes the kinematic aspects satisfactorily, but it leads to an awkward relationship between the stress in the bulk of the element and the tractions across the discontinuity line. Optimal performance is achieved with the nonsymmetric SKON formulation, which uses a very natural stress continuity condition and reasonably represents complete separation at late stages of the fracturing process. This is the formulation to be described next. In section 6, we will present a different approach to the modeling of discontinuities, based on the concept of partition of unity [MEL 96] and refered to as the *extended finite element method* [MOË 99, SUK 00, DAU 00, MOË 02].

2. Triangular element with embedded displacement discontinuity

The optimal combination of static and kinematic equations for elements with embedded discontinuities first appeared in [DVO 90], even though their exact nature is

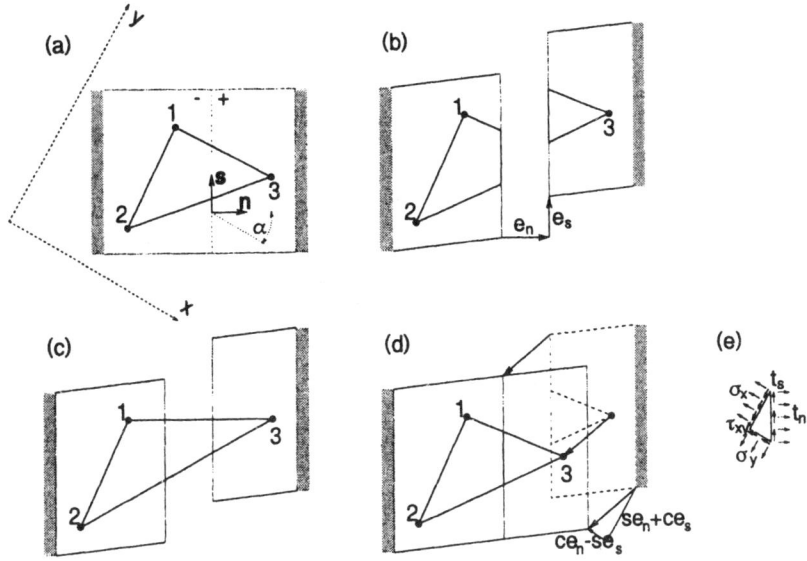

Figure 1. *Constant-strain triangle with an embedded displacement discontinuity*

not easy to understand from that paper. A very similar quadrilateral element based on simple and instructive physical considerations was constructed in [KLI 91]. The same technique was then applied to a constant-strain triangle [OLO 94]. A general version of the SKON formulation for an arbitrary type of parent element was outlined in [SIM 94] and fully described in [OLI 96].

Consider a triangular element crossed by a discontinuity (Figure 1a). The displacement field can be decomposed into a continuous part and a discontinuous part due to the opening and sliding of a crack (Figure 1b). The same decomposition applies to the nodal displacements of a finite element. Instead of smearing the displacement jump over the area of the element and replacing it by an equivalent inelastic strain, as is done by standard smeared crack models (Figure 1c), the discontinuity can be represented by additional degrees of freedom, e_n and e_s, corresponding respectively to the normal (opening) and tangential (sliding) component of the displacement jump and collected in a column matrix $e = \{e_n, e_s\}^T$. The contribution of the displacement jump is then subtracted from the nodal displacement vector, d, and only the part of the nodal displacements produced by the continuous deformation serves as input for the evaluation of strains in the bulk material, ε (Figure 1d). This leads to kinematic equations in the form

$$\varepsilon = B(d - He) \qquad [1]$$

where $\varepsilon = \{\varepsilon_x, \varepsilon_y, \gamma_{xy}\}^T$ is the column matrix of engineering strain components, B is the standard strain-displacement matrix, and H is a matrix reflecting the effect of the displacement jump on the nodal displacements.

In general, the displacement jump could be approximated by a suitable function, for example a polynomial one. This approximation need not be continuous on inter-element boundaries. For triangular elements with a linear displacement interpolation, the strains and stresses in the bulk are constant in each element, and so it is natural to approximate the displacement jump also by a piecewise constant function. This is why we describe the jump in each element by only two parameters, e_n and e_s. These additional degrees of freedom have an internal character and can be eliminated on the element level, which yields the global equilibrium equations written exclusively in terms of the standard unknowns–nodal displacements. From Figure 1d it is clear that if the discontinuity line separates node 3 from nodes 1 and 2 (in local numbering), the crack-effect matrix is given by

$$H = \begin{bmatrix} 0 & 0 \\ 0 & 0 \\ 0 & 0 \\ 0 & 0 \\ c & -s \\ s & c \end{bmatrix} \qquad [2]$$

where $c = \cos\alpha$, $s = \sin\alpha$, and α is the angle between the normal to the crack (discontinuity line) and the global x-axis; see Figure 1a.

Strains in the bulk material generate certain stresses, $\sigma = \{\sigma_x, \sigma_y, \tau_{xy}\}^T$, which are here computed from the equations of linear elasticity,

$$\sigma = D_e \varepsilon \qquad [3]$$

where D_e is the elastic stiffness matrix (for plane stress or plane strain). Note that, in general, the constitutive law for the bulk material could be nonlinear. The tractions transmitted by the crack, t, are linked to the separation vector (displacement jump) by another constitutive law that describes the gradual development of a stress-free crack. One specific form of this law will be presented in Section 3.

The stresses in the bulk and the tractions across the crack must satisfy certain conditions that express internal equilibrium and serve as static equations corresponding to the internal degrees of freedom, e. The most natural requirement is that the traction vector be equal to the stress tensor contracted with the crack normal, similar to static boundary conditions. This internal equilibrium (traction continuity) condition can be derived from equilibrium of an elementary triangle with one side on the discontinuity line; see Figure 1e. In the engineering notation it reads

$$P^T \sigma = t \qquad [4]$$

where

$$P = \begin{bmatrix} c^2 & -cs \\ s^2 & cs \\ 2cs & c^2 - s^2 \end{bmatrix} \qquad [5]$$

is a stress rotation matrix. For linear triangles with a constant displacement jump, both t and σ are constant in each element, and so condition [4] can be satisfied exactly. In general it would have to be enforced in a weak sense. Finally, the nodal forces are evaluated from the standard relation

$$f_{int} = \int_{A_e} B^T \sigma \, \mathrm{d}A = A_e B^T \sigma \qquad [6]$$

where A_e is the area of the element.

For the constant-strain triangle, the kinematic relations and the traction continuity condition follow quite naturally from simple physical considerations. Development of more complicated elements with embedded discontinuities is often done within the framework of enhanced assumed strain (EAS) methods, and such elements are sometimes refered to as EAS elements (which can be somewhat confusing).

3. Traction-separation law in damage format

The basic equations presented in the preceding section must be completed by a law that links the traction transmitted by the discontinuity to the displacement jump. One possible type of such a law was proposed in [JIR 01a] in the form

$$\gamma t = \hat{D} e \qquad [7]$$

where γ is a dimensionless scalar compliance parameter evolving from zero to infinity and

$$\hat{D} = \begin{bmatrix} D_{nn} & 0 \\ 0 & D_{ss} \end{bmatrix} \qquad [8]$$

is a stiffness matrix corresponding to a reference intermediate stage of the degradation process. Before crack initiation, the value of γ is zero. For simplicity, it is assumed here that crack initiation is controlled by the Rankine criterion of maximum principal stress. This means that the discontinuity line is inserted perpendicular to the direction of maximum principal stress, and the shear traction at the instant of crack initiation is zero.

The evolution of γ is described by the loading-unloading conditions in the Karush-Kuhn-Tucker form,

$$\dot{\gamma} \geq 0, \quad f \leq 0, \quad \dot{\gamma} f = 0 \qquad [9]$$

The loading function f characterizing the elastic domain is defined as

$$f(e, \gamma) \equiv F(\bar{e}(e)) - \gamma \qquad [10]$$

where $\bar{e}$ is a scalar measure of the separation vector e, called the equivalent separation (analogous to the equivalent strain in continuum damage mechanics), and F

is a suitable function describing the dependence of the compliance parameter on the equivalent separation during monotonic loading. In [JIR 01a] it was proposed to set

$$\tilde{e} = \sqrt{\frac{e^T \hat{D} e}{D_{nn}}} \qquad [11]$$

and

$$F(\tilde{e}) = \frac{D_{nn}\tilde{e}}{g(\tilde{e})} \qquad [12]$$

where g is a scalar function describing the traction-separation curve for Mode-I cracking.

Alternatively, the traction-separation law could be formulated within the framework of plasticity; see e.g. [OLO 94]. This is especially useful for the description of cohesive zones in metals or shear bands in soils. Due to space limitations, the details cannot be presented here.

4. Evaluation of internal forces and tangent stiffness

In an incremental-iterative analysis of a structure discretized by finite elements with embedded discontinuities, the nodal displacements are computed iteratively from the global equilibrium equations, and the main tasks on the level of one finite element are to evaluate the internal forces and the tangent stiffness matrix for a given increment of nodal displacements.

Substituting [3] and [1] into the traction continuity condition [4], we obtain a useful expression for the traction vector in terms of the kinematic variables,

$$t = P^T D_e B(d - He) = A(d - He) \qquad [13]$$

where we have denoted

$$A = P^T D_e B \qquad [14]$$

Expression [13] for the traction vector substituted into the traction-separation law [7] yields the equation

$$(\gamma AH + \hat{D})e = \gamma Ad \qquad [15]$$

that links the nodal displacements d, separation vector e, and compliance parameter γ.

If d is known, the separation vector e can be solved from [15], taking into account that the compliance parameter γ may change during the step according to [9]–[10]. Adopting the usual numerical scheme, equation [15] is first solved under the assumption of constant damage (unloading), i.e., with γ kept equal to its value at the end of the previous step. If the computed separation vector e satisfies the condition $f(e, \gamma) \leq 0$, then the solution is admissible, otherwise it is necessary to solve [15] with γ replaced by $F(\tilde{e}(e))$. The conditions under which this problem has a unique solution were studied in [JIR 00b], where it was shown that uniqueness may be lost not only for elements that are too large but also for badly shaped elements. Once the separation vector has been determined, the strain, stress and internal forces are easily evaluated by substituting into equations [1], [3] and [6].

The tangent stiffness matrix can be derived by combining the rate forms of the basic equations and is given by the formula [JIR 01a]

$$K = K_e - \gamma K_e H (\hat{D} + \gamma AH - tf^T)^{-1} A \qquad [16]$$

where $K_e = A_e B^T D_e B$ is the elastic element stiffness matrix and $f = \partial F / \partial e$ is the gradient of the loading function with respect to the separation vector. Recalling the definition [14] of matrix A, expression [16] can be rewritten in the equivalent form

$$K = A_e B^T [D_e - \gamma D_e BH (\hat{D} + \gamma P^T D_e BH - tf^T)^{-1} P^T D_e] B \qquad [17]$$

This formula has the same structure as the standard expression $K = A_e B^T DB$ for the tangent element stiffness of a conventional constant-strain triangle with a nonlinear material. It is interesting to note that the "equivalent" tangent material stiffness

$$D = D_e - \gamma D_e BH (\hat{D} + \gamma P^T D_e BH - tf^T)^{-1} P^T D_e \qquad [18]$$

depends not only on the constitutive parameters (elastic bulk stiffness D_e, reference cohesive stiffness $\hat{D}$, compliance parameter γ), but also on the matrices H and P that reflect the orientation of the discontinuity and on matrix B that reflects the size and shape of the finite element. The dependence on orientation (crack direction) means that crack-induced anisotropy is taken into account, and the dependence on element size is a typical feature of smeared models based on the fracture energy concept; see Section 3.2 in [JIR 02]. However, the present model contains additional information on the kinematics of the failure process, because matrix H is affected by the position of the discontinuity with respect to the element and matrix B depends on the element shape.

For a conventional element, the element stiffness K is symmetric if and only if the material stiffness D is symmetric. For the present element with an embedded discontinuity, this happens only if (i) f is a scalar multiple of t and (ii) BH is a scalar multiple of P. The first condition is related to the traction-separation law and is equivalent to symmetry of the tangent stiffness that links the rates of the separation vector and of the traction vector. For the equivalent separation defined by formula [11], the gradient vector $f = \partial F / \partial e = (dF/d\bar{e})(\partial \bar{e}/\partial e) = F' \hat{D} e / (D_{nn} \bar{e})$ is indeed colinear with the traction vector $t = \hat{D} e / \gamma$. The second condition is related exclusively to geometrical properties. It is satisfied only if the discontinuity line is parallel to one of the element sides [JIR 00a]. Therefore, even if the material stiffness is symmetric, symmetry of the structural stiffness is disturbed by the kinematic and static equations. This is why the present SKON formulation is called nonsymmetric. There exist two types of symmetric formulations (KOS and SOS), in which only one group of equations (either kinematic or static) is postulated and the dual group is derived from the principle of virtual work. However, the equations obtained in this way are not "natural" and the resulting formulations have severe drawbacks; see [JIR 00a] for a detailed analysis.

5. Tracing of the discontinuity path

The elements described in the preceding sections can easily accommodate pre-existing discontinuities, e.g., rock joints or delaminating interfaces, but the principal interest lies in modeling of evolving discontinuities such as propagating cracks. In this latter case, the position of the discontinuity is not known in advance and the simulation starts with all elements in their "virgin" state. A discontinuity segment is inserted in an element only when a certain initiation criterion is satisfied. The simplest approach is to formulate this criterion in terms of the stress in the bulk material, e.g., as the Rankine criterion of maximum principal stress, and to place the discontinuity segment in the element center, perpendicular to the maximum principal stress direction. However, if this is done in each element independently of the others, numerical difficulties often appear. They are caused by the finite size of the incremental steps and by the limited accuracy of the constant-strain triangular elements.

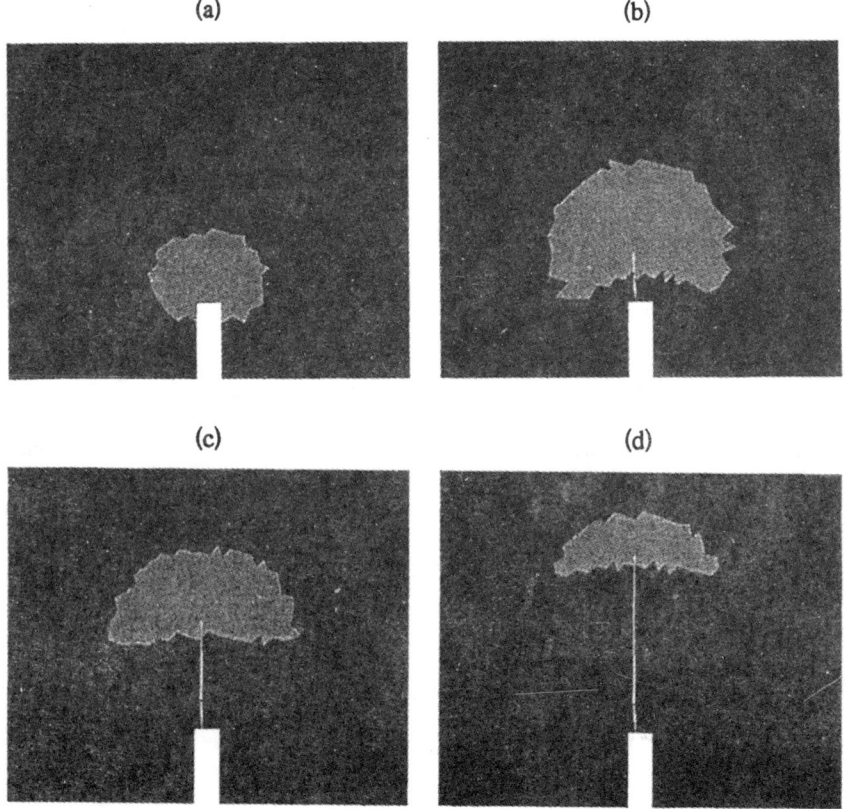

(a) (b)

(c) (d)

Figure 2. *Evolution of the fracture process zone in the central part of a notched beam under three-point bending*

One possible remedy was proposed in [JIR 01b]. The embedded discontinuity model was combined with a smeared (continuum-based) description of inelastic processes. It was argued that early stages of the fracture process in quasibrittle materials are adequately described by distributed damage while the macroscopic crack that forms at later stages is naturally treated as a displacement discontinuity. The best results are obtained with a nonlocal formulation of the continuum part of the combined model. A strong discontinuity is inserted in an element when the damage parameter attains a critical level. This is of course an *ad hoc* criterion that cannot be derived in a rigorous way, but it seems to work reasonably well. The traction-separation law governing the discontinuity must be adjusted so as to ensure that the overall energy dissipation remains correct. The progressive failure of a notched beam under three-point bending is shown in Figure 2. The grey color marks the region in which the damage parameter keeps growing (active part of the process zone). The material in the wake of the propagating crack with decreasing cohesive tractions is unloading, and so in this region the damage parameter does not grow any more.

(a)   (b)

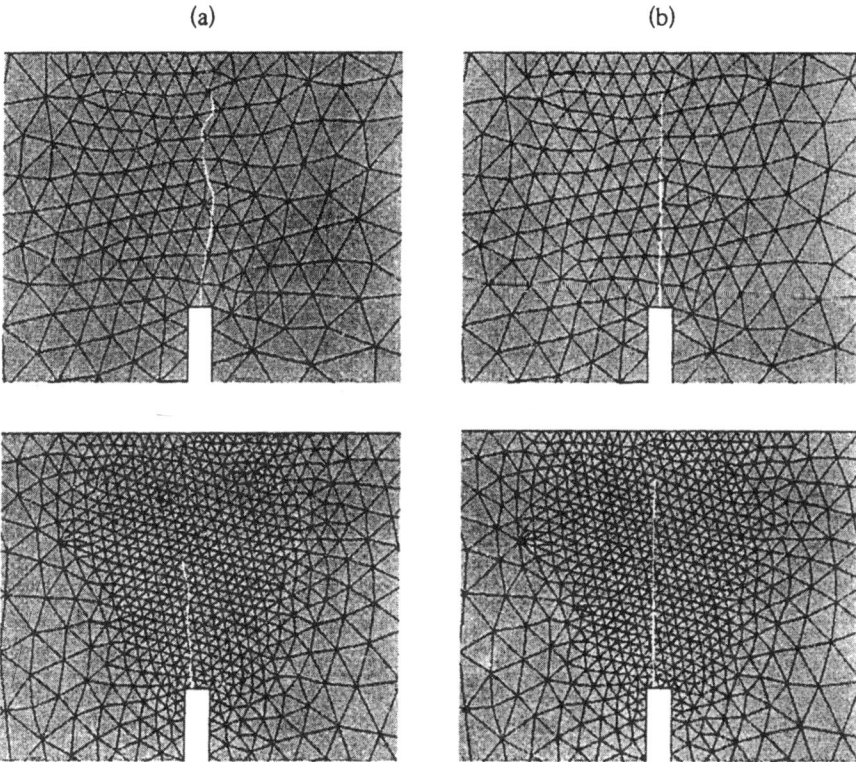

Figure 3. *Embedded crack trajectory for a model with enforced continuity of the crack path, with direction determined (a) from the local strain, (b) from the nonlocal strain*

To make sure that the propagating discontinuity cleanly separates the finite element model into two parts, it is good to enforce continuity of the crack path (note that the interpolation of the crack opening is still only piecewise continuous, with jumps at element boundaries). When a new discontinuity segment is inserted in an element that meets the initiation criterion, it is checked whether one of its neighboring elements is already crossed by a discontinuity and, if it is the case, the newly inserted discontinuity is placed such that it passes through the intersection of the neighboring discontinuity segment with the edge shared by both elements. If the orientation of the discontinuity line is determined from the local orientation of principal strain axes in each element separately, the resulting crack trajectory is tortuous (Figure 3a). Real cracks in quasibrittle materials are not perfectly smooth, but their roughness is controled by the material microstructure while the present tortuosity is purely numerical, dependent on the structure of the finite element mesh. An efficient remedy is to determine the direction of the discontinuity from the principal directions of the *nonlocal* strain. This approach gives a correct crack trajectory even on highly biased meshes, such as that in the bottom part of Figure 3b.

6. Resolution of discontinuities by extended finite elements

Elements with embedded discontinuities provide a better kinematic description of discontinuous displacement fields than pure continuum models that smear the displacement jumps uniformly over the entire element, but they still have certain limitations. Their main disadvantage is that the strain approximations in the two parts of the element separated by a discontinuity are not independent. For instance, using constant-strain triangle, the strains in these two parts are approximated by the same constant tensor. Of course, one could use a higher-order formulation with a spatially variable strain approximation, but there always remains a certain constraint that does not allow modeling of the behavior of two separated material bodies in full generality.

A new class of methods that overcomes this drawback emerged very recently, even though a similar idea could be traced back to the so-called manifold method, developed in the context of discontinuous deformation analysis [SHI 92]. Conceptually, this new approach to modeling of discontinuities can be considered as a particular case of the partition-of-unity method [MEL 96]. The general idea of that method is that the approximation space spanned by a standard basis (e.g., by the standard finite element shape functions) is enriched by products of the standard basis functions with special functions selected by the user and constructed, e.g., from the analytical solution of the problem under some simplifying assumptions. This permits the incorporation of *a priori* knowledge about the character of the problem and its solutions. The enriched displacement approximation is written in the form

$$u(x) = \sum_{i=1}^{N_{nod}} N_i(x) \left(d_i + \sum_{j \in \mathcal{L}_i} G_j(x) e_{ij} \right) \qquad [19]$$

where N_{nod} is the number of nodes of the finite element model; N_i, $i = 1, 2, \ldots N_{nod}$, are the standard shape functions; d_i, $i = 1, 2, \ldots N_{nod}$, are the standard displacement DOFs; G_j, $j = 1, 2, \ldots m$, are the enrichment functions; $\mathcal{L}_i \subset \{1, 2, \ldots m\}$ is the set of integers that indicate which enrichment functions are activated at node i; and e_{ij} are the additional DOFs associated with node i and enrichment function j. The key trick is that the global enrichment functions G_j are multiplied by the nodal shape functions N_i. The products $N_i G_j$ inherit from G_j good approximation properties and from N_i a limited support. Consequently, the enrichment has a local character and the resulting stiffness matrix is sparse (with a proper renumbering it remains banded). The standard approximation is usually enriched only locally in a certain region of interest, e.g., in a localization zone, and the newly added degrees of freedom e_{ij} can be associated with nodes of the existing mesh, without the need for changing the topology. Owing to the partition-of-unity property of the standard shape functions (their sum is equal to one at any point x), the enrichment functions G_j can be reproduced exactly.

This general idea was adapted for linear elastic fracture mechanics, with the enrichment constructed using the singular near-tip asymptotic fields and simple Heaviside functions [MOË 99]. The method was later called the eXtended Finite Element Method (X-FEM). It can efficiently handle three-dimensional cracks [SUK 00] and even branching and intersecting cracks [DAU 00]. A big advantage of this technique is that the displacement interpolation is conforming, with no incompatibilities between elements, and that the strains on both sides of a stress-free crack are fully decoupled.

The partition-of-unity concept is also applicable to cohesive crack models. Wells and Sluys [WEL 01] enriched the interpolation functions by products of the Heaviside function with standard finite element shape functions that correspond to the nodes of those elements that are intersected by the crack. Moës and Belytschko [MOË 02] added special non-singular enrichments around the crack tip, motivated by asymptotic analysis of the strain field at the tip of a cohesive crack. These elegant formulations seem to overcome the difficulties associated with the piecewise constant interpolation of the displacement jump used by embedded discontinuity models, and they even restore the symmetry of the stiffness matrix. Their implementation is, however, somewhat more difficult, because it is necessary to add new global degrees of freedom during the simulation and to refine the integration scheme in the enriched area around the crack.

The improved resolution of discontinuities by the extended finite element method is illustrated by the schematic tests in Figure 4. A square piece of material divided into two parts by a vertical stress-free crack (top row) is first subjected to a relative motion of the two parts (middle row), and then the right hand part is compressed in the direction parallel to the crack (bottom row). Figure 4a depicts the actual physical process. Figure 4b shows the approximation obtained with a standard bilinear finite element. The relative motion of the two parts is transformed into normal and shear strain, and the forces imposed on the right hand part of the body influence the deformation of the left hand part. An element with an embedded discontinuity (Figure 4c) can cleanly reproduce the rigid-body separation but forces parallel to the crack are still transmit-

(a)   (b)   (c)   (d)   (e)

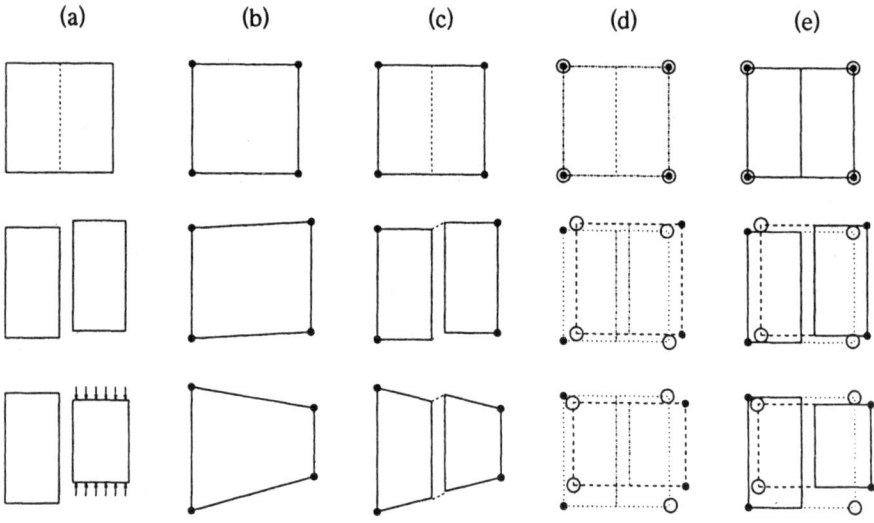

Figure 4. *Illustration of separation tests: (a) real body split into two independent parts, (b) standard finite element, (c) element with embedded discontinuity, (d-e) extended finite element*

ted from the right hand part to the left hand one. This is because the formulation allows a displacement jump but the strain in the bulk material is still interpolated in a continuous manner. The approximation obtained with the extended finite element method can be thought of as using two independent overlayed elements (Figure 4d). The edges of these elements are plotted by dotted and dashed lines, respectively. Solid circles mark standard nodes while empty circles mark enriched nodes. The displacement interpolation constructed with the "dotted" element is valid to the left of the crack, and that constructed with the "dashed" element is valid to the right of the crack (Figure 4e). In this way, both the separation and the deformation of one part can be reproduced exactly.

The foregoing example considered only one single element. Figure 5a shows an assembly of extended triangular finite elements modeling a partially cracked body. All the elements that are crossed by the crack can be thought of as doubled. Each of the "child elements" provides an approximation valid only on one side of the crack and is connected to the standard nodes on this side and to special enrichment nodes (marked by empty circles) on the other side. The enrichment nodes are introduced at the same initial locations as the standard nodes but their displacements are completely independent of the standard ones (Figure 5b). The nodes connected by the edge at which the crack tip is located are not enriched, to make sure that the displacement interpolation along this edge is continuous. As is clear from Figure 5c, the displacement jump is interpolated in a continuous, piecewise linear manner. The deformation on one side of the crack is fully independent of the deformation on the other side.

(a)   (b)   (c)

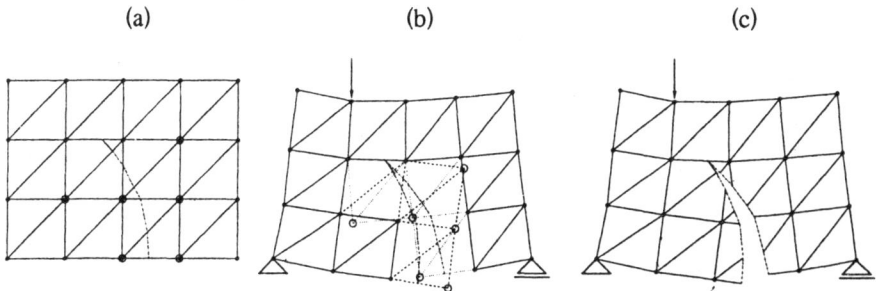

Figure 5. *Extended finite element method: (a) finite element mesh with added degrees of freedom around a crack, (b) displacement approximation using doubled elements and nodes, (c) resulting discontinuous displacement approximation*

7. Concluding remarks

This short paper could provide only an elementary introduction into the wide and complex subject of numerical approximations with built-in discontinuities. This field of research has been evolving very fast in recent years and many interesting applications are currently being developed. For the sake of simplicity, we considered the strong discontinuity as a crack, but it could be as well a slip line or an interface between two different materials. The main idea of the extended finite element method can be adapted for analysis of weak discontinuities and propagating fronts that appear, e.g., in phase transformation, solidification, dessication or leaching problems.

8. References

[BEL 88] BELYTSCHKO T., FISH J., ENGELMANN B. E., "A finite element with embedded localization zones", *Computer Methods in Applied Mechanics and Engineering*, vol. 70, 1988, p. 59–89.

[ČER 94] ČERVENKA J., "Discrete crack modeling in concrete structures", PhD thesis, University of Colorado, Boulder, Colorado, 1994.

[DAU 00] DAUX C., MOËS N., DOLBOW J., SUKUMAR N., BELYTSCHKO T., "Arbitrary branched and intersecting cracks with the extended finite element method", *International Journal for Numerical Methods in Engineering*, vol. 48, 2000, p. 1741–1760.

[DVO 90] DVORKIN E. N., CUITIÑO A. M., GIOIA G., "Finite elements with displacement interpolated embedded localization lines insensitive to mesh size and distortions", *Computer Methods in Applied Mechanics and Engineering*, vol. 90, 1990, p. 829–844.

[JIR 98] JIRÁSEK M., ZIMMERMANN T., "Analysis of rotating crack model", *Journal of Engineering Mechanics, ASCE*, vol. 124, 1998, p. 842–851.

[JIR 00a] JIRÁSEK M., "Comparative study on finite elements with embedded cracks", *Computer Methods in Applied Mechanics and Engineering*, vol. 188, 2000, p. 307–330.

[JIR 00b] JIRÁSEK M., "Conditions of uniqueness for finite elements with embedded cracks", *Proceedings of the Sixth International Conference on Computational Plasticity*, Barcelona, 2000, CD-ROM.

[JIR 01a] JIRÁSEK M., ZIMMERMANN T., "Embedded crack model: I. Basic formulation", *International Journal for Numerical Methods in Engineering*, vol. 50, 2001, p. 1269–1290.

[JIR 01b] JIRÁSEK M., ZIMMERMANN T., "Embedded crack model: II. Combination with smeared cracks", *International Journal for Numerical Methods in Engineering*, vol. 50, 2001, p. 1291–1305.

[JIR 02] JIRÁSEK M., "Objective modeling of strain localization", *Revue française de génie civil*, 2002, in this issue.

[KLI 91] KLISINSKI M., RUNESSON K., STURE S., "Finite element with inner softening band", *Journal of Engineering Mechanics, ASCE*, vol. 117, 1991, p. 575–587.

[MEL 96] MELENK J. M., BABUŠKA I., "The partition of unity finite element method: Basic theory and applications", *Computer Methods in Applied Mechanics and Engineering*, vol. 39, 1996, p. 289–314.

[MOË 99] MOËS N., DOLBOW J., BELYTSCHKO T., "A finite element method for crack growth without remeshing", *International Journal for Numerical Methods in Engineering*, vol. 46, 1999, p. 131–150.

[MOË 02] MOËS N., BELYTSCHKO T., "Extended finite element method for cohesive crack growth", *Engineering Fracture Mechanics*, vol. 69, 2002, p. 813–833.

[OLI 96] OLIVER J., "Modelling strong discontinuities in solid mechanics via strain softening constitutive equations. Part 1: Fundamentals. Part 2: Numerical Simulation", *International Journal for Numerical Methods in Engineering*, vol. 39, 1996, p. 3575–3624.

[OLO 94] OLOFSSON T., KLISINSKI M., NEDAR P., "Inner softening bands: A new approach to localization in finite elements", MANG H., BIĆANIĆ N., DE BORST R., Eds., *Computational Modelling of Concrete Structures*, Pineridge Press, 1994, p. 373–382.

[ORT 87] ORTIZ M., LEROY Y., NEEDLEMAN A., "A finite element method for localized failure analysis", *Computer Methods in Applied Mechanics and Engineering*, vol. 61, 1987, p. 189–214.

[SAO 81] SAOUMA V. E., "Interactive finite element analysis of reinforced concrete: A fracture mechanics approach", PhD thesis, Cornell University, Ithaca, New York, 1981.

[SHI 92] SHI G., "Modeling rock joints and blocks by manifold method", *Rock Mechanics, Proceedings of the 33rd U.S. Symposium*, Santa Fe, New Mexico, 1992, p. 639–648.

[SIM 94] SIMO J. C., OLIVER J., "A new approach to the analysis and simulation of strain softening in solids", BAŽANT Z. P., BITTNAR Z., JIRÁSEK M., MAZARS J., Eds., *Fracture and Damage in Quasibrittle Structures*, London, 1994, E & FN Spon, p. 25–39.

[SUK 00] SUKUMAR N., MOËS N., MORAN B., BELYTSCHKO T., "Extended finite element method for three-dimensional crack modeling", *International Journal for Numerical Methods in Engineering*, vol. 48, 2000, p. 1549–1570.

[WEL 01] WELLS G. N., SLUYS L. J., "A new method for modelling cohesive cracks using finite elements", *International Journal for Numerical Methods in Engineering*, vol. 50, 2001, p. 2667–2682.

Chapter 14

Practical Aspects of the Finite Element Method

Manuel Pastor, Pablo Mira and José Antonio Fernandez Merodo
CEDEX, CETA, Ing. Computacional, Madrid, Spain

1. Introduction

The finite element method has become one powerful tool of analysis which is being used all over the world both in the industry and in research. Many young (and not so young) engineers are nowadays familiar with commercial codes such as ANSYS, ABAQVS, COSMOS..., just to mention a few. Beginners face some questions such as (i) Which element should I use? (ii) How many elements and how big (or small)?, (iii) Which material model? (iv) Is there any saving in using reduced integration?, and so on.

While the number of questions and doubts is huge, there are some important aspects of which all of us should be aware when running finite element codes. The purpose of this Chapter is to provide a pocket guide for travellers in this unknown country. Of course, it is just a pocket guide. We do not pretend otherwise! And they have been described in detail in classical guides, such as those of Bathe, Hughes [HUG 87], Irons and Shrive , Zienkiewicz and Taylor [ZIE 2000],

Therefore, we will describe some pitfalls and difficulties in doing finite element computations, such as poor bending behaviour, locking and modes of zero energy when using reduced integration. We will also deal with topics such as Babuska-Brezzi restrictions (the light version), and will comment on which solvers are easy to program and are efficient.

2. The Mysteries of Bending

Not all elements perform as we wish when dealing with problems in which bending is important. For instance, linear triangles give much stiffer responses to the structure than they should. Moreover, when obtaining natural frequencies of vibration, we will get higher values because of this extra stiffness. To understand the problem, we will consider the simple case of a square $[-1, 1]x[-1, 1]$ under pure bending conditions. We will discretize the domain with one bilinear quadrilateral, as depicted in Figure 1.

The solution of this plane stress problem is the displacement field

$$u = -\frac{M}{EI}xy \quad v = -\frac{M}{EI}\left(1 - x^2\right) \tag{1}$$

with horizontal displacements at the nodes given by $\pm u_0 = \frac{M}{EI}$.

$$u = -\frac{M}{EI}xy \quad v = -\frac{M}{EI}\left(1 - x^2\right) \tag{2}$$

The origin of coordinate axes has been taken at the centre of the square. The strain field is given by

$$\varepsilon_x = -\frac{M}{EI}y \quad \varepsilon_y = 0 \quad \gamma_{xy} = 0 \tag{3}$$

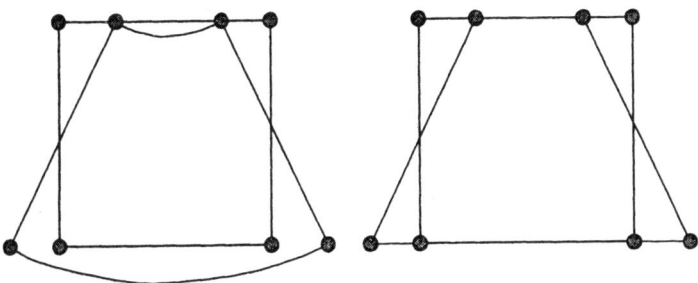

Figure 1. *Bending of a bilinear quadrilateral*

Should we reproduce the displacement field with the bilinear quadrilateral, we would obtain a constant strain field

$$\varepsilon_x^h = -\frac{M}{EI}y \quad \varepsilon_y^h = 0 \quad \gamma_{xy}^h = -\frac{M}{EI}x \tag{4}$$

It is important to note that the discretization has introduced a spurious shear strain which is only zero at the centre. Therefore, the element will be stiffer, and the deformation under a given moment M will be smaller than it should. The reader can verify that if we discretize the square with two linear elements, the situation is the same.

3. Risks of Reduced Integration

Reduced integration consists in using an integration rule of smaller degree of precision than required with less integration points. In this way, we get two advanteges (i) The cost of computation -and therefore the time- is reduced. We can analyze larger problems in the same time or we reduce the computer time. (ii) We obtain better performance (sometimes) when computing limit loads.

Two popular reduced integration rules are: (i) One point for bilinear quadrilaterals (ii) Two by two points for 8 noded quadrilaterals. Let us consider the first case. The bilinear quadrilateral has 8 degrees of freedom, and the dimension of the stiffness matrix is 8x8. The matrix has eight eigenvectors which are sketched in Figure 2. The eigenvalues of the two translation and rotation modes are zero, but this mode of deformation with zero energy is prevented by boundary conditions avoiding rigid solid motions.

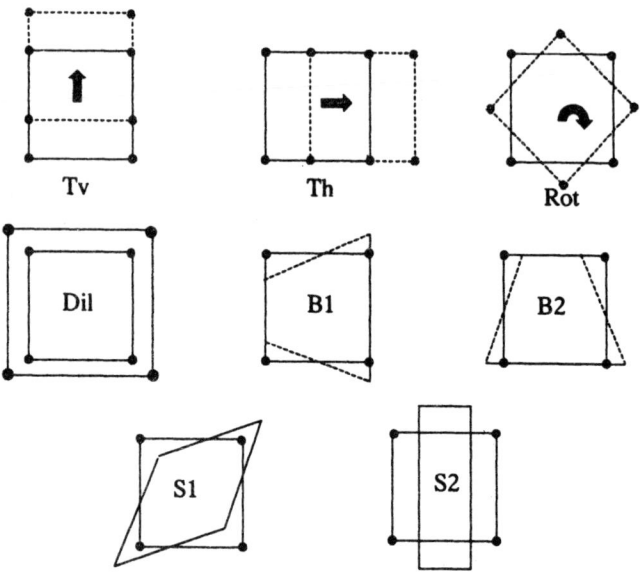

Figure 2. *Eigenvectors of the stiffness matrix*

If we integrate the stiffness matrix using just one point of integration (the centre), the situation changes. The stiffness matrix is given by

$$K = \int_\Omega B^T.D.B \, d\Omega \qquad (5)$$

which is computed as:

$$K = B_0^T.D.B_0 \, .W_0 \qquad (6)$$

where the subindex 0 refers to the integration point. Dimensions of matrix B are $8x3$, and D is a $3x3$ matrix. Therefore, **the rank of** K **is 3**. This means that, in addition to the 3 free energy modes, there are two more now. These modes are the "bending" modes B1 and B2. A finite element mesh can in some conditions (for instance, poor conditioning of the equations system) exhibit an spurious mode called "hourglassing" because of the shape of the deformed elements (Figure 3).

Reduced integration of 8 noded quadrilaterals produce also another hourglass mode. The shape is sketched in Figure 4. The advantages of this under-integrated element are important, specially in computation of failure loads, and some finite element codes incorporate an "hourglass control" to warn the user that this spurious mechanism is present.

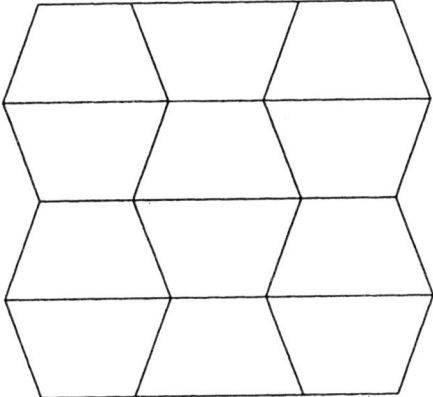

Figure 3. *Hourglassing in a mesh of bilinear quadrilaterals*

4. Volumetric locking and failure loads

We will consider now the case of an incompressible material under plane strain conditions. The problem is sketched in Figure 5. Boundary conditions are: (i) Pre-scribed zero horizontal and vertical displacements at the left side and the bottom, (ii) traction free right side and top, with a point load applied at the corner. If the material is incompressible, the node 4 cannot move in the vertical direction as it belongs to tri-angle 124 which otherwise would change its volume, and it cannot move horizontally as it belongs to triangle 143. Therefore, the node cannot move. This can be repeated for node 6, which belongs to triangles 256 and 264, and for all the remaining nodes in the mesh. All the nodes are "blocked", which is unrealistic. This example can be reproduced with plane strain finite elements using Poisson ratios approaching 0.5 (for instance, 0.49, 0.499, 0.4999 and so on).

The reader should be aware of this problem when trying to model the behaviour of saturated soils under fast loading or undrained conditions, where values of Poisson ratio close to 0.5 are usually chosen.

All displacement based finite elements present this problem to a certain extent. The elements performing better are higher order triangles (15 nodes). Another popular

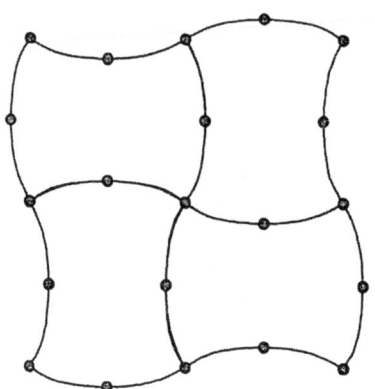

Figure 4. *Hourglassing of 8 noded quadrilaterals*

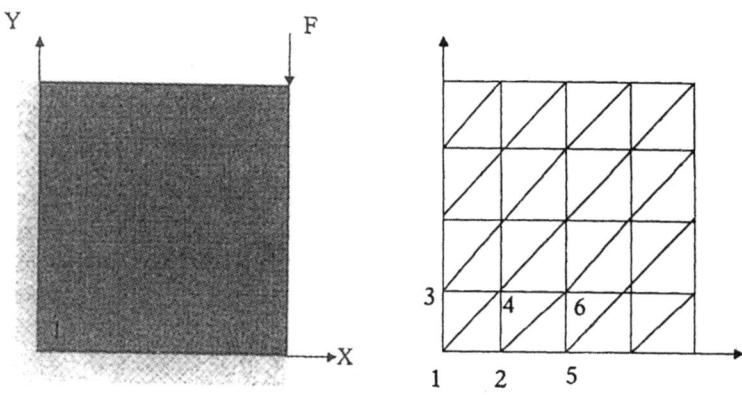

Figure 5. *Volumetric Locking*

alternative is to use quadratical eight noded quadrilaterals with a reduced integration rule of $2x2$ points.

However, the best choices are assumed strain elements (Simo Rifai, for instance), or mixed displacement-pressure formulations. These techniques will be described later on in this book.

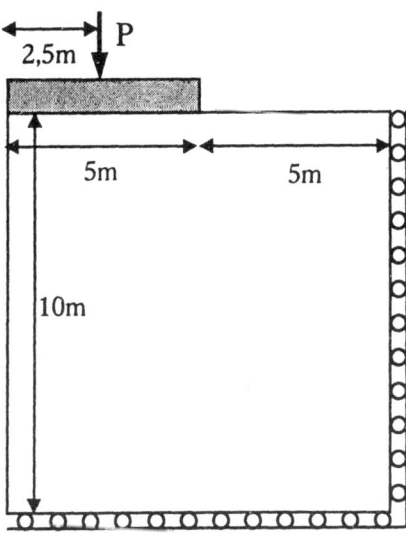

Figure 6. *Footing on vertical slope*

The reader should be aware that this problem is also exhibited in plasticity, because at failure the flow rule imposes an additional condition on the rate of plastic strain (which can be zero). To ilustrate this problem let us analyse the problem of a footing on a vertical slope as sketched in Figure 6. The material models used in the analysis are presented in the following table:

	Material type	E(Pa)	v	σ_y (Pa)
Soil	Von Mises	1.0E5	0.35	200.0
Footing	Linear elastic	1.0E8	0.35	

The objective of the analysis is to obtain a failure mechanism and a value for the limit load using different element types and different meshes. The different element types used in the analysis are listed in the following table:

Standard 3 node displacement triangle	3st0
Standard 4 node displacement quadrilateral	4st0
B 4 node quadrilateral	4bb0
Simo&Rifai 4 node quadrilaterals with 4 internal modes	4sr0
Simo&Rifai 4 node quadrilaterals with 5 internal modes	4sr1
Simo&Rifai 4 node quadrilaterals with 6 internal modes	6sr4
Standard 6 node displacement triangle	6st0
B 6 node triangle	6bb2
Standard 7 node displacement triangle	7st0
B 7 node triangle	7bb2

A theoretical solution for this problem may be obtained based on limit analysis. The failure mechanism would be a shear band descending leftwards from the bottom right corner of the footing in a 45° angle. Accordingly for each element type two mesh orientations were tested :

1) A so called right orientation (r) in which mesh alignment ran parallel to the direction of the expected failure mechanism

2) A so called wrong orientation in which mesh alignment ran perpendicular to the direction of the expected failure mechanism. (w)

The different meshes used are presented in Figure 7.

As will be shown, this test appears to be very demanding. Poor performing elements exhibit not only significant differences in the load-displacement curves and different shear band widths but also significant changes in the failure mechanism direction depending on the mesh orientation. Load displacement diagrams are presented in Figures 8 and 9. Failure mechanisms are presented in Figures 10 and 11.

The worst performers are as always standard displacement elements. The best performers are Simo-Rifai and 7 node elements.

In the case of the poor performers a wrong mechanism appears which significantly prevails over the correct one. In the case of the good performers both mechanisms appear but the correct one prevails over the wrong one.

As we can see from the graphs the failure mechanism appears to be more important than the load-displacement curve since good performers appear to have more problems in the first one than in the second one.

5. Why we cannot choose all elements for geotechnical analysis: the Zienkiewicz-Taylor patch test

Users of finite element codes for geotechnical analysis often wonder why the choice of elements is so limited. For instance, it is not possible to choose bilinear quadrilaterals or enhanced strain elements like those of Simo-Rifai, which are excellent for bending, quasi incompressibility, failure loads, etc. The reason is that in mixed

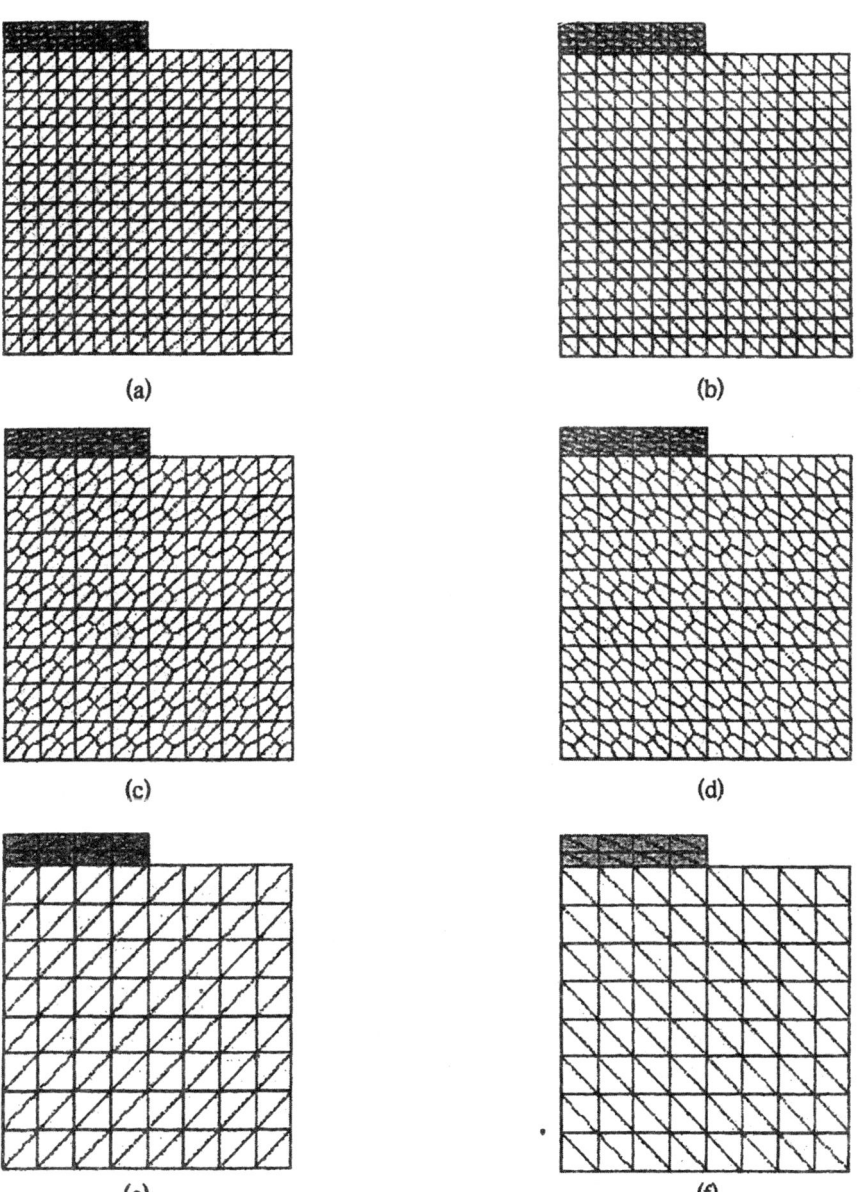

Figure 7. *Meshes used to analyse footing on vertical slope depending on the number of nodes per element: (a) 3 nodes right orientation (b) 3 nodes wrong orientation (c) 4 nodes right orientation (d) 4 nodes wrong orientation (e) 6 and 7 nodes right orientation (f) 6 and 7 nodes wrong orientation*

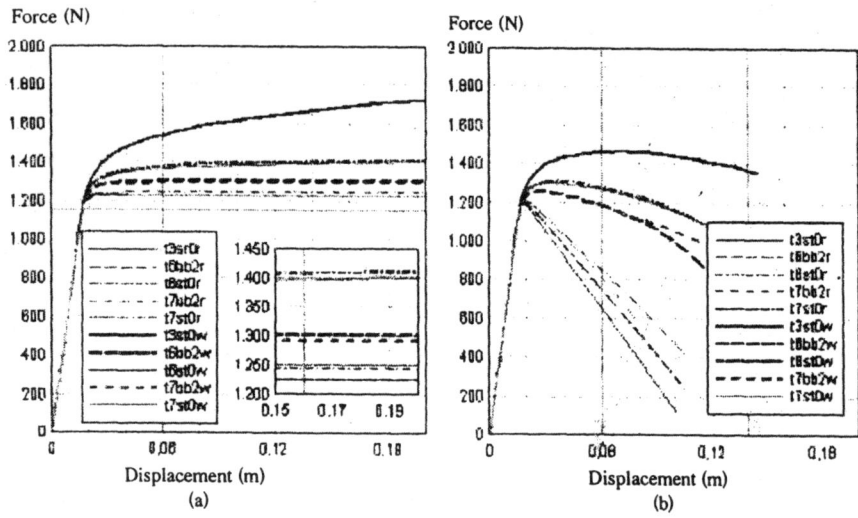

Figure 8. *Force displacement diagrams for triangles (a) H=0.0 (b) H=-0.01E*

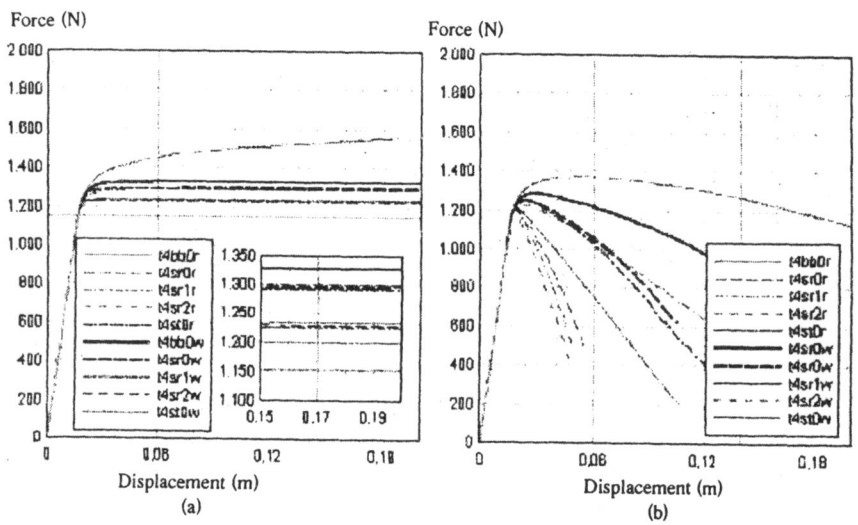

Figure 9. *Force displacement diagrams for quadrilaterals (a) H=0.0 (b) H=-0.01E*

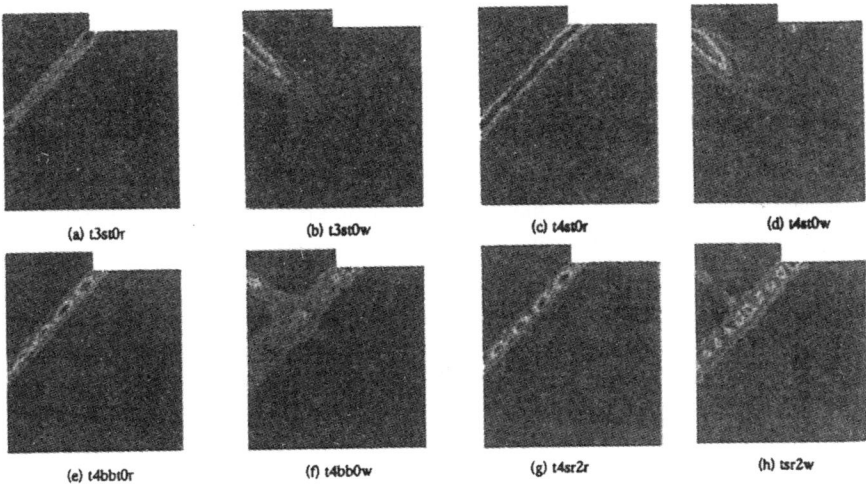

Figure 10. *Failure mechanisms for footing on vertical slope using 3 and 4 node elements and a perfectly plastic material*

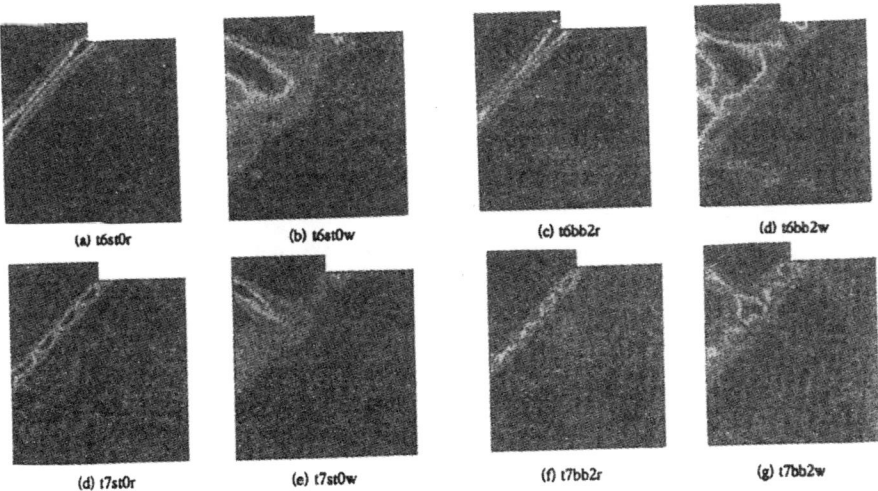

Figure 11. *Failure mechanisms for footing on vertical slope using 6 and 7 node elements and a perfectly plastic material*

displacement-pore pressure problems we cannot use the same shape functions for interpolation of both fields, unless we develop special stabilized elements. Well, we can indeed, but if the permeability is very small (undrained conditions), the system of equations is of the form

$$\begin{bmatrix} \mathbf{A} & -\mathbf{B} \\ -\mathbf{B}^T & 0 \end{bmatrix} \begin{bmatrix} \xi \\ \phi \end{bmatrix} = \begin{bmatrix} f_\xi \\ f_\phi \end{bmatrix} \tag{4.1}$$

This is in detail similar to those found in mixed formulations of incompressible solid mechanics and fluid dynamics problems.

It will be demonstrated next that the system will be singular and present pressure oscillations whenever the number of ξ variables $n_\xi \le n_\phi$,i.e., the number of ϕ variables. Although this condition is not sufficient for stability (and solvability) it is necessary and it generally excludes an equal order of interpolation spaces for pressure and displacements.

A complete mathematical treatment of the problem can be found in Refs. [BAB 73] and [BRE 73]. However, a much simpler explanation is provided by the patch test for mixed formulations proposed by Zienkiewicz *et al.* [ZIE 86], which will be described later.

To illustrate the problem, let us consider a layer of saturated soil of infinite length and depth L depicted in Figure 12, subjected to a harmonic distributed load on its surface.

The problem has been discretized using a column of 1 m. width with the following boundary conditions:

(i) At the top of the layer $y = L$:

Prescribed pore pressure $p_w = 0$

Precribed vertical traction $\bar{t}_y = 100 \exp(-i\omega t) \ (Pa)$ (4.2)

(ii) on the vertical sides

$$\frac{\partial p_w}{\partial x} = 0 \tag{4.3a}$$

$$\bar{t}_y \quad = 0 \quad \text{and} \quad u_x = 0 \tag{4.3b}$$

(iii) on the bottom

$$\frac{\partial p_w}{\partial y} = 0 \tag{4.4a}$$

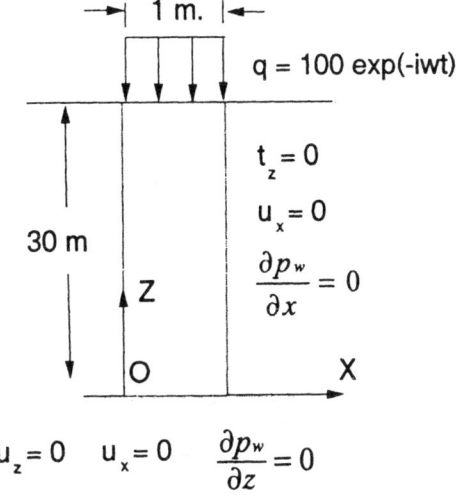

Figure 12. *Boundary conditions for saturated soil column with periodic surface load* ($q = t_y = 100 \exp(-i\omega t)$)

$$u_x = u_y = 0 \tag{4.4b}$$

The material parameters chosen for the analysis are the following:

k_w	$10^{-7} m/s$
n	0.333
E	$7.492\ 10^8$ (Pa)
ν	0.2
ρ_s	$2.0 \times 10^3\ (N/m^3)$
ρ_w	$1.0 \times 10^3\ (N/m^3)$

The height of the column has been taken as $L = 30\ m$ and the excitation frequency chosen is $\omega = 3.379$ rad/s.

The problem was discretized using 20 four node quadrilateral elements with bilinear shape functions for both pressure and displacements. As can be seen in the table, permeability is $10^{-7}\ ms^{-1}$, so the column is close to undrained conditions. Two different compressibilities of water and solid grains will be considered: $10^4 MPa$ and $10^9\ MPa$. The results have been plotted in Figure 13 where it can be seen how spurious oscillations grow as compressibility (i.e. $1/Q^*$) decreases. If we choose now quadratic polynomials for displacements and bilinear for pressure, no oscillations appear, as it can be seen in Figure 14.

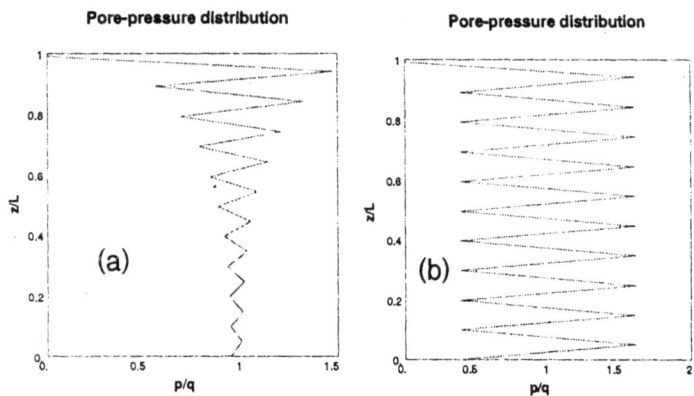

Figure 13. *Amplitude of pore pressure for the soil column problem using 20 Q4P4 elements with* $k = 10^{-7} m/s$ *and (a)* $Q^* = 10^4 MPa$ *(b)* $Q^* = 10^9 MPa$

Therefore, even if some compressibility exists, it is not possible to use any combination of shape functions for pressure and displacements, as oscillations can appear as the undrained-incompressible limit is approached.

In the case of quadrilaterals, allowable interpolation functions are bilinear for pressure and quadratic for displacements. Similarly, quadratic displacement triangles with linear pressure are admissible (Figure 15).

To provide a rational explanation of why some combinations work while some others do not, we will follow that given in Ref [ZIE 86] and begin by rewriting system (4.1) as

$$\begin{bmatrix} \mathbf{K} & -\mathbf{Q} \\ -\mathbf{Q}^T & 0 \end{bmatrix} \begin{bmatrix} \mathbf{u} \\ \mathbf{p} \end{bmatrix} = \begin{bmatrix} \mathbf{f}_u \\ \mathbf{f}_p \end{bmatrix} \tag{4.5}$$

from which we obtain, using the first equation of (4.5)

$$\mathbf{u} = \mathbf{K}^{-1}\mathbf{Q}.\mathbf{p} + \mathbf{K}^{-1}\mathbf{f}_u \tag{4.6}$$

Substituting this into the second equation of (4.5), we arrive at

$$-\mathbf{Q}^T \left(\mathbf{K}^{-1}\mathbf{Q}.\mathbf{p} + \mathbf{K}^{-1}\mathbf{f}_u \right) = 0$$

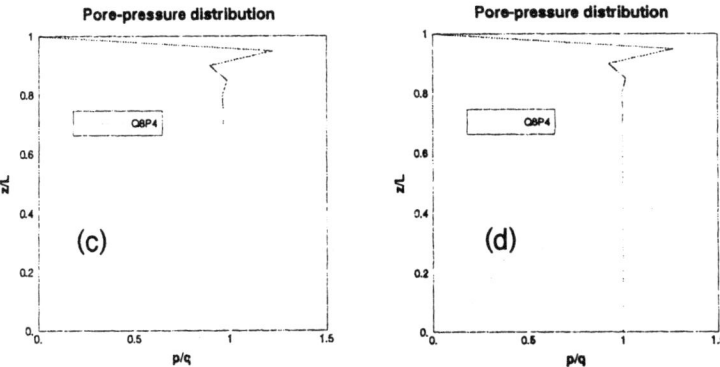

Figure 14. *Amplitude of pore pressure for the soil column problem using 20 Q8P4 elements with* $k = 10^{-7} m/s$ *and (a)* $Q^* = 10^4 MPa$ *(b)* $Q^* = 10^9 MPa$

or

$$(\mathbf{Q}^T \mathbf{K}^{-1} \mathbf{Q})\mathbf{p} = -\mathbf{Q}^T \mathbf{K}^{-1} \mathbf{f}_u \qquad (4.7)$$

which is a system of equations with a matrix of coefficients of dimension $(n_p \mathbf{x} n_p)$ obtained by multiplication of three matrices of dimensions $(n_p \mathbf{x} n_u)$, $(n_u \mathbf{x} n_u)$ and $(n_u \mathbf{x} n_p)$ where n_u and n_p are the number of degrees of freedom of displacements and pressures once suitable boundary conditions have been applied.

If $n_u < n_p$, the system is singular, and spurious oscillations in the pore pressure field will always appear. Therefore, the condition to be fulfilled is

$$n_u \geq n_p \qquad (4.8)$$

for any assembly (or patch) of elements.

It is important to note that this is a necessary but not a sufficient condition, and singularity has to be tested in all cases.

In order to illustrate the application of this condition, we will consider next the case of a quadrilateral with bilinear shape functions for both pressure and displacement (Figure 16).

We shall first use a single element patch. If the displacements have been prescribed on the boundary, and pore pressure has been fixed at one point we have zero degrees of freedom for displacements and three for pressures. The element does not pass the

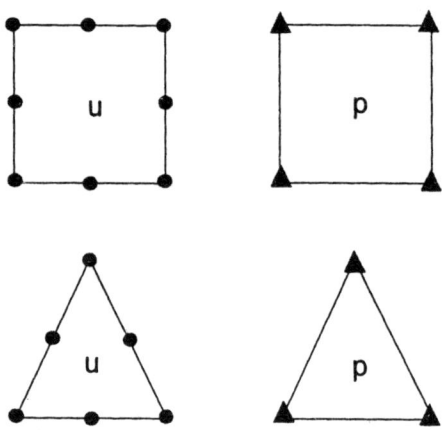

Figure 15. *Some allowed elements*

count conditions of the patch test as $n_u < n_p$ and the element will present oscillations in the undrained-incompressible limit as was shown in previous example. For larger patches obviously the same factors of the count will continue.

Examining the quadratic displacement triangle shown in Figure 17, a first patch consisting in one element alone does not pass the test, as, again, all displacement degrees of freedom have been fixed, and, therefore,

$$(n_u = 0) < (n_p = 2)$$

However, this a rather uncommon situation, and patches incorporating more elements do pass the patch test.

This is illustrated in Figure 18. Now, there are 7 free nodes in the interior of the patch, and 14 degrees of freedom for displacements, and 6 degrees of freedom for the pressure. Therefore,

$$(n_u = 14) > (n_p = 6)$$

and the count of the patch passes the test.

6. Our favourite solver (do we have any?)

Finite elements can give rise to huge systems of equations which we have to solve, perhaps many times if we are analyzing transient problems. If you are thinking

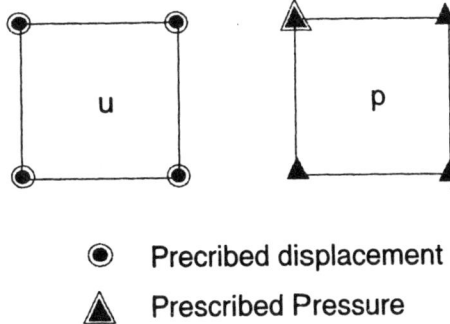

⦿ Precribed displacement
▲ Prescribed Pressure

Figure 16. *Bilinear quadrilateral with equal order of interpolation of displacement and pressure*

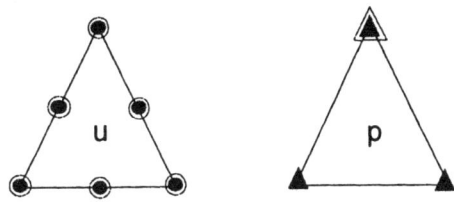

Figure 17. *A single element patch of quadratic displacement - linear pressure triangle*

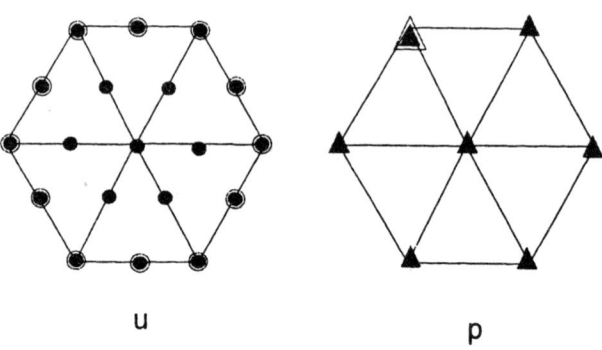

Figure 18. *Patch of six T6P3 elements*

of building your own finite element code, this is a crucial question: which solver should I use. And the answer depends on the problem.

In our opinion, iterative solvers like the preconditioned conjugate gradient or the Jacobi are excellent choices because they are really simple to program, specially if we are using a language like Fortran 90 or C. Another advantage is that they do not require any renumbering to save memory. We will describe the above mentioned iterative solvers in the following section.

The choice of iterative methods mentioned above is in our opinion the best for really large problems (tens or hundreds of thousands of degrees of freedom). For large but not so large problems (just thousands of degrees of freedom) a direct method with a special storage scheme for sparse problems is also a good choice and in general requires fewer numerical operations. Direct methods such as LU Gauss-Jordan fac-torization perform very satisfactorily under these circumstances. Additionally, they present the advantage of requiring relatively few changes to solve non symmetric systems, while conjugate gradient methods require special schemes such as GMRES to solve this type of problem. When using direct methods it is decisive to use sparse storage schemes, otherwise the computational cost and the required storage space are not affordable and round off errors reach unacceptable levels. A well known sparse storage scheme is the skyline method which will be presented in section 6.2.

6.1. *Conjugate Gradient Method with Preconditioning*

One of the most effective and simple iterative methods (when used with pre-conditioning) for solving $\mathbf{A}\mathbf{x} = \mathbf{b}$ is the conjugate gradient algorithm. The algorithm is based on the idea that the solution of $\mathbf{A}\mathbf{x} = \mathbf{b}$ minimizes the total potential $\Pi = \frac{1}{2}\mathbf{x}^T\mathbf{A}\mathbf{x} - \mathbf{x}^T\mathbf{b}$. Hence, the task in the iteration is, given an approximate $\mathbf{x}^k$ to x for which the potential is Π^k, to find an improved approximation $\mathbf{x}^{k+1}$ for which $\Pi^{k+1} < \Pi^k$. However, not only do we want the total potential to decrease in each iteration but we also want $\mathbf{x}^{k+1}$ to be calculate efficiently and the decrease in the total potential to occur rapidly. Then the iteration will converge fast.

In the conjugate gradient method, we use in the kth iteration the linearly independent vectors $\mathbf{p}^1, \mathbf{p}^2, \mathbf{p}^3, ..., \mathbf{p}^k$ and calculate the minimum of the potential in the space of the potential in the space spanned by these vectors. This gives $\mathbf{x}^{k+1}$. Also, we establish the additional basis vector $\mathbf{p}^{k+1}$ used in the subsequent iteration.

The algorithm can be summarized as follows.

Choose the starting iteration vector $\mathbf{x}^1$ (frequently $\mathbf{x}^1$ is the null vector).

Calculate the residual $\mathbf{r}^1 = \mathbf{b} - \mathbf{A}\mathbf{x}^1$. If $\mathbf{r}^1 = 0$, quit.

Else:

Set $\mathbf{p}^1 = \mathbf{r}^1$.

Calculate for $k = 1, 2, ...,$

$$\alpha^k = \frac{\mathbf{r}^{k^T}\mathbf{r}^k}{\mathbf{p}^{k^I}\mathbf{A}\mathbf{p}^k}$$

$$\mathbf{x}^{k+1} = \mathbf{x}^k + \alpha^k\mathbf{p}^k$$

$$\mathbf{r}^{k+1} = \mathbf{r}^k - \alpha^k\mathbf{A}\mathbf{p}^k \qquad (7)$$

$$\beta^k = \frac{\mathbf{r}^{k+1^T}\mathbf{r}^{k+1}}{\mathbf{r}^{k^T}\mathbf{A}\mathbf{r}^k}$$

$$\mathbf{p}^{k+1} = \mathbf{p}^{k+1} + \beta^k\mathbf{p}^k$$

We continue iterating until $\|\mathbf{r}^k\| \le \varepsilon$, where ε is the convergence tolerance. A convergence criterion on $\|\mathbf{x}^k\|$ could also be used.

The conjugate gradient algorihm satisfies two important orthogonality properties regarding the direction vectors $\mathbf{p}_i$ and the residual $\mathbf{r}_i$, namely, we have

$$\mathbf{p}_i^T\mathbf{A}\mathbf{p}_j^k = 0 \qquad (8)$$

$$\mathbf{P}_j^T\mathbf{r}^{k+1} = 0 \qquad (9)$$

where $\mathbf{P}_j = [\mathbf{p}_1, ..., \mathbf{p}_j]$.

Convergence to the solution **x**, in an exact arithmetic, is achieved in at most n iterations. Of course, in practice, we want convergence to be reached in much fewer than n iterations.

The rate of covergence of the conjugate gradient algorithm depends on the condition number of matrix **A**, defined as $\text{cond}(\mathbf{A}) = \lambda_n/\lambda_1$, where λ_1 is the smallest eigenvalue and λ_n is the largest eigenvalue of **A**. the larger the condition number, the slower the convergence, and in practice, when the matrix is ill-conditioned, convergence can be very slow.

To increase the rate of convergence of the solution algorithm, preconditioning is used. The basic idea is that instead of solving $\mathbf{Ax} = \mathbf{b}$, we solve

$$\widetilde{\mathbf{A}}^{-1}\mathbf{Ax} = \widetilde{\mathbf{A}}^{-1}\mathbf{b} \tag{10}$$

where $\widetilde{\mathbf{A}}$ is called the preconditioner. The objective with this transformation is to obtain a matrix $\widetilde{\mathbf{A}}^{-1}\mathbf{A}$ with a much improved conditioned number choosing an easy inverting matrix $\widetilde{\mathbf{A}}$. Various preconditioners have been proposed, the choose of the diagonal part of **A** results in the Jacoby conjugate gradient method (JCG).

The new algorithm introduces an additional set of vectors $\mathbf{z}^k$ defined by:

$$\mathbf{z}^k = \widetilde{\mathbf{A}}^{-1}\mathbf{r}^k \tag{11}$$

which modifies the definitions of α^k, β^k, $\mathbf{p}^k$:

$$
\begin{aligned}
\alpha^k &= \frac{\mathbf{z}^{k^T}\mathbf{r}^k}{\mathbf{p}^{k^T}\mathbf{Ap}^k} \\
\beta^k &= \frac{\mathbf{z}^{k+1^T}\mathbf{r}^{k+1}}{\mathbf{z}^{k^T}\mathbf{r}^k} \\
\mathbf{p}^{k+1} &= \mathbf{z}^{k+1} + \beta^k\mathbf{p}^k
\end{aligned}
\tag{12}
$$

We considered in this section only the case of a symetric coefficient matrix, and should note that also of much interest is the iterative solution of equations with non-symmetric coefficient matrices. Here the benefits of savings in storage and computing time can be even more significant. For nonsymmetric coefficient matrices, the conjugate gradient method has been generalized and other iterative schemes, notably, the generalized minimal residual (GMRes) method, have been developped and researched.

6.2. Skyline storage scheme

The skyline scheme consists of storing the stiffness matrix in a vector including the diagonal terms and the off diagonal terms between the non zero off diagonal term which is farthest from the diagonal in each row (or column) and the corresponding diagonal term. Additionally it will be necessary to use an N component integer vector to store the direction in the stiffnes vector where the last component of the ith row (or column) is stored. Let us assume our stiffness matrix is the following one:

$$\begin{bmatrix} 2 & -2 & 0 & 0 & -1 \\ -2 & 3 & -2 & 0 & 0 \\ 0 & -2 & 5 & -3 & 0 \\ 0 & 0 & -3 & 10 & 4 \\ -1 & 0 & 0 & 4 & 10 \end{bmatrix}$$

Taking advantage of the symmetry of the matrix and storing the upper half would produce the following vector:

i	1	2	3	4	5	6	7	8	9	10	11	12
$a(i)$	2	-2	3	-2	5	-3	10	-1	0	0	4	10

The integer vector storing the positions of diagonal terms would be:

i	1	2	3	4	5
$jdiag(i)$	1	3	5	7	12

Once the matrix has been stored in this fashion, all the operations leading to the solution of the linear equation system are performed on vector a. Factorization will change the contents of **a** and back substitution will use these new contents in conjunction with the load vector to produce the solution.

7. Conclusions

The finite element method is a powerful tool that has been used in many fields of engineering analysis. But, this 'weapon' should be used with care. We have discussed some practical aspects that could introduce important error in the results of computations. One fundamental aspect is the choice of the kind of element to do the analysis. The element type chosen strongly affects the results, and can also be the source of unsatisfactory performance in bending and incompressible situations. For mixed formulations, the element type should also verify some important restrictions (patch test). Concerning the type of algorithm that should be used to solve the system of equation obtained with the FEM the discussion is open. The choice of iterative methods is in our opinion the best for really large problems.

8. References

[HUG 87]   T.J.R.Hughes, **The Finite Element Method: Linear Static and Dynamic Finite Element Analysis,** Prentice-Hall, Inc., Englewood Cliffs, New Jersey, 1987.

[BAB 73] I.Babŭska, "The finite element method with Lagrange multipiers", **Num. Math. 20**, 179-192, 1973.

[BRE 73] F. Brezzi, "On the existence, uniqueness and approximation of saddle point problems arising from lagragian multipliers", **RAIRO 8-R2**, 129-151, 1974.

[BAT 96] K.J. Bathe, Finite element Procedures, Prentice Hall, New Jersey, 1996.

[IRO 83] B.Irons and N.Shrive, **Finite element primer**, Ellis Horwood Limited, 1983.

[ZIE 83] O.Z.Zienkiewicz and K.Morgan, **Finite Element and Approximation,** John Wiley and Sons, New York, 1983.

[ZIE 2000] O.Z.Zienkiewicz and R.L.Taylor, **The Finite Element Method 5th edition** (3 Vols.), Butterword-Heinemann, Oxford, 2000.

[ZIE 86] O.C. Zienkiewicz, S. Qu, R.L. Taylor and S. Nakazawa
"The patch test for mixed formulations".**Int.J.Numer.Meth.Eng. 23**, 1873-1883, 1986.

Chapter 15

Mechanical Effects of Chemical Degradation of Bonded Geomaterials in Boundary Value Problems

Riccardo Castellanza and Roberto Nova
Department of Structural Engineering, Milan University of Technology, Italy

Claudio Tamagnini
Dipartimento di Ingegneria Civile e Ambientale, Università di Perugia, Italy

1. Introduction

The process of rock weathering can have an influence on the behaviour of engineering structures, especially when soft rocks are involved. For instance foundation settlements and even failures were observed in saline and gypsiferous rocks in Russia because of the progressive degradation caused by prolonged soaking (Petruckhin, 1993). The stability of rock slopes can be reduced by the degradation of the mechanical properties of the material (Clough et al., 1981). Weathering processes, acting from the free surface of the rock mass, induce a reduction of shear strength that, under constant gravity load, brings the slope to an unstable condition. The process continues on the debris material and, after a sufficient period of time, an initial steep slope can be transformed into a gentle one (Chigira et al., 1999). The collapse of the pillars in some abandoned mines in Lorraine (France), induced a subsidence of the superficial soil layers of more than one meter, causing severe damage in adjacent buildings (Deck et al., 2002). The origin of this process has been found in the changes in the hydrological regime after the end of the mining works, which, in turn, induced a progressive chemical degradation of the rock in which the pillars have been excavated.

Rock weathering acts at the constitutive level essentially by reducing the strength of the bonds joining the grains together: a bonded geomaterial, like a soft rock, can be progressively reduced into a (residual) granular soil. In order to reproduce the effects of this constitutive transformation at the macroscopic level, an elastoplastic strain-hardening model has been recently proposed in which the loading function depends on some measure of weathering in which the loading function depends on some measure of weathering (Nova and Castellanza, 1999, 2001). The numerical implementation of this model for its application in nonlinear finite element analysis of practical boundary value problems has been discussed in Tamagnini et al. (2003). In this paper, after a brief presentation of the model, three examples of application to the practical problems previously mentioned are discussed in detail.

2. Extension of plasticity theory to cope with weathering effects

As shown in Nova (1997, 2000), and in Nova and Castellanza (1999, 2001), the same conceptual framework used for describing the behaviour of bonded geomaterials can be exploited to model rock weathering. This is possible by extending the evolution laws for the bonding–related internal variables by including a non–mechanical contribution linked to the time rate of change of a suitable scalar measure of degradation.

As suggested by Nova (1986, 1992), the main difference between an intact soft rock and a granular soil lies in that for soft rocks, contrary to soils, the initial elastic domain is not related to the previous loading history (i.e., preconsolidation) but rather exists per se and its size is linked to the strength of the bonds.

At the microscopic level, weathering causes a progressive destruction of intergranular bonds (see Castellanza, 2002). By collecting qualitatively all the information presented in Figure 1a, b in the axisymmetric ("triaxial") plane, it is possible to conclude that the most relevant consequence of weathering is the progressive reduction of the size of the initial elastic domain (Figure 1c).

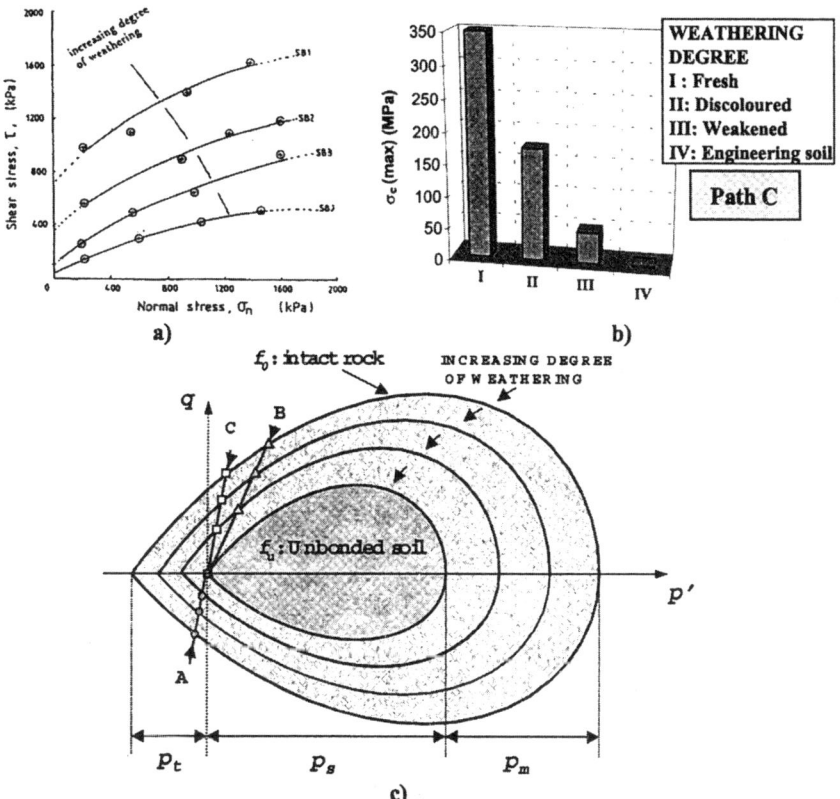

Figure 1. *a) Failure loci in direct shear tests for different degrees of weathering of granite (after Kimmance, 1988); b) Decay of uniaxial strength with respect to weathering degree (after Baynes and Dearman, 1978); c) Yield loci for rock at different degree of weathering (after Castellanza, 2002)*

This is equivalent, in the framework of the theory of plasticity, to a progressive shrinkage of the initial yield surface of the intact rock.

As proposed by Nova (1992), for a bonded geomaterial the size of the (convex) yield surface (of equation $f_0 = 0$) is controlled by three scalar internal variables p_s, p_m and p_t, as shown in Figures 1 and 2. The variable p_s controls the size of the elastic domain for the uncemented material, while the two additional state variables p_m and

p_t are introduced to take into account the macroscopic effects of the intergranular bonds. The first, p_t is linked to the tensile strength of the bonded soil, while the second p_m to the widening of the elastic domain in the compression range. The two quantities p^* and p_c appearing in Figure 2 are defined as:

$$p^* = p' + p_t \qquad [1]$$

$$p_c = p_s + p_m + p_t \qquad [2]$$

In general, p_s, p_m and p_t are assumed to depend on the total plastic strain tensor ε_{ij}^p and, possibly, on a scalar degradation parameter X_d ranging between 0 (intact rock) and 1 (full degradation), according to the following hardening laws:

$$\begin{aligned} p_s &= p_s(\varepsilon_{ij}^p) \\ p_m &= P_m(\varepsilon_{ij}^p)Y(X_d) \\ p_t &= P_t(\varepsilon_{ij}^p)Y(X_d) \end{aligned} \qquad [3]$$

The quantities P_m and P_t are scalar functions decreasing monotonically with ε_{ij}^p. In this way it is possible to describe the effect of mechanical destructuration upon yielding (Lagioia and Nova, 1995). The scalar quantity Y is a normalised, monotonically decreasing function of the degree of weathering X_d, and takes into account the effects of chemical degradation.

The general expression for the yield surface in stress space is thus given by:

$$f[\sigma_{ij}', p_k(\varepsilon_{rs}^p, X_d)] = 0 \qquad [4]$$

where σ_{ij}' is the effective stress tensor and p_k denotes the set of internal variables controlling the size of the elastic domain, whose evolution is given by eq. [3].

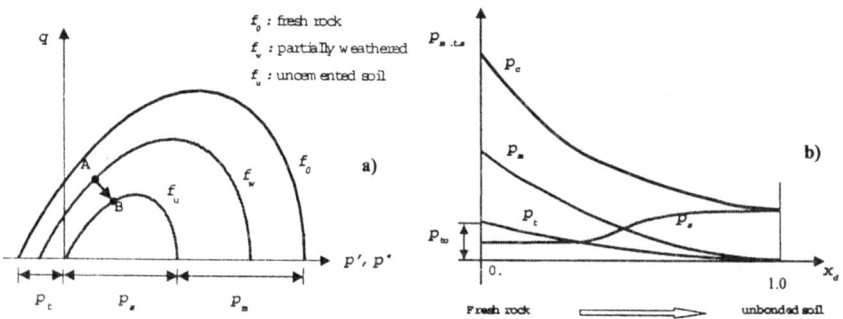

Figure 2. a) Evolution of yield locus with weathering, the initial stress state being represented by point A; b) Evolution of the hardening variables with weathering

As both p_m and p_t are associated with the same physical mechanism at the microscopic level (i.e., interparticle bonding), for the sake of simplicity, in the following p_t is assumed to be equal to $k\,p_m$, with $k=$ const.

A chemical degradation process can be simulated by increasing the value of X_d from 0, when the rock is intact, to 1 for a fully decomposed material (see Figure 2). By means of the function $Y(X_d)$, a progressive shrinkage of the initial elastic domain is forced from f_0 (fresh rock with $Y(X_d)=1$) to f_u corresponding to an unbonded material, for which p_m and p_t are nil ($Y=0$). In Figure 2, f_w represents the yield locus for a partially weathered material.

Consider now a specimen of soft rock in oedometric conditions subject to a given constant axial stress and undergoing weathering. Until the stress state reaches the shrinking yield surface (point A in Figure 2) the behaviour of the material is elastic and no strain or stress changes occur; when the shrinking yield locus touches the point A, the stress state is forced to move towards a higher mean pressure and a lower deviator stress level, since the stress states cannot lie outside the yield locus. This in turn implies an increase of the radial stress. The process stops at point B in Figure 2, when full degradation is achieved (see Nova and Castellanza, 2001). As shown in Figure 2b, the hardening parameter p_s may increase during plastic loading even if the size of the elastic domain is decreasing, due to the occurrence of positive plastic volume strains.

This kind of chemo-mechanical coupling described theoretically by the model (Figure 2) has been observed also experimentally by Castellanza (2002) in a series of oedometric tests in which samples of cemented sand have been degraded by exposing them to an acid flow at constant rate. The experimental results and the model predictions are shown in Figure 3 in terms of stress path (in the "triaxial plane"), radial stress and axial strain as functions of the degree of weathering. The good agreement between predictions and measurements corroborate the proposed model as far as mechanical effects of weathering are concerned.

The analytical expression of the function $Y(X_d)$ employed for the simulations reported in Figure 3 was established on the basis of experimental data on rock specimens tested at different weathering degrees. As reported in Castellanza (2002), a suitable choice for $Y(X_d)$ is provided by the following parabolic equation:

$$Y(X_d) = 1 - rX_d + (r-1)X_d^2 \qquad [5]$$

where X_d has been assumed equal to the ratio t/T between exposure time to acid flow before testing, t, and the total time T required to complete the weathering process.

The complete analytical formulation of the constitutive equation, as well as the details about the role of the different material constants required for the calibration of the model are reported in Castellanza (2002) and in Tamagnini et al. (2003). Here it will be enough to recall that the plastic potential and the yield function are those given by Lagioia et al. (1996), the non linear elastic constitutive equation derives from Borja and Tamagnini (1998) and the hardening laws are derived from Nova (1988) and Lagioia & Nova (1995).

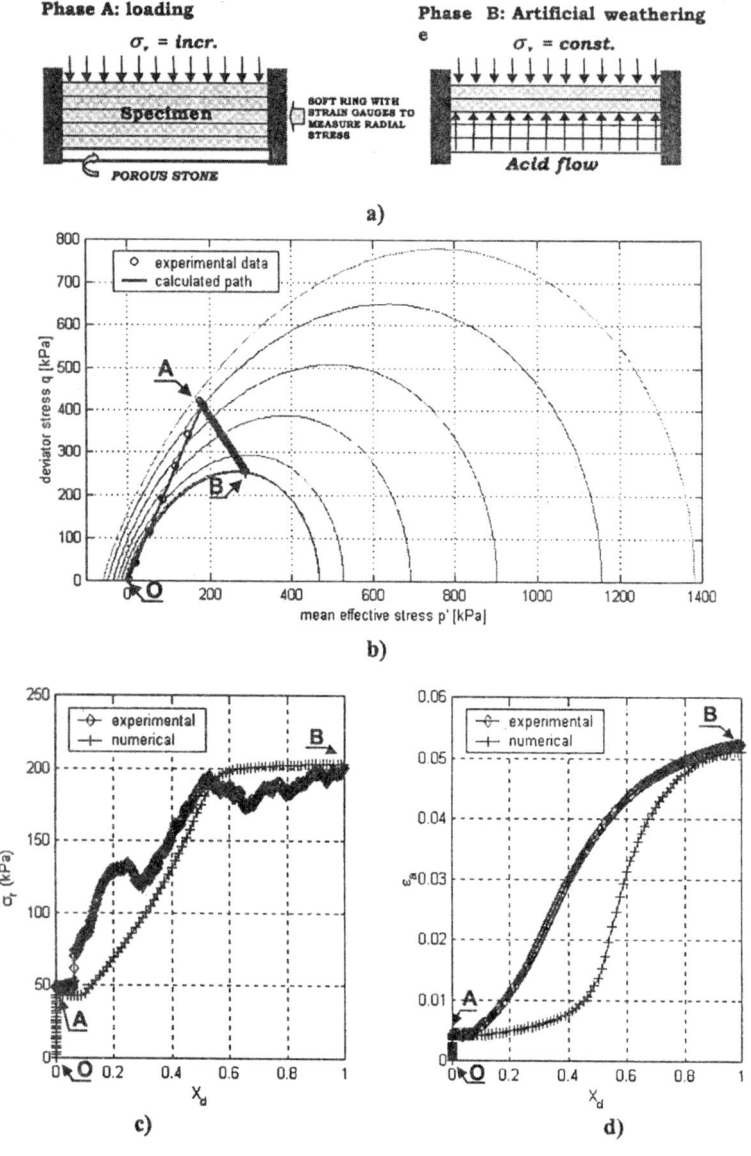

Figure 3. *Oedometric weathering test: a) schematic layout of the testing conditions. Experimental curves and model simulation (after Castellanza, 2002): b) stress path in triaxial plane; c) radial stress vs. degree of weathering; d) axial strain vs. degree of weathering (OA: stress and strain paths in the loading stage; AB: stress and strain paths in the weathering stage)*

3. Weathering effects in boundary value problems: three examples of application

A series of numerical simulations is presented in the following to demonstrate the applicability of the proposed approach for the qualitative and quantitative analysis of some practical situations in which weathering processes may induce — in a time scale comparable to that of typical engineering structures — a relevant alteration of the equilibrium condition of the rock mass, with an adverse impact on the surrounding environment. All the numerical analyses reported in this section have been carried out with a modified version of the code GEHOMADRID (Fernandez Merodo *et al.*, 1999). Details on the numerical implementation of the proposed constitutive model are given in Tamagnini *et al.* (2003).

3.1. *Weathering induced subsidence in a circular foundation*

The geometry of the problem considered is detailed in Figure 4. A perfectly flexible circular footing, with radius $R = 1.3$ m, rests on a layer of soft rock ($H = 5.2$ m thick), underlain by a rigid bedrock. The top surface of the rock layer is overlain by a granular fill modelled as a uniform applied load $Q_0 = 100.0$ kPa. Under working conditions, the foundation is loaded with a uniform pressure $\Delta Q = 300.0$ kPa. Due to the action of weathering, the foundation rock undergoes a process of chemical degradation starting at the top surface and progressing downwards with time.

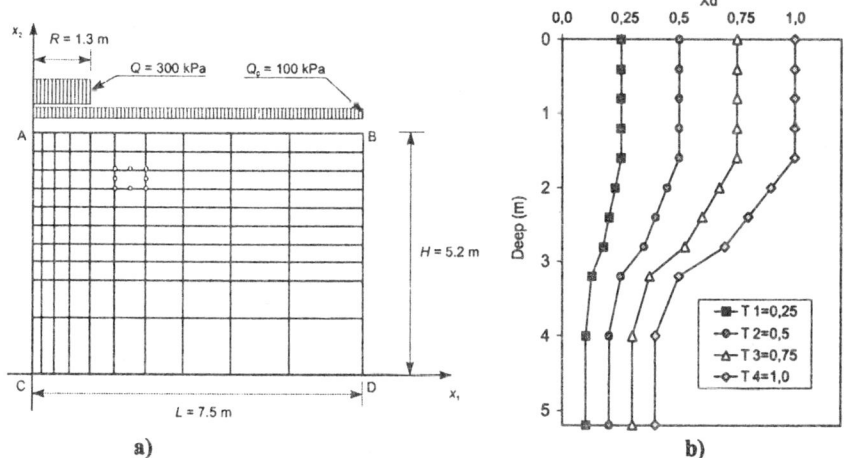

a)   b)

Figure 4. *a) Circular foundation on a weak rock layer: geometry of the problem and FE discretisation; b) Profiles of X_d at different time stations*

The (axisymmetric) finite element mesh adopted for the analyses, also shown in the figure, is composed of 100 eight-noded serendipity elements, underintegrated with a uniform 2×2 Gauss quadrature rule. The material parameters adopted for the rock are listed in Table 1. Note that, in order to highlight the influence of chemical degradation effects alone, mechanical degradation has been ruled out by a convenient choice of the relevant material parameters.

The initial stress state before fill placement and foundation loading is assumed as geostatic, with a unit weight $\gamma = 20.0$ kN/m^3 and a coeffiecient of earth pressure at rest $K_0 := \sigma'_h/\sigma'_v = 0.5$. No pore water pressure is present. Due to the relatively small thickness of the rock layer, the internal variables p_s and p_m have been assumed constant with depth and equal to 250.0 and 1000.0 kPa, respectively. As for the imposed displacement boundary conditions, horizontal displacements have been fixed along the vertical boundaries AC and BD, while both horizontal and vertical displacements have been fixed at the bottom boundary CD (see Figure 4a).

α	k	G_0	p_r	ρ_s	ξ_s	ρ_m	ξ_m	k
(–)	(–)	(kPa)	(kPa)	(–)	(–)	(–)	(–)	(–)
0.0	18.5	20200	1.0e+6	100	0.0	0.0	0.0	0.2
a_g	m_g	M_{gc}	M_{ge}	a_f	m_f	M_{fc}	M_{fe}	r
(–)	(–)	(–)	(–)	(–)	(–)	(–)	(–)	(–)
1.0e-5	2.00	0.95	0.75	0.63	0.95	0.95	0.75	2.0

Table 1. *Material parameters adopted in the finite element analysis*

The numerical simulations have been performed by splitting the loading history into four different loading stages. In the first stage, gravity loads have been applied in two steps, in order to generate the geostatic stresses. Then the uniform surface load and the foundation load have been applied in the next two steps. Due to the assumed initial values of p_s and p_m, no plastic yielding occurs in the foundation rock during fill placement and foundation loading.

The effects of weathering of the foundation rock are simulated in the fourth and last stage of the analysis. Here, a suitable degradation profile inside the layer has been imposed by prescribing the time-histories of X_d in each of the ten layers of elements in the FE discretisation. As the time scale of the degradation process is arbitrary, the duration of the degradation stage has been normalised to a unit time.

The resulting shape of the degradation profile inside the layer is shown in Figure 4b in terms of the function $X_d(x_2,t)$ at four different time stations ($t = 0.25, 0.50, 0.75,$ and 1.0, respectively). It is worth noting that the assumed average rate of degradation with depth can be interpreted qualitatively as a progressive downward motion of a horizontal degradation front from the top surface. This is in qualitative agreement with the observations reported by Fookes *et al.* (1988) on a weathered granite rock and by Chigira and Oyama (1999) on a weathered sedimentary rock.

The response of the rock-foundation system is shown in Figure 5, where the time-history of the additional vertical displacement of point A on the foundation axis (see Figure 4) is portrayed. The two displacement histories obtained with 400 and 20 steps are displayed in the figure, to show the effects of the step size on the numerical solution. It can be observed that, at the beginning of the degradation process (i.e., up to $t \cong 0.55$), the gradual reduction in size of the yield locus implied by the assumed hardening law for p_m has no effects on the equilibrium state of the system, since the current stress state is, at this stage, still inside the yield locus at all integration points. However, as soon as the actual stress point touches the shrinking yield locus, plastic loading occurs and plastic strains develop in order to restore the consistency condition. This, in turns leads to a drastic change in the stress and elastic strain fields. As the resulting elastic and plastic strains beneath the foundation are compressive, point A settles more rapidly at first, and then more slowly, up to a final maximum settlement $\Delta u_2 = -15.3$ mm.

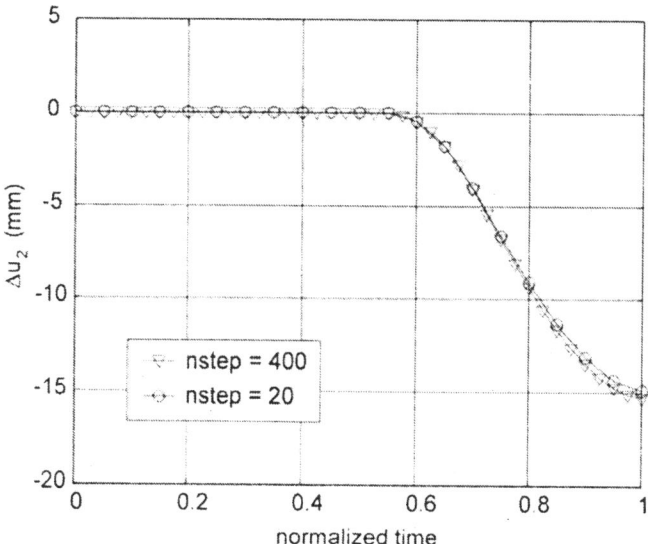

Figure 5. *Additional vertical displacement of point A* vs *normalised time during the chemical degradation stage*

It is interesting to note that a qualitatively similar behaviour has been reported by Petruckhin (1993) who observed weathering-induced settlements in a similar foundation resting on a gypsiferous rock layer. It is also worth noting that the final settlement increment due to weathering is comparable to the settlement observed during fill placement and subsequent foundation loading (-17.1 mm). As for the accuracy of the numerical procedure adopted, the results in Figure 5 indicate that the largest difference between the most and the least accurate numerical solution occurs at the end of the process ($t = 1.0$) and is only about 2.6% of the largest total displacement increment obtained with 400 steps.

The plastic volumetric strain fields computed at four different time stations (t = 0.625, 0.75, 0.875, and 1.0, respectively) are shown in Figure 6. By looking at how plastic strains develop in time beneath the foundation, it can be noticed that, as the degradation proceeds, the initially dilative plastic flow (Figure 6a) changes rapidly into a compressive one, with strong positive plastic volumetric strains. This is a consequence of the particular set of material parameters adopted and of the assumed initial state.

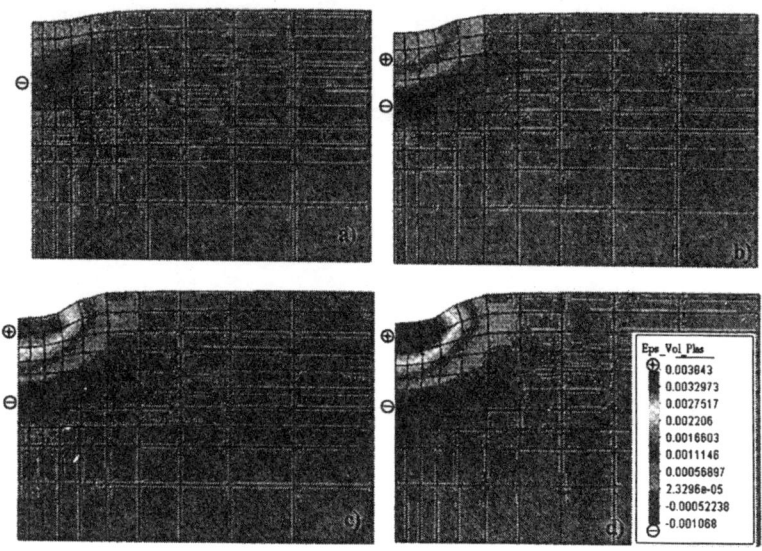

Figure 6. *Contours of plastic volumetric strains at different time steps: a) t = 0.625; b) t = 0.75; c) t = 0.875; d) t = 1.0*

In fact, by observing the stress paths at two Gauss points located at depths equal to B/2 and 2B, respectively, along the foundation axis (Figure 7), the stress state at the inception of yielding lies slightly at the left of the top of the yield surface (point D in Figures 7a and 7b), where the volumetric component of the plastic strain rate is negative (note that a non-associative flow rule is considered in this case). However, as the vertical stress component changes only very slightly during the degradation stage - being controlled primarily by the equilibrium condition along the vertical direction - the stress paths following plastic yielding (path DE in Figures 7a and 7b) are such that the mean stress p' is increased while the deviator stress q is reduced. The stress states at all points beneath the foundation are thus dragged towards the right side of the yield surface, where the plastic flow direction is compressive. In Figure 7 the path ABC refers to the application of the uniform load and the foundation load. Path CD represents the stress redistribution that occurs, in elastic conditions, at the beginning of the weathering phase as a consequence of the stress variation of the upper Gauss points (already in plastic state).

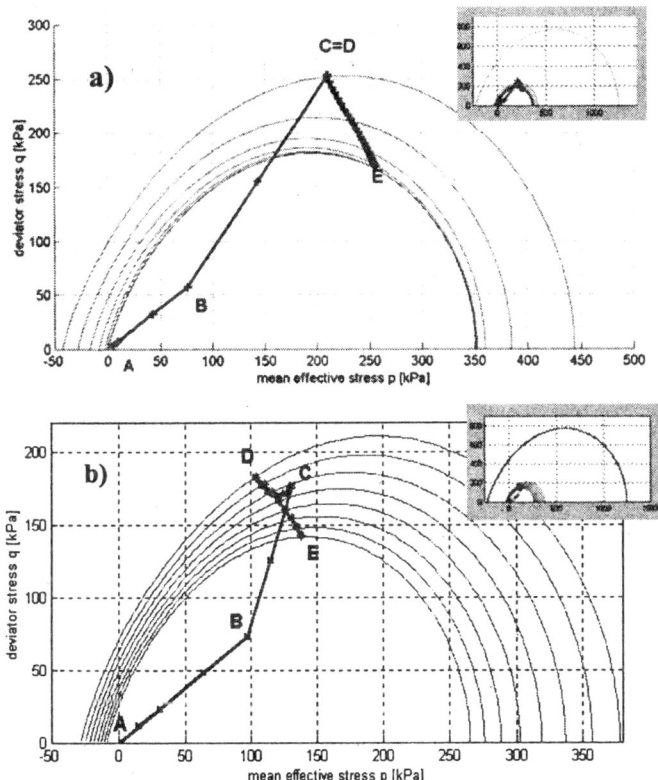

Figure 7. *a) Stress path for a Gauss point at a depth equal to B/2 under the foundation; b) Stress path for a Gauss point at a depth 2B under the foundation. The insets show the initial yield surfaces of the intact soft rock*

The overall variation in the stress field is clearly visible in Figures 8 and 9, in terms of the invariants p' and q. As observed previously for the two Gauss points considered, the weathering of the foundation rock induces a marked increase in the hydrostatic component of the stress state and a drastic reduction of its deviatoric component. It is interesting to note that since the bond strength is nullified by weathering (soft rock is transformed into a non cohesive soil), the second invariant of deviator stress tensor, characterised by higher values below the foundation, is strongly reduced when the weathering process occurs (Figure 9b).

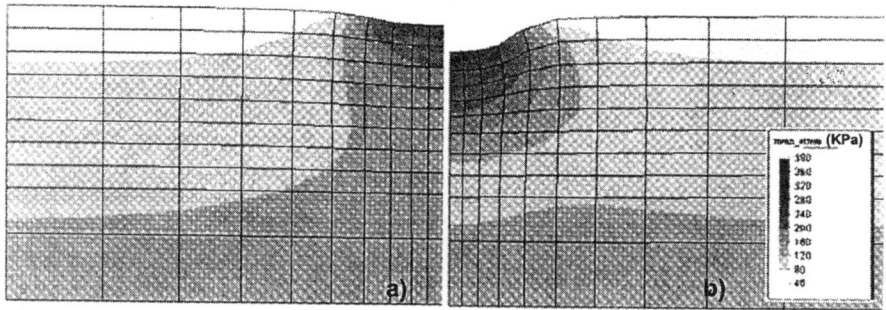

Figure 8. *Mean stress p' (kPa): a) before degradation stage; b) at the end of degradation stage*

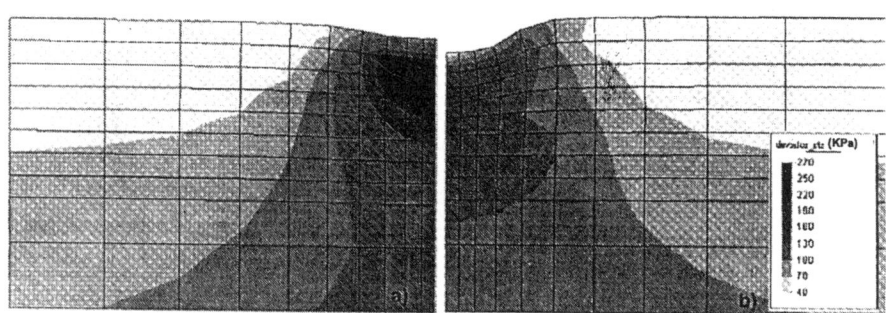

Figure 9. *Deviatoric stress q (kPa): a) before degradation stage; b) at the end of degradation stage*

From an engineering point of view, it is interesting to see how weathering affects the horizontal component of the in-situ stress state. A consistent increase in the horizontal effective stress σ'_h occurs beneath the foundation; such a change corresponds quite well to the observed evolution of the coefficient $K = \sigma'_h/\sigma'_v$, since the vertical stress remains almost constant throughout the weathering process. The biggest increment is reached along the foundation axis at a point 1.2 m deep (more or less the radius length), where the horizontal stress doubles the initial geostatic value (see Figure 10).

3.2. Weathering effects on a slope

In the second example considered, the model is applied to the analysis of the effects of weathering on the equilibrium and stability conditions of a rock slope. Consider, for instance, the two rock slopes in Figure 11a, both affected by weathering progressing in time from the ground surface to the deeper portions of the rock mass, according to a mechanism similar to the one depicted in Figure 11b. As discussed in Clough *et al.* (1981), in such conditions weathering represents a risk for the housing developments on the crest.

This situation is schematized here as shown in Figure 12, detailing the slope geometry and the finite element discretization employed in the analysis. The initial stress state before weathering is in equilibrium with the gravity load resulting from a unit weight of the rock $\gamma = 20$ kN/m^3 and a vertical, uniform strip load $Q = 100$ kPa, 12 m wide, placed at the crest of the slope, as shown in Figure 12. The (plane strain) finite element mesh is composed by 256 eight-noded serendipity elements, under-integrated with a uniform 2×2 Gauss quadrature rule.

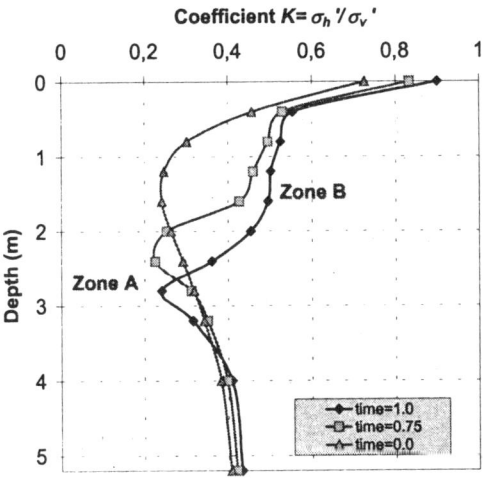

Figure 10. *Variation with depth of the principal stress ratio K under foundation centre at different time steps*

The material parameters of the rock have been (arbitrarily) assumed equal to those adopted in the previous analysis (Table 1). For the sake of simplicity and considering the relatively small height of the slope, the soft rock has been considered homogeneous; and the internal variables p_s and p_m have been assumed as constant, and equal to 180 kPa and 1200 kPa, respectively.

The numerical simulation has been performed by splitting the loading history in two stages. In the first one, the gravity loads and the external loads Q are applied in one step. During this stage, no plastic yielding occurs and the slope response is elastic. In the second stage, the weathering process has been simulated by simply imposing a linear increase of the parameter X_d from 0 to 0.75 in a portion of the rock mass close to the ground surface, identified by the element group W in Figure 12.

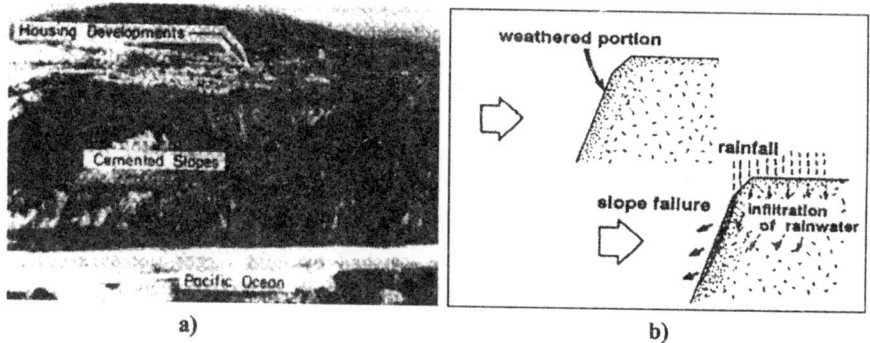

a)   b)

Figure 11. *Weathering process in a slope: a) Daly City, California (after Clough* et
al., 1981); b)schematic mechanism (after Yokota and Iwamatsu, 1999)

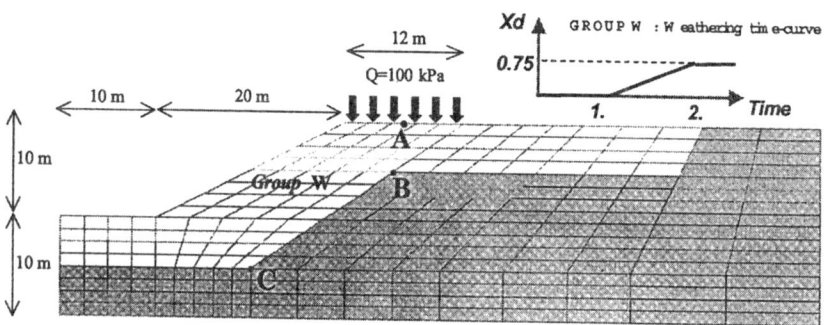

Figure 12. *Finite element discretisation and external load (Group W: zone
subjected to weathering)*

The response of the slope to the imposed weathering process is shown in Figure
13, in terms of displacement field, and in Figure 14, in terms of (additional) vertical
displacements of point A, located at the center of the strip load, as a function of the
degradation parameter X_d. It can be observed how, at the beginning of the
degradation process and up to $X_d = 0.55$, the reduction in size of the yield locus has
no effect on the equilibrium conditions of the slope, as the current stress state is still
inside the yield locus at all the integration points; and no additional displacements
occur. Then, when at least one stress point touches the shrinking yield locus, plastic
yielding occurs, causing additional plastic strains and an overall stress redistribution
to restore consistency. This process stops when the degradation parameter reaches
its maximum value ($X_d = 0.75$ in this case). The response of the slope to weathering
can be summarised as follows:

i) the vertical displacement of point A increases with the weathering degree (see Figure 13);

ii) plastic strains induced by weathering appear to concentrate at the interface between the stiff, intact rock and the softer weathered rock (line BC in Figure 12). Total volumetric and deviatoric plastic strain contours at the end of the weathering process are plotted in Figure 15;

iii) a marked increase of shear stresses is observed in the weathered slope, as shown in Figure 16.

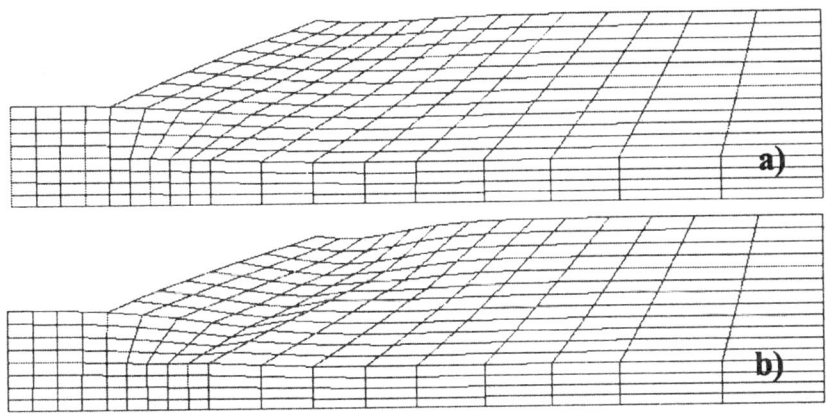

Figure 13. *Deformed mesh (magnification factor equal to 60) a) before weathering b) after weathering*

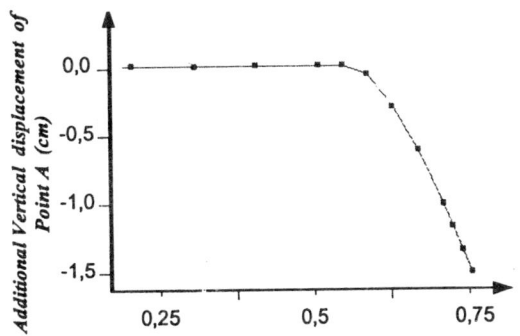

Figure 14. *Additional vertical displacement of a point A vs. weathering degree X_d*

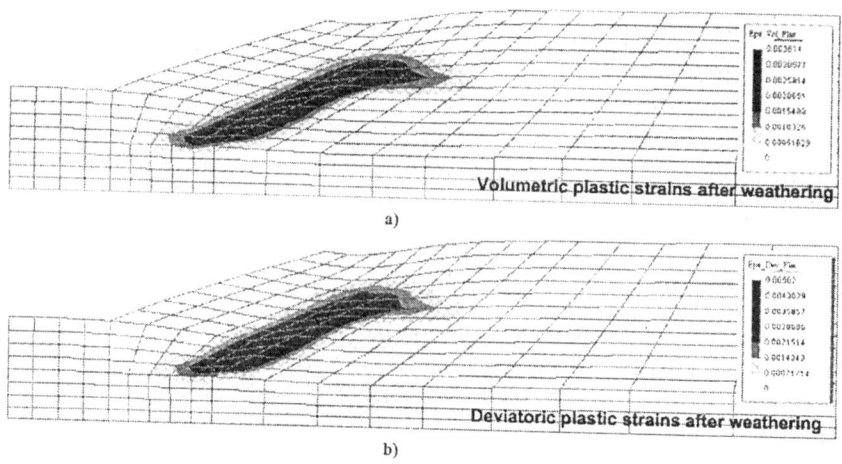

Figure 15. *a) volumetric plastic strain after weathering process in group W; b) deviator plastic strain after weathering process in group W*

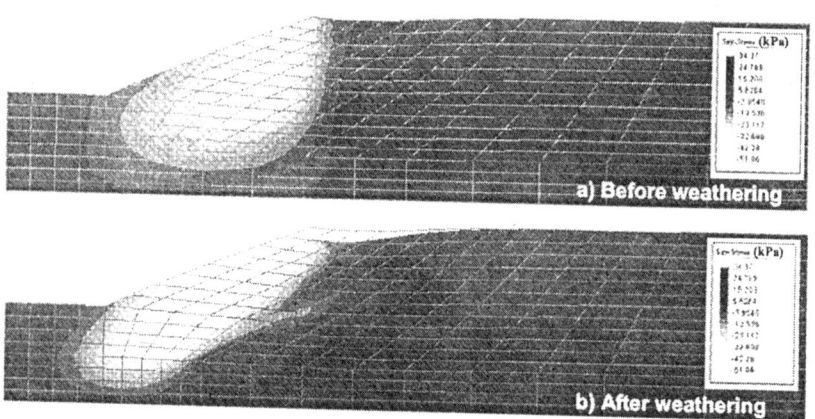

Figure 16. *Shear stress component* τ_{xy} *before (a) and after (b) weathering (magnification factor for the deformed mesh equal to 60)*

This analysis should be considered as a preliminary example to demonstrate the applicability of the model in the analysis of weathering-induced slope movements and instability. From a qualitative point of view, the results show that weathering is associated with the development of an additional displacement field, at constant external loading, which is the result of yielding processes and subsequent stress redistribution induced by the progressive reduction of the shear strength of the rock. Furthermore, the development of a possible slip surface at the interface between the intact and the weathered material (line BC) is highlighted. Of course, the assumed boundary and initial conditions, as well as the particular weathering profile adopted in this case, may have a strong impact on the computed result from a quantitative point of view. In a complete study of this complex problem, the advection/diffusion processes which control the concentration fields for the chemical agents responsible for rock degradation, and the nature of their reactions with the solid phase should be properly taken into account in the balance equations. In this respect, this preliminary investigation can be seen, at least, as a springboard for a more complete study to be conducted in the future.

3.3. Weathering effects in abandoned mines

The last example considered shows an application of the proposed model for describing the effects of the debonding process in abandoned underground mines. In Lorraine (France) this phenomenon has led to the collapse of the pillars in some abandoned mines which has caused a subsidence at ground surface of more than one meter (Homand et al., 2001), with associated damage or even collapse in adjacent structures, and, in some cases the appearance of sinkholes on the ground surface. This process is schèmatised in Figure 17.

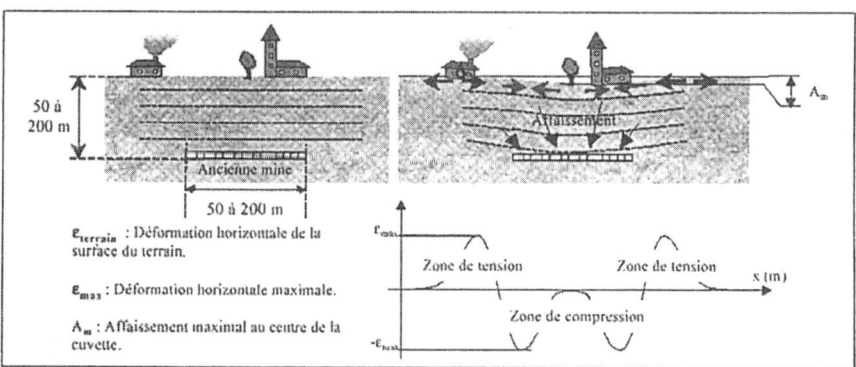

Figure 17. Schematic process of mine collapses in Lorraine (France) after Deck et al. (2001)

The collapse of the pillars occurs mainly after the flooding of the underground cavities, after the end of the mining works. As discussed by Homand et al. (2001), the cause of pillar instability is a rapid decay of the mechanical properties of the pillar rock, due to chemical degradation. This degradation process is associated with the change in the hydrological conditions of the subsoil, which promote the activity of some bacteria, responsible for the changes in the mineralogical composition of the rock mass (Grgic et al., 2001).

Figure 18. *a) Building failure and sinkhole induced by the mining collapse in Auboue (Lorraine, France)*

The initiation of pillar collapse is considered here as the result of a progressive debonding of the pillar rock, caused by the weathering process. In order to define a representative numerical model for the actual site conditions in Lorraine, an idealised room-and-pillars mining excavation located at 100 meter deep has been considered. The geometry of the problem and the finite element discretisation adopted are shown in Figure 19. The (plane strain) finite element mesh is composed of 616 eight-noded serendipity elements, underintegrated with a uniform 2×2 Gauss quadrature rule. To limit somehow the size of the problem, only a limited portion of the rock mass surrounding the mining works has been included in the FE discretization. In particular, the upper boundary of the mesh corresponds to a horizontal layer located at 67 m below the ground surface. The gravity loads due to the overburden have been simulated by a uniform vertical pressure, as shown in Figure 19. The parameters adopted in the finite element analysis have been extrapolated from triaxial test results reported by Grgic *et al.* (2001), and are given in Table 2. In absence of more detailed experimental data, the internal variables p_s and p_m have been assumed constant with depth and equal to 3000 kPa and 20000 kPa, respectively.

α	k	G_0	p_r	ρ_s	ξ_s	ρ_m	ξ_m	k
(–)	(–)	(kPa)	(kPa)	(–)	(–)	(–)	(–)	(–)
0.0	22.7	5.0e+4	1.0e+6	2.52	10.	4.0	2.0	0.05
a_g	m_g	M_{gc}	M_{ge}	a_f	M_f	M_{fc}	M_{fe}	r
(–)	(–)	(–)	(–)	(–)	(–)	(–)	(–)	(–)
3.0e-2	1.03	2.22	2.15	1.7	0.99	0.90	0.78	2.0

Table 2. *Material parameters for the rock mass*

In the first step of the numerical simulation, the gravity load ($\gamma = 20$ kN/m^3) and the vertical pressure $Q = 1520$ kPa equivalent to the overburden have been applied. For the assumed initial conditions, no plastic yielding occurred in this stage.

In the second step, the degradation process occurring after the end of the mining activity is simulated by prescribing a linear increase of the weathering parameter up to $X_d = 1.0$ (complete debonding) for the element group W1 in Figure 19, and up to $X_d = 0.5$ for the element group W2 (partial debonding). The assumed evolution of the weathering parameter with (fictitious) time for the two element groups is also shown in Figure 19.

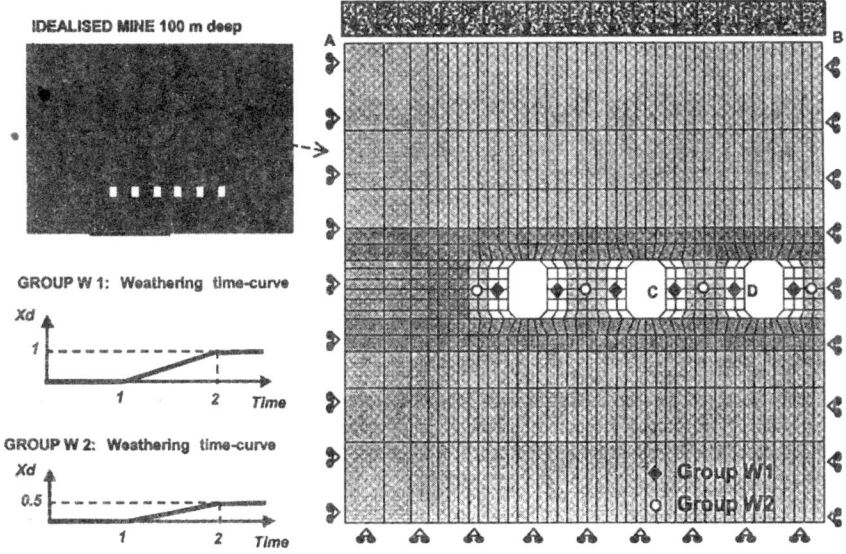

Figure 19. *Boundary value problem: FEM mesh, loading and weathering profile*

The response of the rock mass to the imposed degradation process is given in Figures 20-23. Figure 20 shows the vertical displacements computed along the boundary AB (top of discretized zone). After the degradation of the pillars, an additional settlement is observed at constant external load; with the highest value (35 cm) at point B, on the axis of symmetry. Since at the end of the simulation none of the pillars has yet collapsed, the estimated max. subsidence is small as compared to the figures reported in actual case-histories. However, the predicted trend of behavior is qualitatively similar to the one observed in the field.

As in the previous cases, the additional strain and displacement fields are a consequence of the plastic yielding which occurs in the external portions of the pillars, as a consequence of the reduction in strength caused by weathering. The mechanical degradation of the rock mass is associated with both the developments of plastic strains, and to a stress redistribution in the zones of the rock mass where the state of stress is still inside the yield locus.

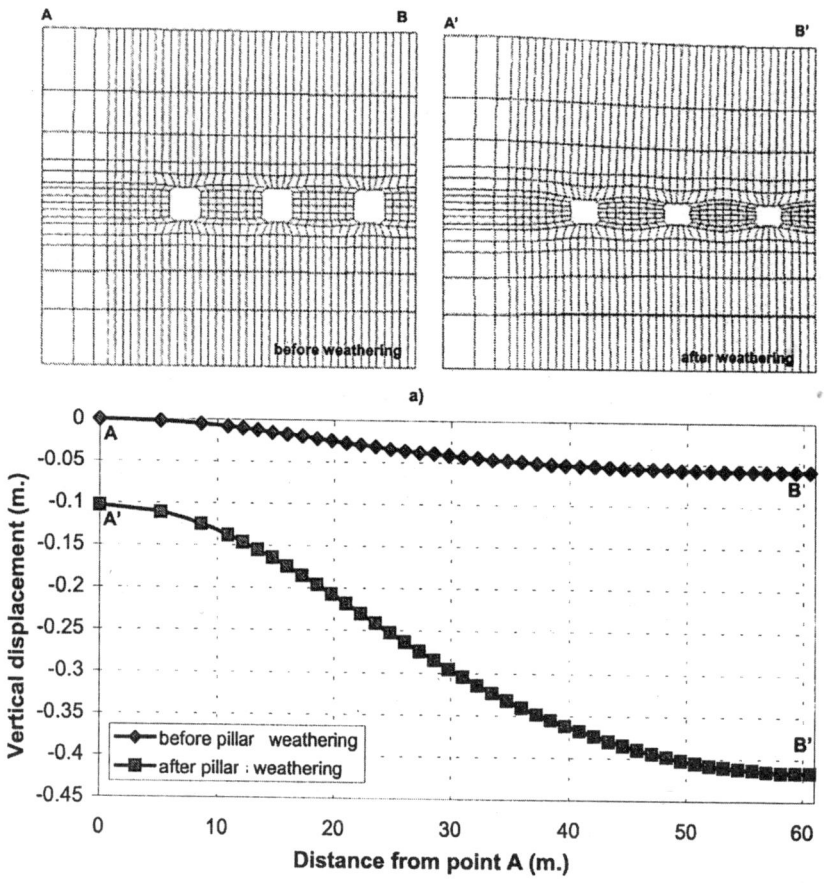

Figure 20. *a) deformed mesh before and after pillar weathering (magnification factor equal to 5); b) Vertical displacement of the boundary AB before and after pillar weathering*

Figure 21 shows the changes in the total strain field induced by weathering in the portion of the rock mass surrounding one of the rooms. It is interesting to note the marked convergence of the pillar boundaries towards the center of the excavation (negative horizontal strains denote extension). This trend indicates the onset of an imminent pillar collapse; in fact by looking at the changes in the vertical stress component induced by weathering — in Figure 22 — a marked reduction in vertical load can be observed in the outermost part of the pillar, while a definite increase in vertical load occurs in the pillar core.

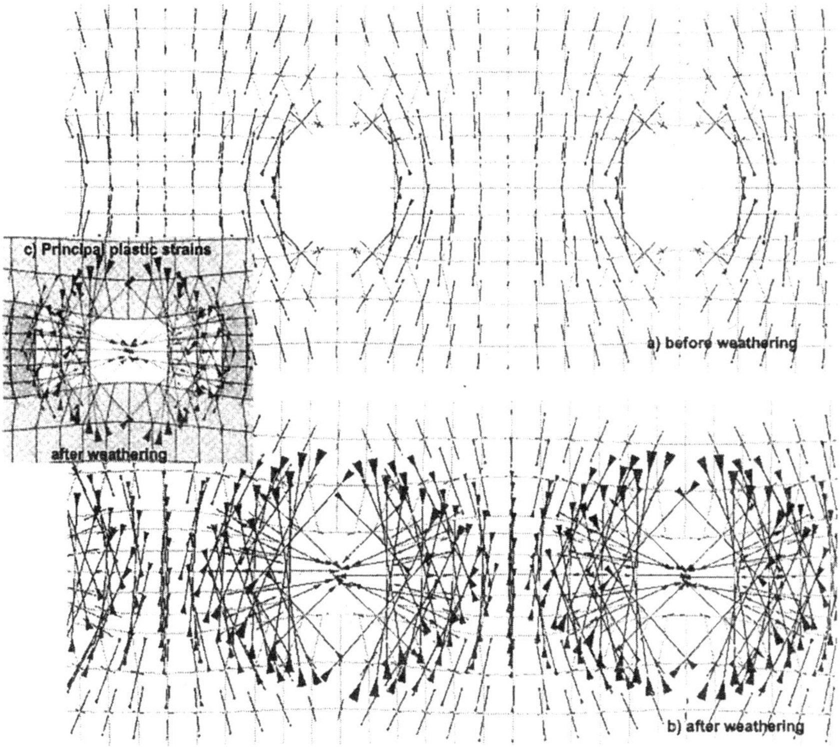

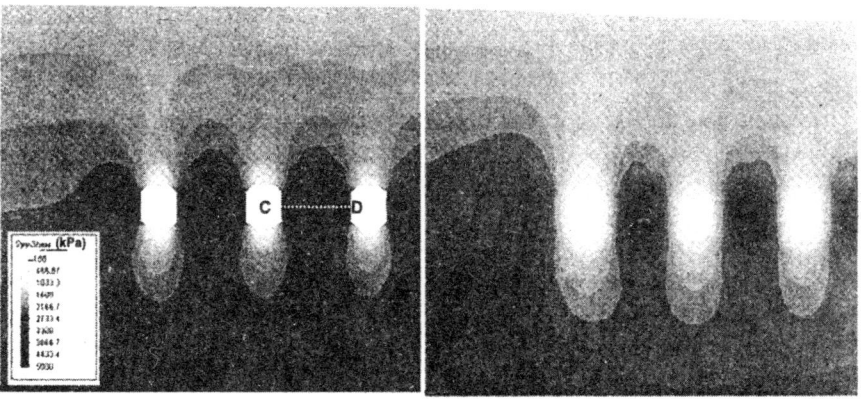

Figure 21. *Principal total strains around a room: a) before weathering; b) after weathering. c) Principal plastic strains around a room at the end of the weathering process (after Carboni, 2002)*

Figure 22. *Vertical stress contours: a) before weathering; b) after weathering (after Carboni, 2002)*

Such a redistribution is apparent from the data reported in Figure 23, where the vertical stress component is plotted along the horizontal line CD (see Figure 22), at different stages of the weathering process.

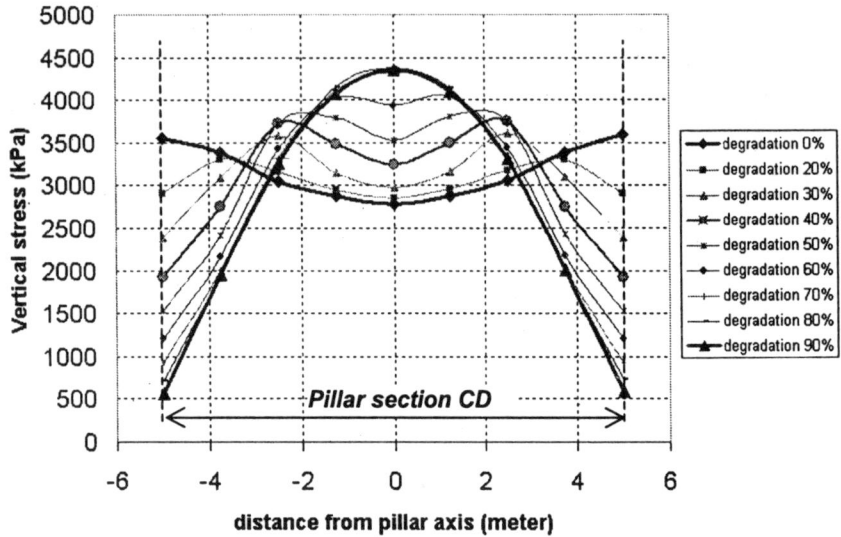

Figure 23. *Weathering induced vertical stress variation in the pillar (line CD)*

4. Conclusions

Despite the fact that the three cases considered represent only preliminary studies, the results demonstrate that the theory of plasticity, suitably extended to account for non-mechanical degradation processes, can be successfully employed to deal with the effects of weathering on engineering structures.

In particular, it is shown that the mechanical response of the rock markedly changes when weathering occurs. Although the external loads remain constant, in all the three boundary value problems considered, additional strains and displacements occur as weathering progresses, as a result of the yielding phenomena associated with the reduction in size of the yield locus. Plastic yielding is also responsible for a marked stress redistribution in the rock mass, from the more weathered regions to adjacent, intact or less weathered ones. It is remarkable that for all the cases considered such variations in the displacement and in the stress fields appear qualitatively similar to those observed in real case-histories.

Acknowledgements

The authors would like to thank Prof. M. Pastor, Dr. P. Mira and Dr. J. A. Fernandez Merodo for permitting to implement the model in the GEHOMADRID code and Mr. R. Carboni for the significant contribution in the last case shown.

5. References

Baynes, F.J., Dearman, W.R. (1978), "The relationship between the microfabric and the engineering properties of weathered granite", *Bulletin of the International Association of Eng. Geol.*, n. 18, 191-197.

Borja, R.I., Tamagnini, C. (1998), "Cam-Clay plasticity, part III: Extension of the infinitesimal model to include finite strains", *Mech. Cohes.-Frict. Mater.*, 155, 73-95.

Carboni, R., (2002), *Modellazione numerica di cedimenti indotti dalla degradazione di miniere abbandonate*, Degree Thesis, Milan University of Technology (Politecnico) (in Italian).

Castellanza, R. (2002), *Weathering effects on the mechanical behaviour of bonded geomaterials: an experimental, theoretical and numerical study*, PhD Thesis, Milan University of Technology (Politecnico).

Chigira, M., Oyama, T. (1999), "Mechanism and effect of chemical weathering of sedimentary rocks", *Engineering Geology*, 55, 3-14.

Clough, G.W., Sitar, N., Bachus, R.C., Rad, N.S. (1981), Cemented sands under static loading, *J. Geot. Eng. Div., ASCE*, 107, n. GT6, 799-817.

Deck, O., Al Heib, M., Homand, F. (2001), Etude des conséquences des affaissements miniers sur le bâti par la modélisation numérique, *Proc. XVe Congrès Français de Mécanique*, 2-5 Nancy, septembre 2001.

Fernandez Merodo, J.A., Mira, P., Pastor, M. and Li, T. (1999), *GeHoMadrid User Manual*, Internal Report, CEDEX, Madrid.

Fookes, P. G., Gourley, C. S., Ohikere, C. (1988), Rock weathering in engineering time, *Quarterly Journal of Engineering Geology*, 21, 33-57

Grgic, D., Homand, F., Hoxha, D. (2001), Instabilités des mines de fer abandonnées de lorraine: approche hydromecanique, *Proc. XVe Congrès Français de Mécanique*, 2-5 Nancy, septembre 2001.

Homand, F., Feuga, B., Kouniali, S., Josien, J.P. (2001), Les instabilités des mines de fer abanonnées de Lorraine, *Proc. XV ^m Congrès Français de Mécanique*, 2-5 Nancy, septembre 2001.

Kimmance, G.C. (1988), Computer aided risk analysis of open pit mine slopes in kaolin mined deposit, Ph.D. Thesis, Univ. of London.

Lagioia, R., Nova, R. (1995), "An experimental and theoretical study of the behavior of a calcarenite in triaxial compression", *Géotechnique*, 45, n. 4, 633-648.

Lagioia, R., Puzrin, A.M., Potts, D.M. (1996), A new versatile expression for yield and plastic potential surfaces, *Comp. & Geotechnics*, 19, 171-191.

Nova, R. (1986), "Soil Models as a Basis for Modelling the Behaviour of Geophysical Materials", *Acta Mechanica*, 64, 31-44.

Nova, R. (1988), "Sinfonietta Classica: an exercise on classical soil modelling", *in Constitutive Equations for Granular Non-Cohesive Soils*, Cleveland, Saada and Bianchini (eds.), Balkema, Rotterdam, 501-519.

Nova, R. (1992), "Mathematical modelling of natural and engineered geomaterials", *Europ. J. Mech. A Solids*, 11 (special issue), 135-154.

Nova, R. (1997), "On the modelling of the mechanical effects of diagenesis and weathering", *ISRM News Journal*, 4, n. 2, 15-20.

Nova, R. (2000), "Modelling the weathering effects on the mechanical behaviour of granite", *in Constitutive Modelling of Granular Materials*, Kolymbas, D. eds, Springer, Berlin, 397-412.

Nova, R., Castellanza, R. (1999), "The effect of rock weathering on the geostatic stress state", *Proc. J.-P. Boehler Memorial Symposium*, F. Darve, B. Loret eds., Grenoble, France, p. 79-84.

Nova, R., Castellanza, R. (2001), "Modelling weathering effects on the mechanical behaviour of soft rocks", *Proc. Int. Conf. on Civil Engineering* (ICCE 2001) Bangalore, India, Interline Publishing, p. 157-167.

Park, H.D. (1996), "Assessment of the geotechnical properties of the weathered rocks at historical monuments in Korea", *Eurock '96*, 1413-1416.

Petruckhin, V.P. (1993), *Construction of structures on saline soils*. Oxford & IBH, New Delhi.

Tamagnini, C., Castellanza, R., Nova, R. (2003), "A Generalized Backward Euler algorithm for the numerical integration of an isotropic hardening elastoplastic model for mechanical and chemical degradation of bonded geomaterials", accepted for publication, *Int. J. Num. Anal. Meth. Geomech.*

Yokota, S., Iwamatsu, A. (1999), "Weathering distribution in a steep slope of soft pyroclastic rocks as an indicator of slope instability", *Eng. Geol.*, 55, 57-68.

Chapter 16

Finite Element Modelling of Landslides

José Antonio Fernandez Merodo and Manuel Pastor
CEDEX, CETA, Ing. Computacional, Madrid, Spain

1. Introduction

Slope stability analysis has always been a problem of great importance in geotechnical engineering either in the form of limit load analysis or, alternatively in the framework of safety factor computation. Traditionally this type of analysis has been performed using a set of numerical techniques based on limit analysis, which in a first stage were conceived and applied as hand calculations and at a later stage were implemented in computer codes. However, these techniques, maybe due to their original conception as hand computations, assume a series of important simplifications in the mechanical model which may be applied to a great number of typical cases in engineering practice but are not applicable to many of the most complex cases.

The finite element method is a very powerful and flexible analysis tool that overcomes the limitations associated with traditional methods, allowing the user to introduce sophisticated constitutive models (classical plasticity, generalized plasticity, etc.) and complicated geometrical configurations and boundary conditions (contact problems, rainfall as a loading option, etc.). Besides, the finite element method allows coupled formulations which introduce the effect of fluid pore pressure in a very general fashion, unlike the traditional techniques which introduce this effect in a very simplified way. Slope failure occurs naturally through the zones in which the shear strength of the soil is insufficient to resist the shear stresses.

Several mechanisms are present and follow one another during a landslide. It is difficult to classify landslides according to the size, the shape or the kind of mass movement. This movement is induced by the gravity force and it is caused by changes in the effective stresses, variation of material properties or changes in the geometry. Changes of effective stresses can be induced either directly, as consequence of variation of the external forces (earthquakes, human action), or indirectly through pore pressures (rainfall effects). Variations in material properties can be caused by processes of degradation (weathering and chemical attack). Finally, geometry can change because of natural causes (erosion) or human action (excavation, construction, reshaping...).

It is possible to subdivide the whole of mass movement into three major classes [BRO 86][DIK 96]: slides, flows and falls. The major differences between these three types are in the way in which the movement takes place. In a slide, the moving material remains largely in contact with the parent or underlying rocks during the movement, which takes place along a discrete boundary shear surface. The term flow is used when the material becomes disaggregated and can move without the concentration of displacement at the boundary shear. Lastly, falls normally take place from steep faces in soil or rocks, and involve immediate separation of the tailling material from the parent rock or soil mass with movement involving only infrequent or intermittent contact thereafter. This kind of movement will not be studied here because a discrete approximation seems to be a more suitable method for the analysis.

In some cases, the failure mechanism consists of a clearly defined surface where shear strain concentrates. From a mathematical point of view, the inception of the shear band is characterized by a discontinuity in the strain field, which can evolve towards a discontinuity in the displacements at a later stage. Much effort has been devoted during the past years to better understand this phenomenon. The problem is ill posed for elastoplastic materials, and the results obtained using numerical models depend on the mesh size and alignment. The interested reader is addressed to the text [VAR 95] where a detailed explanation is provided. Depending on the mechanical behaviour of the material, a local collapse can be produced, with an important increase in pore pressures and a corresponding decrease in the effective stresses which can lead to liquefaction on the failure surface [SAS 00].

In other cases, the failure mechanism involves a much larger mass of soil, as in the case of the liquefaction of a sand layer induced by an earthquake. This mechanism of failure can be referred to as "diffuse", and it is characteristic of soils presenting very loose or metastable structures with a strong tendency to compact under shearing. One paramount feature is that effective stresses approach zero, and the material behaves like a viscous fluid in which buildings can sink, as happened during the 1966 earthquake of Niigata in Japan. When this failure mode takes place in a slope, the mass of mobilized soil can propagate downhill, evolving into flow slides or mudflows.

Generally a landslide can be split into an initiation phase, when the material arrives at a non equilibrium state or failure and into a propagation phase, when the movement of material takes place. The purpose of the paper is to describe the initiation mechanism using the models which are more suitable. After describing the model for the coupling of pore fluid and the solid structure, we will propose a discretization technique based on the finite element method. Then we will concentrate on the numerical difficulties of obtaining the failure conditions and mechanisms, providing some solutions. We will present two application cases of landslides representing two kinds of failures:

– localised failure of a slope due to heavy rain

– diffuse failure of a slope due to an earthquake.

2. The mathematical model

Geomaterials (soil, rocks, concrete and other similar materials) are porous materials with voids which can be filled with water, air and other fluids. They are, therefore, multiphase materials, exhibiting a mechanical behaviour governed by the coupling between all the phases. Pore pressure plays a paramount role in the behaviour of a soil structure and, indeed, its variations can induce failure.

The first mathematical model describing this coupling was proposed by Biot [BIO 41, BIO 55] for linear elastic materials. This work was followed by further development at Swansea University, where Zienkiewicz and coworkers [ZIE 80, ZIE 84, ZIE 90a, ZIE 90b, ZIE 00] extended the theory to non-linear materials and large deformation problems. The mathematical model of solid skeleton and pore fluid interaction which will be recalled here is that proposed by Zienkiewicz et al. and described in [ZIE 00].

In what follows we will assume that the voids are filled with air and water, and introduce S_w and S_a, the degrees of saturation of water and air, defined as the volume percent of each phase. It is possible to define an intersticial pressure $\bar{p}$ as $\bar{p} = S_w p_w + S_a p_a$, where p_w and p_a are the pressures of the water and air respectively.

2.1. The effective stress tensor

The effective stress σ' is defined from

$$d\sigma' = d\sigma + \mathbf{m} d\bar{p} \tag{1}$$

where $\mathbf{m}$ is the vector $\mathbf{m} = [1, 1, 1, 0, 0, 0]^T$. We have used positive values of the stress for tension and negative for extension.

We will assume also that the air is at atmospheric pressure, i.e., $p_a = 0$ and, therefore, $\bar{p} = S_w p_w$. This approach is valid for many engineering cases of interest. A more complete description can be found in [LEW 98].

To account for the compressibility of the solid grains a modification can be introduced, by defining a "real effective stress"

$$d\sigma'' = d\sigma + \alpha \mathbf{m} d\bar{p} \tag{2}$$

where α is a corrective coefficient.

The effective stress causes all the significant deformation of the solid skeleton. The constitutive relationship may be written as

$$d\sigma' = \mathbf{D}_{ep} \left(d\varepsilon - d\varepsilon^p - d\varepsilon^T - d\varepsilon^0 \right) \tag{3}$$

where the matrix $\mathbf{D}_{ep}$ is the elasto-plastic matrix which in general depends on σ', its history and the direction of stress increment, $d\varepsilon$ represents the local strain of the solid, $d\varepsilon^p$ represents the volumetric strain due to the uniform compression of the solid grains

$$d\varepsilon^p = -\mathbf{m} \frac{d\bar{p}}{3K_s} \tag{4}$$

K_s is the bulk modulus of the solid grains, $d\varepsilon^T$ represents the thermal strain

$$d\varepsilon^T = \mathbf{m} \frac{\beta_s}{3} dT \tag{5}$$

T is the temperature and β_s is the thermal expansion coefficient of the solid phase, $d\varepsilon^0$ represents the other 'initial' strains and they are not considered here.

We introduce now the 'real' or 'modified' effective stress σ'' to account for the deformation of solid grains. This allows us to determine Biot's constant α.

Substituting equation (4) into equation (3) we obtain:

$$d\sigma' = \mathbf{D}_{ep} \left(d\varepsilon - d\varepsilon^T \right) + \mathbf{D}_{ep} \mathbf{m} \frac{d\bar{p}}{3K_s} \tag{6}$$

Applying equation (1) to equation (6) leads to:

$$d\sigma = \mathbf{D}_{ep}\left(d\varepsilon - d\varepsilon^T\right) - \left(\mathbf{m} - \mathbf{D}_{ep}\mathbf{m}\frac{1}{3K_s}\right)d\overline{p} \tag{7}$$

or

$$d\sigma = d\sigma'' - \left(\mathbf{m} - \mathbf{D}_{ep}\mathbf{m}\frac{1}{3K_s}\right)d\overline{p} \tag{8}$$

where

$$d\sigma'' = \mathbf{D}_{ep}\left(d\varepsilon - d\varepsilon^T\right) \tag{9}$$

in which the constitutive relation is defined.

We can rewrite equation (8) as:

$$d\sigma = d\sigma'' - \left(1 - \frac{\mathbf{m}^T\mathbf{D}_{ep}\mathbf{m}}{9K_s}\right)\mathbf{m}d\overline{p} \tag{10}$$

or

$$d\sigma = d\sigma'' - \alpha\mathbf{m}d\overline{p} \tag{11}$$

where in the case of an elastic isotropic material

$$\alpha = 1 - \frac{\mathbf{m}^T\mathbf{D}_{ep}\mathbf{m}}{9K_s} \approx 1 - \frac{K_t}{K_s} \tag{12}$$

K_t is the bulk modulus of the porous medium.

For most soils α is very close to unity and the difference between the stress σ'' and the effective stress σ' disappear. For rock or concrete, however, α can be in the range of 0.4-0.6 and in such cases it is important to take the effect of solid grain compressibility into account.

2.2. Kinematics

We will describe the displacement of soil skeleton by the vector field $\mathbf{u}(\mathbf{x}, t)$ and the displacement of the intersticial water by $\mathbf{U}(\mathbf{x}, t)$, which will be decomposed into two parts, corresponding to the displacements of the soil skeleton and the relative movement with respect to it as:

$$\dot{\mathbf{U}} = \dot{\mathbf{u}} + \frac{\dot{\mathbf{w}}}{n} \tag{13}$$

where n is the material porosity. Above, $\dot{\mathbf{w}}$ is a fictitious velocity which provides the same debit when we consider flow over the total section (and not only the pores).

2.3. Balance of mass

The purpose of this section is to relate the change of volume of an infinitesimal element of the mixture to the changes of volume of the constituents and the net balance of fluid flowing through it.

We will characterize the rates of volume change caused by:

(i) deformation of soil skeleton $\dot{\theta}_1$,

(ii) volumetric deformation of solid particles caused by changes in pore pressure $\dot{\theta}_2$,

(iii) volumetric deformation of pore fluid caused by changes in pore pressure $\dot{\theta}_3$,

(iv) deformation of solid particles caused by changes in effective stresses $\dot{\theta}_4$ and

(v) water storage induced by change of degree of saturation $\dot{\theta}_5$.

The first term is given by:

$$\dot{\theta}_1 = S_r \text{tr}\,(\dot{\varepsilon}) \tag{14}$$

The second can be written as:

$$\dot{\theta}_2 = \frac{1-n}{K_s}\dot{\bar{p}} = \frac{1-n}{K_s}\left\{ S_w + p_w \frac{C_s}{n} \right\} \dot{p}_w \tag{15}$$

where we have introduced C_s, the specific storage coefficient, as

$$C_s = n\frac{\partial S_w}{\partial t} \tag{16}$$

To derive the above equation, we have used the following relationships:

$$\frac{\partial \bar{p}}{\partial t} = \frac{\partial}{\partial t}\,(S_w p_w) = S_w \frac{\partial p_w}{\partial t} + p_w \frac{\partial S_w}{\partial t} \tag{17}$$

$$\frac{\partial S_w}{\partial t} = \frac{\partial S_w}{\partial p_w}\frac{\partial p_w}{\partial t} = \frac{C_s}{n}\frac{\partial p_w}{\partial t}$$

The volume change due to fluid compressibility is given by:

$$\dot{\theta}_3 = n\frac{S_w}{K_w}\frac{\partial p_w}{\partial t} \tag{18}$$

where K_w is the volumetric stiffness of the fluid.

The fourth term, $\dot{\theta}_4$, accounts for the effect of the effective stress on the solid particles:

$$\dot{\theta}_4 = -\frac{S_w}{3K_s}\text{tr}\,(\dot{\sigma}') \tag{19}$$

If we substitute in the above the value of $\dot{\sigma}'$ given by (6), we obtain:

$$\dot{\theta}_4 = -\frac{S_w}{3K_s}\text{tr}\left\{ \mathbf{D}_{ep} : \dot{\varepsilon} - \alpha \mathbf{m} d\dot{\bar{p}} \right\} \tag{20}$$

Making use of (17) and of the definition of α we finally arrive at:

$$\dot{\theta}_4 = -S_w (1 - \alpha) \operatorname{tr}(\dot{\varepsilon}) - \frac{S_w}{K_s} (1 - \alpha) \left\{ S_w + p_w \frac{C_s}{n} \right\} \frac{\partial p_w}{\partial t} \tag{21}$$

Finally, the term $\dot{\theta}_5$ is

$$\dot{\theta}_5 = n \frac{\partial S_w}{\partial t} = C_s \frac{\partial p_w}{\partial t} \tag{22}$$

The sum of all five contributions should be equal to the rate of fluid entering (or exiting) the domain:

$$-\nabla^T \dot{w} = \dot{\theta}_1 + \dot{\theta}_2 + \dot{\theta}_3 + \dot{\theta}_4 + \dot{\theta}_5 \tag{23}$$

from where

$$\nabla^T \dot{w} + S_w \alpha \operatorname{tr}(\dot{\varepsilon}) + C_s \dot{p}_w + \frac{1}{Q} \dot{p}_w = 0 \tag{24}$$

where we have introduced the mixed volumetric stiffness Q as

$$\frac{1}{Q} = \left\{ n \frac{S_w}{K_w} + \frac{\alpha - n}{K_s} \left(S_w + p_w \frac{C_s}{n} \right) \right\} \tag{25}$$

2.4. Balance of linear momentum

The balance of linear momentum for the mixture can be written as:

$$S^T \sigma + \rho b - \rho \, \ddot{u} - n \rho_w S_w \frac{D}{Dt} \left(\frac{\dot{w}}{n S_w} \right) = 0 \tag{26}$$

where the first three terms account for the joint movement of the solid and fluid phases and the last for the relative displacement of the fluid with respect to the solid. The transpose of the strain operator S coincides with the vector representation of the divergence operator:

$$S = \begin{bmatrix} \dfrac{\partial}{\partial x} & 0 & 0 \\[2mm] 0 & \dfrac{\partial}{\partial y} & 0 \\[2mm] 0 & 0 & \dfrac{\partial}{\partial z} \\[2mm] \dfrac{\partial}{\partial y} & \dfrac{\partial}{\partial x} & 0 \\[2mm] 0 & \dfrac{\partial}{\partial z} & \dfrac{\partial}{\partial y} \\[2mm] \dfrac{\partial}{\partial z} & 0 & \dfrac{\partial}{\partial x} \end{bmatrix} \tag{27}$$

where we have represented the stress tensor in vector form:

$$\sigma = \left(\sigma_{xx}, \sigma_{yy}, \sigma_{zz}, \sigma_{xy}, \sigma_{yz}, \sigma_{zx} \right)^T \tag{28}$$

In the above, $\mathbf{b}$ are the body forces, ρ_m the density of the mixture and $\ddot{\mathbf{u}}$ the acceleration of the solid skeleton.

The balance of momentum of the fluid phase is given by the expression

$$-\nabla p_w + \rho_w \mathbf{b} - k_w^{-1}\dot{\mathbf{w}} - \rho_w \left[\ddot{\mathbf{u}} + \frac{D}{Dt}\left(\frac{\dot{\mathbf{w}}}{nS_w}\right)\right] = 0 \qquad (29)$$

2.5. A note on permeability and degree of saturation

In the proposed approach we have assumed that the air is always at atmospheric pressure. Concerning the pore pressure and the degree of saturation, it can be assumed that they are related by a given empirical law. Further simplification can be made by assuming that the evolution of both variables does not present hysteresis effects. Still, it is important to recall that degree of saturation will depend on pore pressure.

Concerning the permeability tensor, it will also depend on the degree of saturation and will change also with the volumetric strains induced in the soil.

2.6. The u-p_w formulation

Under certain conditions [ZIE 80], we can neglect the relative accelerations of the fluid phase and eliminate the relative displacements of the fluid phase w. This can be done by substituting the value of $\nabla^T\dot{\mathbf{w}}$ obtained from (29)

$$\nabla^T\dot{\mathbf{w}} = \nabla^T\left[k_w\left(-\nabla p_w + \rho_w \mathbf{b} - \rho_w \ddot{\mathbf{u}}\right)\right] \qquad (30)$$

into the balance of mass of the fluid phase (24)

$$\nabla^T\dot{\mathbf{w}} + S_w\alpha\,\mathrm{tr}\,(\dot{\varepsilon}) + C_s\dot{p}_w + \frac{1}{Q}\dot{p}_w = 0 \qquad (31)$$

from where we obtain

$$\nabla^T\left\{k_w\left(-\nabla p_w + \rho_w \mathbf{b} - \rho_w \ddot{\mathbf{u}}\right)\right\} + \left(C_s + \frac{1}{Q}\right)\dot{p}_w + S_w\alpha\,\mathrm{tr}\,(\dot{\varepsilon}) = 0 \qquad (32)$$

The main advantage of this formulation is that the field variables reduce to two, displacements $\mathbf{u}$ and pressures p_w.

Therefore, the basic equations of this $\mathbf{u}-p_w$ formulation are (26) and (32), together with the constitutive equation (11) and the kinematic relations between strain and displacement.

2.7. Boundary and initial conditions

The equations presented so far have to be complemented by suitable boundary and initial conditions for the problem variables. In the $u - p_w$ the conditions are the following:

(i) u prescribed on Γ_u;

(ii) tractions prescribed on Γ_t, $\sigma.n = \bar{t}$, $\Gamma_u \cup \Gamma_t = \Gamma$, $\Gamma_u \cap \Gamma_t = \{\emptyset\}$;

(iii) p_w prescribed on Γ_{pw};

(iv) flux $-k_w \nabla p_w n$ prescribed along Γ_q, $\Gamma_{pw} \cup \Gamma_q = \Gamma$, $\Gamma_{pw} \cap \Gamma_q = \{\emptyset\}$.

In the case of a non-saturated slope under rain, the boundary conditions to be applied can be simplified to a prescribed pressure (atmospheric) on the surface, but care should be taken as the inflow cannot be larger than the amount of percolating water. In this way, the saturation will increase within the material and the effective stresses will decrease.

The initial conditions will be:

(i) solid displacements and velocities at $t = 0$,

(ii) pore pressures at $t = 0$.

2.8. A note on the incompressible undrained limit

The $u - p_w$ version of Biot equations can be further simplified according to whether accelerations are small, leading to what is known as "consolidation", or "slow consolidation phenomena", which in the case of saturated materials are:

$$S^T \sigma + \rho b = 0 \tag{33}$$

and

$$\nabla^T \{k_w (-\nabla p_w + \rho_w b)\} + \frac{1}{Q}\dot{p}_w + \text{tr}(\dot{\varepsilon}) = 0 \tag{34}$$

If we now consider a material of very small permeability and very large volumetric stiffness, the above equations can be written as:

$$S^T (\sigma' + m p_w) + \rho b = 0$$
$$\nabla^T \dot{u} = 0$$

which are similar to the equations found in solid mechanics for incompressible materials. Here, important difficulties regarding both the interpolation spaces which can be used for displacements and pressures and simulation of failure processes are found, and will be discussed later.

3. The numerical model

3.1. Discretization of the u-p_w model

The mathematical model consists on the following system of PDE's:

$$\mathbf{S}^T \boldsymbol{\sigma} + \rho \mathbf{b} - \rho \ddot{\mathbf{u}} - n\rho_w S_w \frac{D}{Dt}\left(\frac{\dot{\mathbf{w}}}{nS_w}\right) = 0 \tag{35}$$

$$\boldsymbol{\nabla}^T \{k_w(-\boldsymbol{\nabla}p_w + \rho_w\mathbf{b} - \rho_w\ddot{\mathbf{u}})\} + \left(C_s + \frac{1}{Q}\right)\dot{p}_w + S_w\alpha\,\mathrm{tr}(\dot{\boldsymbol{\varepsilon}}) = 0 \tag{36}$$

or

$$S_w\alpha\,\mathrm{tr}(\dot{\boldsymbol{\varepsilon}}) - \boldsymbol{\nabla}^T(k_w\boldsymbol{\nabla}p_w) + \left(\frac{1}{Q_*}\right)\dot{p}_w + \boldsymbol{\nabla}^T(k_w\rho_w\mathbf{b}) - \boldsymbol{\nabla}^T(k_w\rho_w\ddot{\mathbf{u}}) = 0 \tag{37}$$

where

$$\frac{1}{Q_*} = C_s + \frac{1}{Q}$$

The system of partial differential equations can be discretized using standard Galerkin techniques. After approximating the fields $\mathbf{u}$ and p_w as: $\mathbf{u}(\mathbf{x},t) = \mathbf{N}_u(\mathbf{x})\cdot\mathbf{U}(t)$, $p_w(\mathbf{x},t) = \mathbf{N}_p(\mathbf{x})\cdot\mathbf{P}_w(t)$, it results in two ordinary differential equations:

$$\mathbf{M}\ddot{\mathbf{U}} + \int_\Omega \mathbf{B}^T\sigma''d\Omega - \mathbf{Q}\mathbf{P}_w = \mathbf{f}^u \tag{38}$$

$$\mathbf{Q}^T\dot{\mathbf{U}} + \mathbf{H}\mathbf{P}_w + \mathbf{S}\dot{\mathbf{P}}_w = \mathbf{f}^p \tag{39}$$

where $\mathbf{M}$ is the mass matrix :

$$\mathbf{M} = \int_\Omega \mathbf{N}_u^T\rho\mathbf{N}_u d\Omega \tag{40}$$

$\mathbf{Q}$ is the coupling matrix :

$$\mathbf{Q} = \int_\Omega \mathbf{B}^T\alpha S_w\mathbf{m}\mathbf{N}_p d\Omega \tag{41}$$

$\mathbf{H}$ is the permeability matrix :

$$\mathbf{H} = \int_\Omega (\nabla\mathbf{N}_p)^T\mathbf{k}(\nabla\mathbf{N}_p)d\Omega \tag{42}$$

$\mathbf{S}$ is the compressibility matrix :

$$\mathbf{S} = \int_\Omega \mathbf{N}_p^T\frac{1}{Q_*}\mathbf{N}_p d\Omega \tag{43}$$

and

$$\mathbf{f}^u = \int_\Omega \mathbf{N}_u^T \rho \mathbf{b} d\Omega + \int_\Gamma \mathbf{N}_u^T \mathbf{t} d\Gamma \qquad (44)$$

$$\mathbf{f}^p = \int_\Omega (\nabla \mathbf{N}_p)^T k\rho_w \mathbf{b} d\Omega - \int_\Gamma \mathbf{N}_p^T q d\Gamma \qquad (45)$$

The time derivatives of $\mathbf{U}$ and $\mathbf{P}_w$ are approximated in a typical step of computation using the Generalized Newmark GN22 scheme for displacements and a GN11 for the water pressure [KAT 85, ZIE 91].

If we introduce the following notation [KAT 85]:

$$\dot{\mathbf{U}}_{n+1} = \dot{\mathbf{U}}_n + \Delta t \ddot{\mathbf{U}}_n + \beta_1 \Delta t \Delta \ddot{\mathbf{U}}_n \qquad (46)$$

$$\mathbf{U}_{n+1} = \mathbf{U}_n + \Delta t \dot{\mathbf{U}}_n + \frac{1}{2}\Delta t_n^2 \ddot{\mathbf{U}} + \frac{1}{2}\beta_2 \Delta t^2 \Delta \ddot{\mathbf{U}}_n \qquad (47)$$

where $\Delta \ddot{\mathbf{U}}_n = \ddot{\mathbf{U}}_{n+1} - \ddot{\mathbf{U}}_n$ and

$$\mathbf{P}_{n+1} = \mathbf{P}_n + \Delta t \dot{\mathbf{P}}_n + \theta \Delta t \Delta \dot{\mathbf{P}}_n \qquad (48)$$

where $\Delta \dot{\mathbf{P}}_n = \dot{\mathbf{P}}_{n+1} - \dot{\mathbf{P}}_n$

we obtain the discretized system of equation in which only $\Delta \ddot{\mathbf{U}}_n$ and $\Delta \dot{\mathbf{P}}_n$ remain as unknowns:

$$\mathbf{G}_{n+1}^u = \mathbf{M}\Delta \ddot{\mathbf{U}}_n + \left(\int_\Omega \mathbf{B}^T \sigma'' d\Omega \right)_{n+1} - \mathbf{Q}\theta \Delta t \Delta \dot{\mathbf{P}}_n - \mathbf{F}_{n+1}^u = 0 \quad (49)$$

$$\mathbf{G}_{n+1}^p = \mathbf{Q}^T \beta_1 \Delta t \Delta \ddot{\mathbf{U}}_n + \mathbf{H}\theta \Delta t \Delta \dot{\mathbf{P}}_n + \mathbf{S}\Delta \dot{\mathbf{P}}_n - \mathbf{F}_{n+1}^p = 0 \qquad (50)$$

where

$$\mathbf{F}_{n+1}^u = \mathbf{f}_{n+1}^u - \mathbf{M}\ddot{\mathbf{U}}_n + \mathbf{Q}(\mathbf{P}_n + \Delta t \dot{\mathbf{P}}_n) \qquad (51)$$

$$\mathbf{F}_{n+1}^p = \mathbf{f}_{n+1}^p - \mathbf{Q}^T(\dot{\mathbf{U}}_n + \Delta t \ddot{\mathbf{U}}_n) - \mathbf{H}(\mathbf{P}_n + \Delta t \dot{\mathbf{P}}_n) - \mathbf{S}\dot{\mathbf{P}}_n \qquad (52)$$

If this system is non-linear, it can be solved by using a Newton-Raphson method with a suitable jacobian matrix:

$$\begin{bmatrix} \frac{\partial \mathbf{G}^u}{\partial \Delta \ddot{\mathbf{U}}} & \frac{\partial \mathbf{G}^u}{\partial \Delta \dot{\mathbf{P}}} \\ \frac{\partial \mathbf{G}^p}{\partial \Delta \ddot{\mathbf{U}}} & \frac{\partial \mathbf{G}^p}{\partial \Delta \dot{\mathbf{P}}} \end{bmatrix}^{(i)} \begin{bmatrix} \delta\left(\Delta \ddot{\mathbf{U}}_n\right) \\ \delta\left(\Delta \dot{\mathbf{P}}_n\right) \end{bmatrix}^{(i+1)} = - \begin{bmatrix} \mathbf{G}^u \\ \mathbf{G}^p \end{bmatrix}^{(i)} \qquad (53)$$

Using equations (49) and (50) we can write the above step as:

$$\begin{bmatrix} \mathbf{M} + \mathbf{K}_T \frac{1}{2}\beta_2 \Delta t^2 & -\mathbf{Q}\theta \Delta t \\ \mathbf{Q}^T \beta_1 \Delta t & \mathbf{H}\theta \Delta t + \mathbf{S} \end{bmatrix}^{(i)} \begin{bmatrix} \delta\left(\Delta \ddot{\mathbf{U}}_n\right) \\ \delta\left(\Delta \dot{\mathbf{P}}_n\right) \end{bmatrix}^{(i+1)} = - \begin{bmatrix} \mathbf{G}^u \\ \mathbf{G}^p \end{bmatrix}^{(i)} \qquad (54)$$

where $\mathbf{K}_T$ is the tangent stiffness matrix: $\mathbf{K}_T = \int \mathbf{B}^T \mathbf{D}_{ep} \mathbf{B} d\Omega$.

3.2. Restrictions on the interpolation spaces for displacements and pressures

The system of equations given in equation (54) becomes in steady state in the undrained-incompressible limit, *i.e.*, with $k_w \to 0$, $Q^* \to \infty$ a system of the type frequently found in constrained problems

$$
\begin{bmatrix} \mathbf{A} & -\mathbf{B} \\ -\mathbf{B}^T & 0 \end{bmatrix} \begin{bmatrix} \Delta\xi \\ \Delta\phi \end{bmatrix} = \begin{bmatrix} \mathbf{f}_\xi \\ \mathbf{f}_\phi \end{bmatrix}. \tag{55}
$$

This is in detail similar to those found in mixed formulations of incompressible solid mechanics and fluid dynamics problems, and it can demonstrated that the system will be singular and present pressure oscillations whenever the number of $\Delta\xi$ variables $n_\xi \leq n_\phi$, *i.e.*, the number of $\Delta\phi$ variables. Although this condition is not sufficient for stability (and solvability) it is necessary and it generally excludes equal orders of interpolation spaces for pressure and displacements.

A complete mathematical treatment of the problem can be found in [BAB 73] and [BRE 74]. However, a much simpler explanation is provided by the patch test for mixed formulations proposed by Zienkiewicz, Taylor and Chan [ZIE 91]. Therefore, the interpolation spaces for displacements and pressures cannot be the same unless stabilization is provided, such as the one proposed by the authors in references [PAS 97, PAS 99, ZIE 98, PAS 99].

4. Applications

The simulations proposed are carried out using a finite element code, GeHoMadrid [FER 01, MIR 01], which implements the Biot-Zienkiewicz coupled $u - p_w$ model and a Generalized Plasticity model able to reproduce sand liquefaction under dynamic conditions.

4.1. Failure of a cut slope under rain action

The first example we will present concerns the progressive failure of a non saturated clay cut slope due to rainfall. The height has been chosen as 10 meters and the slope is 2:1. Initial conditions have been simulated by assuming an initial horizontal soil layer which has been excavated to obtain the cut slope. The finite element mesh can be seen in Figure 1. The interpolation functions are quadratic for displacements and bilinear for pore pressures. A reduced integration rule of 2x2 points has been used.

Concerning boundary conditions, we have used $\partial_n p_w = 0$ and $u.n = 0$, where n is the normal to the artificial boundary. In the surface, we have assumed traction free conditions and a suction of $20\,kPa$.

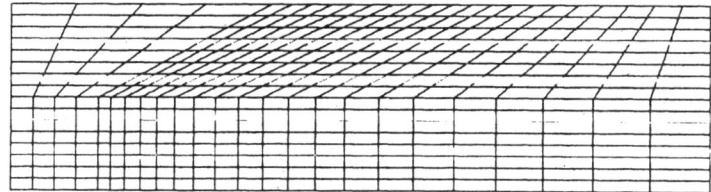

Figure 1. *Finite element mesh*

The material has been assumed elasto-plastic, with a non associated Drucker–Prager yield criterion and a softening law characterized by $\varepsilon^p_{peak} = 0.05$, $\phi_{peak} = 20°$, $c_{peak} = 7$ kPa, $\varepsilon^p_{res} = 0.20$, $\phi_{res} = 13°$ and $c_{res} = 2$ kPa. The Young's modulus varies with the effective confining pressure as

$$E = \min\left(400\ kPa,\ 25\left(p' + 100\right)\right) \qquad \nu = 0.2$$

After checking equilibrium steady state conditions after excavation (Figure 2), the rain has been applied in two simplified ways:

1) Flooding, $p_w = 0$

2) Rain intensity known, flow = 6 mm/day.

In both cases failure is reached. Figure 2 shows the plastic strain contours at different times. The failure occurs naturally and progressively through a localised surface. Figure 3 displays the time evolution of the horizontal displacement of the midslope point.

4.2. Liquefaction failure of a dyke under earthquake action

The case we will consider next is that of an earthquake induced flowslide in very loose saturated sand. The problem consists of a dyke 10 m in height with slope 2:1, founded on a sand layer which extends 10 m in depth and lies on a rigid rockbed. The material of both the dyke and the foundation is a very loose saturated sand.

Initial conditions correspond to geostatic equilibrium under gravity forces. Pore pressure at the surface has been assumed to be equal to -20 kPa. The finite element mesh can be seen in Figure 5 and consists of 500 quadrilaterals with 8 nodes for displacements and 4 for pore pressures. A reduced integration rule has been used in the solid part to avoid locking. The number of nodes is 1611, with 3535 degrees of freedom.

Loading is applied by prescribing horizontal accelerations at the base. We have used the horizontal accelerations of the *NS* component of El Centro earthquake. A simplified absorbing boundary condition has been applied at lateral boundaries. Concerning pore pressures, it has been assumed that no flux occurs at artificial boundaries, and the constant value of -20 kPa has been kept at the surface.

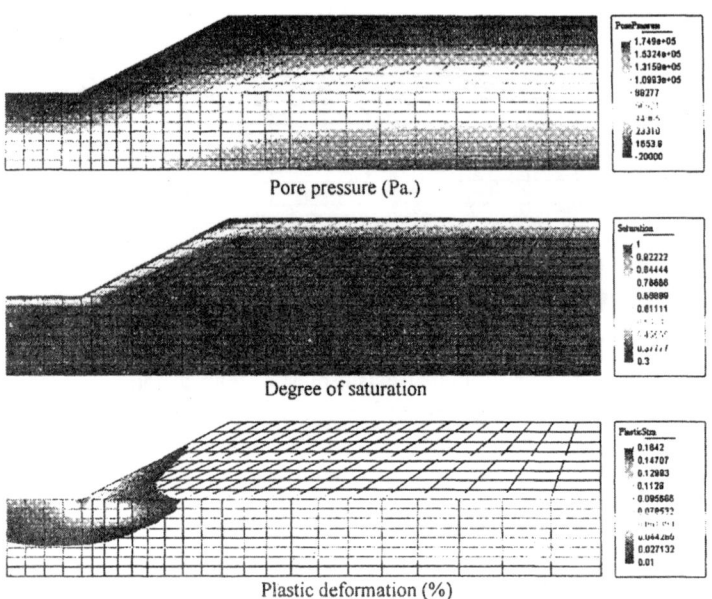

Pore pressure (Pa.)

Degree of saturation

Plastic deformation (%)

Figure 2. *Equilibrium steady state conditions after excavation*

The behaviour of the loose material, $D_R = 27°$, is represented, using the Pastor-Zienkiewicz model for sand. To better understand this behaviour we simulate a loading-unloading-reloading nondrained triaxial test. The material loses progressively its resistant capacity, the effective stress decreases and the pore pressure increases, Figure 6.

The results can be seen in Figures 7, 8, 9, 10, where the contours of pore pressure, plastic strain, p'/p'_0 and displacements are given at different times. Plastic strain accumulates in two zones, with much higher values at the outer. The ratio between the mean effective confining pressure and its initial value p'/p'_0 has been used as an indicator of the extent of the liquefied zones. From these results, it can be concluded that failure of the dyke is caused by liquefaction of the outer liquefied zone.

A note on air liquefaction failure

If the soil is partially saturated, its behaviour depends on the coupling between the solid skeleton and the pore air and water. In the limit case of a dry soil, the air has to flow out from the pores for the material to consolidate, but typical air consolidation times are much smaller than those of the water. Therefore, in practical cases, the role of air pressures is neglected, as the characteristic time of loading is much larger than consolidation time. However, it is possible to imagine situations with much smaller loading times, where coupling between pore air and soil skeleton plays a paramount

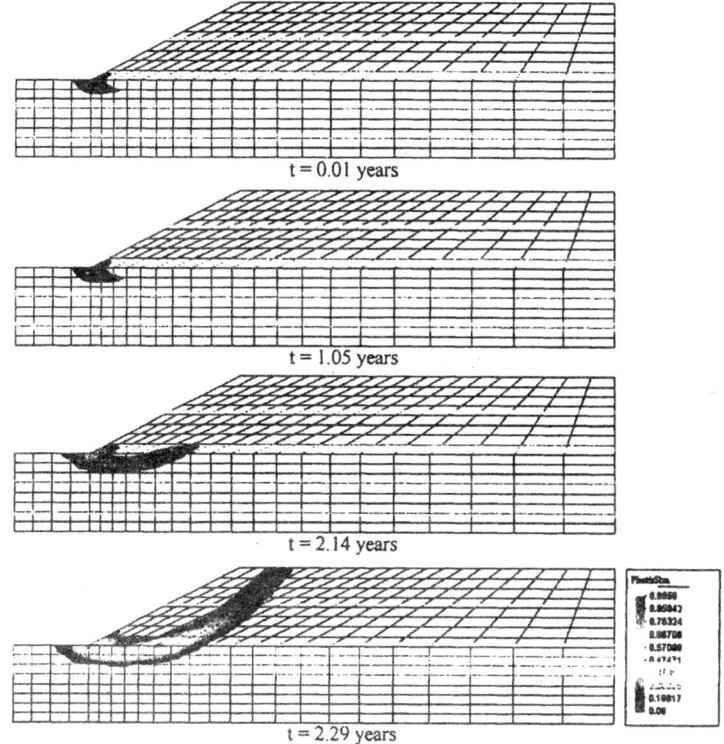

Figure 3. *Plastic strain contour*

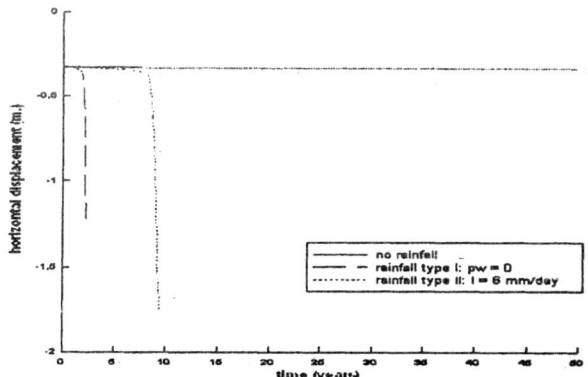

Figure 4. *Horizontal displacement at middle height of slope*

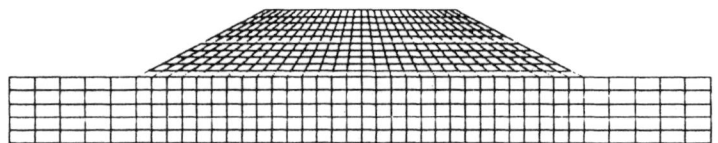

Figure 5. *Finite element mesh*

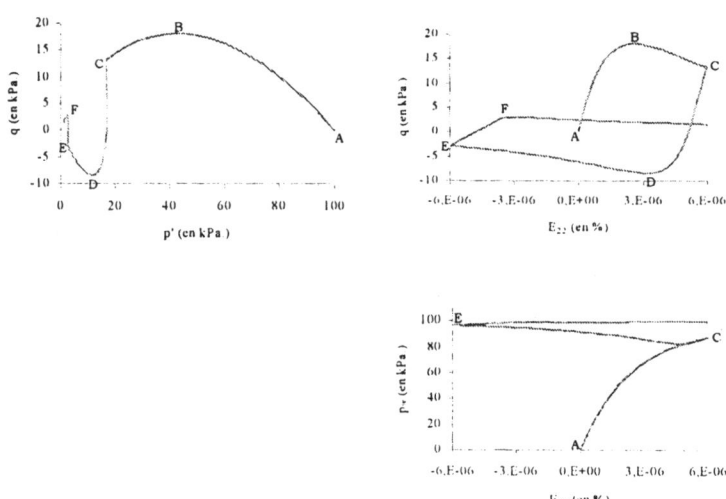

Figure 6. *Nondrained triaxial simulation of the loose sand, loading, unloading, reloading*

role. This is the case of fluidized granular beds, just to mention a particular example of industrial interest.

Bishop (1973) describes the case of Jupille flowslide, which happened in Belgium in February 1961. A tip of uncompacted fly ash located in the upper part of a narrow valley collapsed, and the subsequent flowslide travelled for about 600 m at very high speeds (130 km/h) until it stopped. Triggering mechanism was suggested to be "collapse due to undermining of a steep, partly saturated and slightly cohesive face". Bishop referred to Calembert and Dantinne (1964), who pointed out the role of the entrapped air. The mechanism of pore air consolidation can explain the "sort of fog" which formed above the flowing material, as warmer air met the colder winter air in the exterior. The "honeycomb" like structure of soil in the tip was responsible for its sudden collapse when movement started.

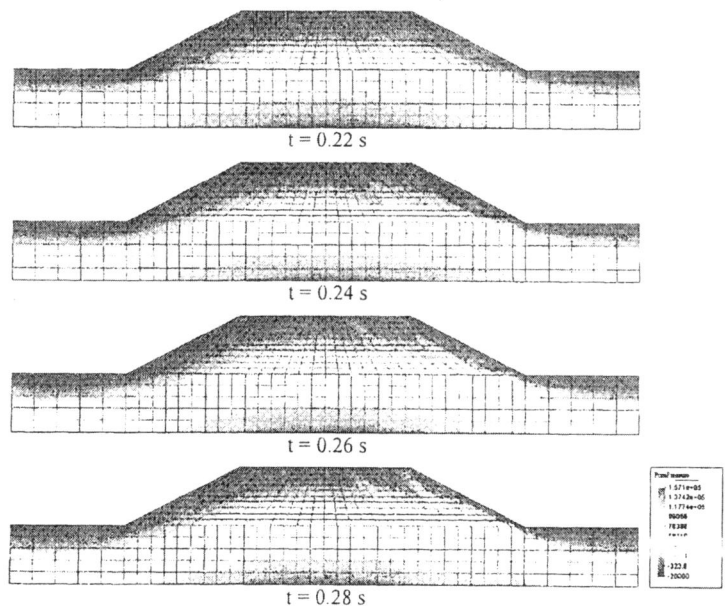

t = 0.22 s

t = 0.24 s

t = 0.26 s

t = 0.28 s

Figure 7. *Pore pressure contour (Pa.)*

5. Conclusions

Landslides are responsible of major damage to lives and properties. We have presented in this paper an approach to reproduce the initiation of landslides. The initiation phase can be modelled using a non linear finite element aproximation of the coupled formulation for the solid skeleton and the pore fluid.

Several failure mechanisms are presented in the examples proposed. A simple Drucker-Prager model incorporating softening is able to trigger the localized failure in the case of the collapse of cut clay slope under rain action. It has to be noted that diffuse failure mechanisms such as occurs when a part of the slope liquefies under earthquake action require more complex constitutive models such as Generalized Plasticity.

The study of the initiation phase is important because it is possible to design the mitigation action if necesary. The final state obtained after this study (volume of soil colapsed, stress and pore pressure state...) can also be used in a following propagation analyse as can be seen in [PAS 02].

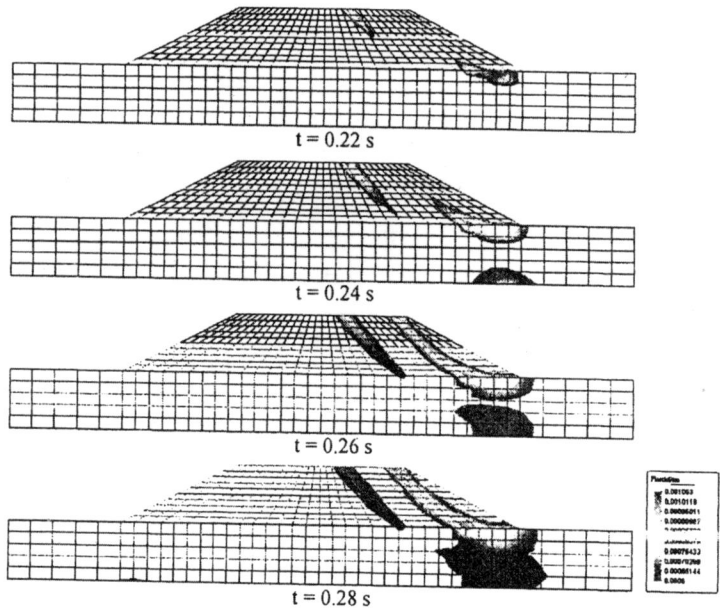

t = 0.22 s

t = 0.24 s

t = 0.26 s

t = 0.28 s

Figure 8. *Plastic deformation contour*

6. References

[BAB 73] I. BABŮSKA: The finite element method with Lagrange multipliers, *Num. Math.* **20**, pp. 179–192, 1973.

[BIO 41] M.A. BIOT: General theory of three-dimensional consolidation, *J. Appl. Phys.* **12**, pp. 155–164, 1941.

[BIO 55] M.A. BIOT: Theory of elasticity and consolidation for a porous anisotropic solid, *J. Appl. Phys.* **26**, pp. 182–185, 1955.

[BRE 74] F. BREZZI: On the existence, uniqueness and approximation of saddle point problems arising from lagrangian multipliers, *RAIRO* **8-R2**, pp. 129–151, 1974.

[BRO 86] Bromhead, E. N. 1986. The stability of slpoes. 2nd edition. Blackie Academic & Professional.

[DIK 96] R. DIKAU, D. BRUNDSEN, L. SCHROTT, M.L. IBSEN: *Landslide Recognition*, John Wiley and Sons, New York, 1996.

[FER 01] J.A. FERNÁNDEZ MERODO: *Une approche à la modélisation des glissements et des effondrements de terrains: initiation et propagation*, Thèse Ecole Centrale Paris, n° 2001-33.

[KAT 85] M.G. KATONA, O.C. ZIENKIEWICZ: A unified set of single-step algorithms. Part 3: the beta-m method, a generalisation of the Newmark scheme, *Int. J. Num. Meth. Eng.* **21**, pp. 1345–1359, 1985.

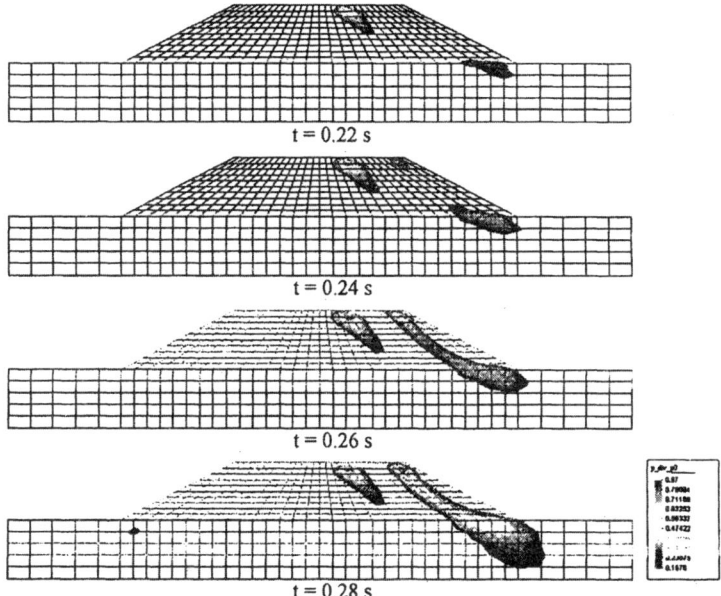

Figure 9. p'/p_0' *contour*

[LEW 98] R.L. LEWIS, B.A. SCHREFLER: *The Finite Element Method in the Static and Dynamic Deformation and Consolidation of Porous Media*, John Wiley and Sons, New York, 1998.

[PAS 97] M. PASTOR, T. LI, J.A. FERNÁNDEZ-MERODO: Stabilized finite elements for harmonic soil dynamics problems near the undrained-incompressible limit, *Soil Dyn. Earthquake Eng.* **16**, pp. 161–171, 1997.

[PAS 99] M. PASTOR, T. LI, X. LIU, O.C. ZIENKIEWICZ: Stabilized low order finite elements for failure and localization problems in undrained soils and foundations, *Comp. Meth. Appl. Mech. Eng.* **174**, pp. 219–234, 1999.

[PAS 99] M. PASTOR, O.C. ZIENKIEWICZ, T. LI, X. LIU, M. HUANG: Stabilized finite elements with equal order of interpolation for soil dynamics problems, *Arch. Comp. Mech.* **6**, pp. 3–33, 1999.

[PAS 02] M. PASTOR: *Flowslides and Debris Flow*, Lecture notes of the ALERT course, RFGC Hermes Science Publications, Paris, 2002.

[MIR 01] P. MIRA: *Análisis por elementos finitos de problemas de rotura de geomateriales*, Tesis Doctoral ETSI. Caminos, Canales y Puertos UPM, 2001.

[SAS 00] SASSA, K. Mechanism of flows in granular soils, pp. 1671–1702, in GeoEng2000, *Int. Conf. on Geotechnical and Geological Engineering*, Melbourne, 19–24 Nov.2000, Technomic Publishing Co.Inc. Lancaster.Basel.

[VAR 95] VARDOULAKIS, I. AND SULEM, J. *Bifurcation Analysis in Geomechanics*, Blakie Academic & Professional, 1995.

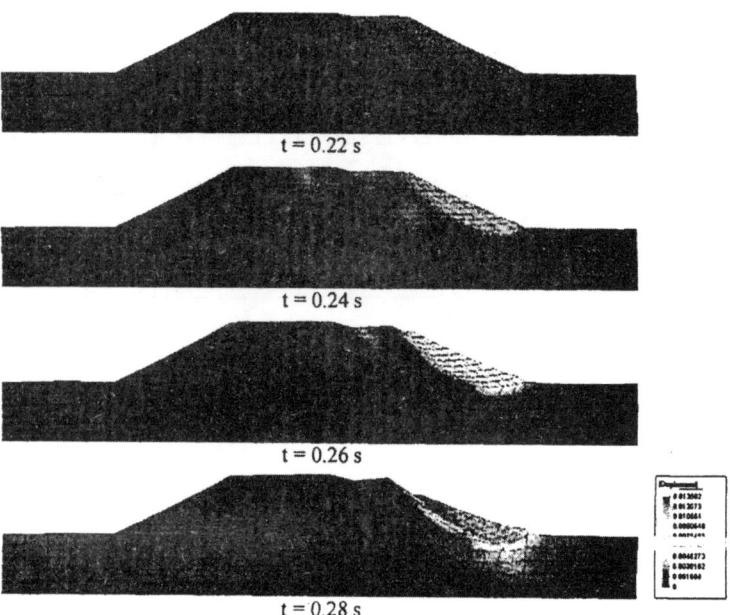

t = 0.22 s

t = 0.24 s

t = 0.26 s

t = 0.28 s

Figure 10. *Deformation contour*

[ZIE 80] O.C. ZIENKIEWICZ, C.T. CHANG, P. BETTESS: Drained, undrained, consolidating dynamic behaviour assumptions in soils, *Géotechnique* **30**, pp. 385–395, 1980.

[ZIE 84] O.C. ZIENKIEWICZ, T.SHIOMI: Dynamic behaviour of saturated porous media: the generalised Biot formulation and its numerical solution. *Int. J. Num. Anal. Meth. Geomech.* **8**, pp. 71–96, 1984.

[ZIE 90a] O.C. ZIENKIEWICZ, A.H.C. CHAN, M. PASTOR, D.K. PAUL, T. SHIOMI: Static and dynamic behaviour of soils: a rational approach to quantitative solutions. I. Fully saturated problems, *Proc. R. Soc. Lond.* **A 429**, pp. 285–309, 1990.

[ZIE 90b] O.C. ZIENKIEWICZ, Y.M. XIE, B.A. SCHREFLER, A. LEDESMA, N. BICANIC: Static and dynamic behaviour of soils: a rational approach to quantitative solutions. II. Semi-saturated problems, *Proc. R. Soc. Lond.* **A 429**, pp. 311–321, 1990.

[ZIE 91] O.C. ZIENKIEWICZ, R.L. TAYLOR: *The Finite Element Method*, Vol. 2, 4th-edition, McGraw-Hill, New York, 1991.

[ZIE 98] O.C. ZIENKIEWICZ, J. ROJEK, R.L. TAYLOR, M. PASTOR: Triangles and tetrahedra in explicit dynamic codes for solids, *Int. J. Num. Meth. Eng.* **43**, pp. 565–583, 1998.

[ZIE 00] O.C. ZIENKIEWICZ, A.H.C. CHAN, M. PASTOR, B. SCHREFLER, T. SHIOMI: *Computational Geomechanics*, John Wiley and Sons, New York, 2000.

Chapter 17

Modelling of Debris Flows and Flow Slides

Manuel Pastor, José Antonio Fernandez Merodo, Maria Isabel Herreros, Elena González and Pablo Mira
Centro de Estudios y Experimentacion de Obras Publicas, Madrid, Spain, and Department of Applied Mathematics, ETS de Ingenieros de Caminos UPM, Madrid, Spain

Manuel Quecedo
Department of Applied Mathematics, ETS de Ingenieros de Caminos UPM, Madrid, Spain, and ENUSA, Madrid, Spain

1. Introduction

The term "landslide" embraces a wide variety of phenomena, from the very slow reptations with velocities in the range of cm/yr, to the fast catastrophic flowslides and debris flows, where velocities can be of the order of 60 Km/h. The causes, origin, and nature of these phenomena are also very different from each other. Among the causes we can mention the following: (i) changes in soil or rock properties due to weathering, (ii) changes of effective stresses induced by variations in pore pressures caused in turn by the rain, (iii) earthquakes, and (iv) loading or change of geometry due to human action. The slopes can be natural, or man made, as in the case of tailing dams and mine waste dumps.

We will focus here in the propagation phase of fast landslides, specially flowslides, debris flow and mudflows. They are dangerous because they can propagate over long distances in very short times without warning, reaching zones which were considered previously "safe".

The main difference between flowslides and debris flows is the amount on origin of the water. In the case of flowslides, they can be of course triggered by heavy rains, but the mechanism of fluidification is the tendency of the soil to compact which causes pore pressure to increase. On the contrary, the origin of debris flows is different. The mixture of mobilized soil and water has a composition which can vary. A typical case (Figure 1) consists of a zone where rainfall erodes the slopes and concentrates into steep, narrow canyons (ravines or gullies). The material is transported along the valley and it is deposited in the fan which develops when the velocity falls below a critical value due to a widening of the valley or a reduction of the slope.

This mechanism can be combined with others. For instance, Figure 2 [DUK 87] shows several landslides which have provided additional material in the ravine.

Depending on topography, rain intensity, etc, debris flows can also be produced in larger valleys, where tributary ravines arrive. In this case, material provided by secondary debris flows can be eroded by the main stream and become part of a larger flow. This is the case of the tributary ravine of the Jiangjia valley in Yunnan (China), where the deposition fan has been eroded extensively by subsequent debris flows in the main valley.

Debris flows are quite often of catastrophic nature. They cause:

– Destruction of farmland and crops

– Losses of human life, houses and equipment in cities close to active ravines, as for instance in Dongchuan (Yunnan, China).

– Partial destruction of lifelines such as roads, railroads, etc. In China, the 2400 km long road from Chengdu (Sichuan) to Lhasa (Tibet) is cut 2 months each year, and in the south province of Yunnan railroads are often interrupted. Figure 3 shows the result of an event which took place on the 9th of September 1999, burying some 600 m of railroad.

Figure 1. *Tributary gully at JiangJia ravine (Yunnan, China)*

Another example of debris flow is that of Venezuela (Dec.1999), where a previous period of rain had softened slopes. New heavy rains caused then general landslides, providing the solid material of the debris flows. Hugh rock blocks were transported, causing in some cases severe damage to buildings. Figures 4 and 5 show some of the damage caused.

As examples of flowslides, we can mention the cases of Aberfan or Jupille, reported by Bishop [BIS 73], and the more recent Santa Tecla in El Salvador.

A paramount aspect is the fluidization or liquefaction process, in which the soil mass is transformed into a fluid like material. Most of the soils involved in flowslides are very loose and metastable, with a strong tendency to compact under shear. If time of loading is much smaller than consolidation time, pore pressures (of water or air)

Figure 2. *Landslides provide material for debris flows (Du et al 1987)*

will make the effective stresses path to approach failure conditions. In the limit, the mean hydrostatic confining pressure can become very close to zero, and the soil will behave as a viscous fluid in which buildings can sink, as it happened in Niigata during the 1966 earthquake.

It has to be mentioned here that this mode of failure can be exhibited also by non-saturated soils such as those of volcanic origin. Indeed, collapse of the material under the loading induced by an earthquake can make the pore air pressure to increase. The phenomenon is controlled by two characteristic time scales, a characteristic time for the consolidation and a characteristic time of loading. If the former is much larger than the latter, there will be not enough time for dissipation of air pore pressures, and the material will arrive to a "dry" liquefaction. Bishop describes the failure of a fly ash tip at Jupille in Belgium and refers to the explanation provided by Calembert and Dantinne [CAL 64]. Some of the catastrophic landslides caused in El Salvador by the 13th of January 2001 earthquake can be explained by this mechanism. Figure 6 shows an aerial view of the dry flowside triggered as a consequence of the 13th February earthquake.

Figure 3. *Railroad cut by debris flow*

Figure 4. *Debris Flow in Venezuela (Photograph by Prof.Zhang Sucheng)*

Figure 5. *Damage to buildings in Venezuela (Photograph by Prof.Zhang)*

Figure 6. *Flow slide at Santa Tecla, El Salvador (2001)*

From an engineering point of view, it is important to:

(i) Assess the risk of flowslide occurrence, providing remedial measures such as soil reinforcement or improvement.

(ii) Predict the characteristics of the flowslide (mass of material involved, velocity, runout distance and path of propagation), in order to design suitable protection and channeling structures.

Unfortunately, classical engineering tools such as slip circles or finite element codes with Mohr-Coulomb like models are unable to provide accurate predictions even for the initiation phase. Which is why more accurate models are needed to describe the propagation phase.

This chapter presents an overview of rheological, mathematical and numerical models which can be used for this purpose.

2. Mathematical Model: Depth Integrated Equations

Once failure has been triggered, the soil mass will move with a velocity which will depend largely on the type of problem. In some cases (flow slides, mudflows,...) the material will flow in a "fluid-like" manner. There exists a strong coupling between the solid and the fluid phases, and the equations describing the movement are similar to those of the initiation phase. However, a first simplification of assuming a single phase can be made in two limit cases (i) very permeable or dry materials, and (ii) materials for which the time scale of consolidation is much larger than that of propagation. Examples of both situations are rock avalanches and mudflows. Care should be taken when selecting constitutive material properties, as the apparent friction angle can be much smaller than the effective, because of the pore pressures. In fact, liquefied material can flow over slopes much smaller than its effective friction angle.

If some assumptions are made about the vertical structure of the flow, it is possible to integrate the balance equations in depth, arriving at the so-called "depth-integrated" equations or "shallow water equations" in coastal and hydraulic engineering. The equations can be further simplified by integrating on cross sections, arriving at simple 1D models.

The depth integrated equations can be cast either in a Eulerian or a Lagrangian form. The second approach has been used mainly in one dimensional situations. The models proposed by Savage & Hutter [SAV 91], Hutter & Koch [HUT 91], Hungr [HUN 95] and Rickenmann & Koch [RIC 97] incorporate features such as introducing a K_0 coefficient relating the horizontal and vertical stress components, or formulating the equations in a curvilinear coordinate systems following the terrain. However, Lagrangian formulations present some important difficulties in the 2D case. For instance, we can mention the problem of bifurcation of the flow when arriving at an obstacle, or merging after it has been passed. Because of this problem, other authors have chosen Eulerian formulations (Vulliet & Hutter [VUL 88]; Laigle & Coussot [LAI 97]; Laigle [LAI 97b]).

The equations of the depth averaged model are obtained integrating the balance of mass and momentum equations along X_3, and using Leibniz's rule. Integration along X_3 of the balance of mass equation results in

$$\frac{\partial h}{\partial t} + \frac{\partial}{\partial x_j}(h\bar{u}_j) = 0 \quad \text{with } j = (1,2) \tag{1}$$

where $\bar{u}_j$ is the depth-averaged velocity along X_j (Figure 7).

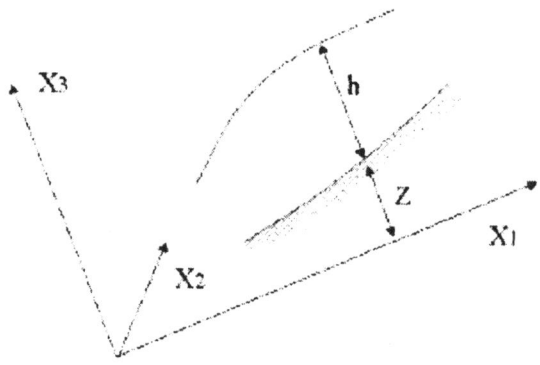

Figure 7. *Notation and reference system*

Next, integration along X_3 of the equations of balance of linear momentum along X_i $(i = 1, 2)$ results in:

$$\rho\frac{\partial}{\partial t}(h\bar{u}_i) + \rho\frac{\partial}{\partial x_j}(h\bar{u}_i\bar{u}_j) = \rho b_i h + \frac{\partial}{\partial x_j}(h\,\bar{\sigma}_{ij}) + t_i^A + t_i^B \quad (i,j = 1,2) \tag{2}$$

where t_i^S and t_i^B are the surface forces at the free surface and the bottom respectively,

$$t_i^S = \sigma_{ij}.n_j|_{Z+h} \tag{3}$$

$$t_i^B = \sigma_{ij}.n_j|_Z \tag{4}$$

and $\bar{\sigma}_{ij}$ is the stress tensor averaged over depth. Surface tractions are caused by interaction with the atmospheric air, and will be not considered in this analysis, $t_i^S = 0$.

3. Depth integrated rheological models

Balance of momentum equations include two terms involving the stress tensor: (i) Divergence of the depth averaged stresses

$$\frac{\partial}{\partial x_j}\left(h\,\bar{\sigma}_{ij}\right) \quad (i,j=1,2) \tag{5}$$

and (ii) tractions acting on the bottom

$$t_i^B = \sigma_{ij}.n_j\big|_Z \tag{6}$$

It is therefore necessary to relate stresses to flow characteristics.

In the initiation phase, the analysis is carried out taking into account the coupling between pore fluid and solid skeleton, using the concept of effective stress and assuming that the strain rate can be related to stresses and their rates by suitable constitutive relations.

Once the flowslide has been triggered, the material fluidizes and a different constitutive approach is needed. The behaviour of fluidized mixtures of soil and water has been described by rheological laws relating total stresses to rate of strains (or deformations):

$$\sigma = \sigma\,(\mathbf{D}) \tag{7}$$

where the rate of deformation tensor $\mathbf{D}$ is given by

$$D_{ij} = \frac{1}{2}\left(\frac{\partial u_i}{\partial x_j} + \frac{\partial u_j}{\partial x_i}\right) \tag{8}$$

Examples of rheological laws are those introduced by Bagnold (1954) in his pioneering work, or the more advanced proposed by Chen and Lin (1996).

If the fluid is isotropic, theorems of representation of isotropic tensor functions provide a general expression for σ :

$$\sigma = -p\,\mathbf{I} + \Phi_1\mathbf{D} + \Phi_2\mathbf{D}^2 \tag{9}$$

where p is a pressure term and Φ_1 and Φ_2 are functions of the invariants of $\mathbf{D}$

$$I_{2D} = \frac{1}{2}\mathrm{tr}\left(\mathbf{D}^2\right) \tag{10}$$

$$I_{3D} = \frac{1}{3}\mathrm{tr}\left(\mathbf{D}^3\right) \tag{11}$$

as we have assumed that fluid is incompressible, and therefore $I_{1D} = 0$.

This "total stresses" approach is valid in two extreme cases: (i) when the permeability is very high and pore pressures dissipate in a time much smaller than propagation time (fully drained behaviour), and (ii) when the permeability is very small, and pore pressures do not change much during the slide.

However, in real situations, pore pressures p_w and effective stresses σ' may change during the flowslide, and therefore, a more rational approach can be obtained using the effective stress principle:

$$\sigma' = -p'\,\mathbf{I} + \Phi_1\mathbf{D} + \Phi_2\mathbf{D}^2 \tag{12}$$

with

$$\sigma = \sigma' - p_w \mathbf{I} = -\left(p' + p_w\right)\mathbf{I} + \Phi_1 \mathbf{D} + \Phi_2 \mathbf{D}^2 \tag{13}$$

The decision as to which approach should be used will depend on the type of problem being solved and the available experimental data as will be shown later.

The models that will be used in this work are a particular class of (9),

$$\sigma = -p\,\mathbf{I} + \Phi_1 \mathbf{D} \tag{14}$$

where p is the hydrostatic pressure, given by:

$$p = -\rho b_3 \left(h - \xi\right) \tag{15}$$

In above ξ is the distance to the bottom along X_3, i.e., $\xi = x_3 - Z$. The scalar function Φ_1 is composed of two parts, static and dynamic. The static component Φ_{1s} is

$$\Phi_{1s} = \frac{\tau_y}{\sqrt{I_{2D}}} \tag{16}$$

where τ_y defines a yield locus in the stress space within which no flow occurs. Depending on the choice of τ_y the model will be (i) a Bingham fluid if τ_y is constant, and (ii) a frictional fluid if $\tau_y = -p' \tan \phi$. The contribution of the static component to the stress tensor is therefore

$$\frac{\tau_y}{\sqrt{I_{2D}}} \mathbf{D} \tag{17}$$

which is rate independent. The dynamic component will be assumed to be of the form

$$\Phi_{1d} = 2\mu \tag{18}$$

which corresponds to a classical Newtonian fluid. Therefore, the total stress tensor is given by:

$$\sigma = -p\,\mathbf{I} + \left(\frac{\tau_y}{\sqrt{\mathbf{I}_{2D}}} + 2\mu\right)\mathbf{D} \tag{19}$$

This expression cannot be used directly in depth integrated models to obtain depth averaged stresses and tractions at the bottom, because there is no available information on how velocity changes along X_3. Indeed, details of the flow structure along X_3 have been lost when averaging the velocities.

A possible solution to this problem consists of assuming that the flow structure is that of a "simple shear" flow, where u_1 and u_2 depend only on x_3 and u_3 is zero. Denoting by θ the angle between velocity and the X_1 axis, components of velocity will be $u_1 = u \cos \theta$ and $u_2 = u \sin \theta$, where u is the velocity modulus. The rate of deformation tensor $\mathbf{D}$ is

$$\mathbf{D} = \begin{pmatrix} 0 & 0 & \frac{1}{2}\frac{\partial u}{\partial x_3}\cos\theta \\ 0 & 0 & \frac{1}{2}\frac{\partial u}{\partial x_3}\sin\theta \\ \frac{1}{2}\frac{\partial u}{\partial x_3}\cos\theta & \frac{1}{2}\frac{\partial u}{\partial x_3}\sin\theta & 0 \end{pmatrix} \tag{20}$$

from where I_{2D} is obtained as:

$$I_{2D} = \frac{1}{2}\frac{\partial u}{\partial x_3}$$
(21)

The components of the total stress tensor are:

$$
\begin{aligned}
&\sigma_{11} = \sigma_{22} = \sigma_{33} = -p \\
&\sigma_{12} = \sigma_{21} = 0 \\
&\sigma_{13} = \sigma_{31} = \left(\tau_y + \mu\frac{\partial u}{\partial x_3}\right)\cos\theta \\
&\sigma_{23} = \sigma_{32} = \left(\tau_y + \mu\frac{\partial u}{\partial x_3}\right)\sin\theta
\end{aligned}
$$
(22)

Now it is possible to integrate along X_3 the stresses to obtain the depth averaged stresses

$$
\begin{aligned}
&\bar{\sigma}_{11} = \bar{\sigma}_{22} = -\bar{p} = \tfrac{1}{2}\rho b_3 h \\
&\bar{\sigma}_{12} = \bar{\sigma}_{21} = 0
\end{aligned}
$$
(23)

We will consider next the bottom tractions for two particular cases of interest: the Bingham fluid and the pure frictional fluid.

3.1. Bingham fluids

In the case of Bingham fluids, shear stress in simple shear flows is given by $\tau_y + \mu\frac{\partial u}{\partial x_3}$ where the yield stress τ_y and the viscosity μ are material constants which do not depend on the stress level nor on the rate of shear strain. If we consider a Bingham fluid initially at rest, and we increase the shear stress, the fluid will continue at rest until the shear stress reaches τ_y. Once the threshold is passed, the rate of deformation will increase with the shear stress. If the external forces decrease, the velocity will decrease until the fluid stops, which will occur at $\tau = \tau_y$. Therefore, this rheological law is able to reproduce both an static resistance to initiation of the flow, and stoppage of it.

One of the main features of Bingham fluids is the formation of "plugs", or zones where velocity is constant and rate of deformation is zero. The relative height of the plug is given by

$$\frac{h_p}{h} = \frac{\tau_y}{\tau_B}$$
(24)

where τ_B is the shear stress at the bottom. It is interesting to note that if $\tau_y = 0$ no plug will exist.

The shear stress τ_B can be related to the depth averaged velocity u by

$$\bar{u} = \frac{\tau_B}{6\mu}h\left(1 - \frac{\tau_y}{\tau_B}\right)^2\left(2 + \frac{\tau_y}{\tau_B}\right)$$
(25)

which can be written as

$$x^3 - (3 + a) x + 2 = 0 \qquad (26)$$

where we have introduced the unknown $x = \tau_y/\tau_B$ and the parameter $a = (6\mu\,\bar{u})\,/\,(h\,\tau_y)$ The solution lies in the interval $[0, 1]$.

Bingham models cast in terms of total stresses have been widely applied to the analysis of flowslides, mudflows and debris flows. Rickenmann and Koch [RIC 97] propose values of τ_y and μ in the ranges 100-800 Pa and 400-800 Pa.s in simulations of debris flows in Kamikamihori (Japan) and Saas (Switzterland) valleys events. From their simulations, it can be concluded that Bingham models predict higher velocities than other models.

3.2. Frictional Fluids: the role of pore pressures

The Bingham fluid model has been written in terms of total stresses. This approximation is valid when the consolidation time scale is much larger than the time of propagation where the behaviour is undrained, or when the permeability is so high that the overall behaviour can be considered as drained. In both cases, material properties can be taken as constant. Care has to be taken when obtaining the properties for the undrained conditions, as the apparent friction angle in total stresses will depend on the initial conditions. For high water contents, this angle is usually taken as zero, as in the case of tailings.

This simplification is not valid when the pore pressures can evolve during propagation, as the apparent friction angle will change, making the mobilized soil stop. This was pointed out by Hutchinson [HUT 86], who proposed a "sliding-consolidation" model to predict runout of flow slides. In that work, he predicted the runout distance of the Aberfan flow slide using a model where the effective friction angle was chosen as $\phi' = 36°$, and the excess pore pressures varied according to consolidation along vertical paths.

It is also worth mentioning the model proposed by Hungr [HUN 95] , where he introduced a bottom friction term depending on the product $(1 - r_u) \tan \phi'$, where the factor r_u accounts for the pore pressures. This is equivalent to substituting the effective friction angle by an "apparent" friction angle (or "total" friction angle) given by $\tan \phi_{ap} = (1 - r_u) \tan \phi'$, where r_u is assumed to be constant. In the limit case of a fully liquefied mass of soil, the angle ϕ_{ap} is zero, and the behaviour of the flowing soil can be described by a Bingham model, for instance.

However, there will be cases where the evolution of the pore pressures within the soil mass will play a paramount role in the propagation. Stopping of the flow slide can be caused both by a reduction on the terrain slope and by a decrease of the excess pore pressures.

The second particular case of the rheological law (19) that will be considered here is the frictional fluid, which is obtained by choosing $\tau_y = -p' \tan \phi'$ and $\mu = 0$.

The shear stress at the bottom is given by:

$$\tau_B = -p'_B \tan\phi' \tag{27}$$

where p'_B is the effective hydrostatic pressure, $p'_B = p_B - p_{wB}$, and p_{wB} the pore pressure at the bottom.

Following Hutchinson [HUT 86] we will assume that there exists a layer of saturated soil of height h_s on the bottom of the flowing material. The decrease in pore pressures is caused by vertical consolidation of this layer. Pore pressures on the top and bottom of this layer can be either estimated from the values of the vertical stresses or obtained directly from the results of finite element computations.

Assuming that the excess pore pressure evolves as

$$p_w(x_3, t) = N(x_3)\,\bar{p}_w(t) \tag{28}$$

it is possible to obtain a closed form solution of the consolidation equation. In the case of $N(x_3) = \sin\left(\frac{\pi}{2}\frac{x_3}{h_s}\right)$, the solution is

$$\bar{p}_w(t) = \bar{p}_w^0 \exp\left(-\frac{t}{T_v}\right) \tag{29}$$

where $T_v = \frac{4h_s^2}{\pi^2 c_v}$ and c_v is the coefficient of consolidation. Finally,

$$p_w(x_3, t) = \bar{p}_w^0\, N(x_3) \exp\left(-\frac{t}{T_v}\right) \tag{30}$$

The coefficient r_u can be estimated as:

$$r_u = r_u^0 \exp\left(-\frac{t}{T_v}\right) \tag{31}$$

In above, we have used "pore pressures" instead of "pore water pressures". The reason is that dry or partially saturated materials can collapse generating high air pressures on their pores, which will cause a similar effect.

4. Initial conditions

The proposed depth-integrated model requires the following initial conditions as input:

(i) Sliding mass, defined by $h(x_i; t = 0)$

(ii) Basal pore pressure factor r_{u0}

(iii) Parameters of the rheological law

Most of the examples presented in the literature deal with flowslides which have been produced in the past. Therefore, the sliding mass is known, and both material parameters and the basal pore pressure factor can be determined by back analysis.

In real situations it is necessary to predict also the sliding mass and to estimate the factor r_{u0}. Moreover, it is important to know whether the slide will evolve or not into a flowslide.

Nowadays, it is difficult to provide accurate answers to these questions. The fundamental difficulty is the "phase transition" between the solid and fluidized states (and viceversa), for which a consistent model is required.

The approximation proposed in this paper consists of using non-linear coupled models implementing suitable constitutive equations to determine the extent of the mobilized mass of soil together with the initial distribution of pore pressures. The modelling method will consist of determining the initial conditions (stresses and pore pressures), and then simulating the failure mechanism (changes in pore pressures induced by rain, external loading, earthquake,..). Even if the external loading is not of dynamic nature, it is highly recommended to perform the analysis taking into account possible accelerations which may develop when failure approaches. Otherwise, loss of convergence of non-linear iterations will stop the process. This is the case, for instance, of an applied load higher than the limit load. Indeed, these computations are usually carried out under displacement control.

5. Discretization

The equations described in previous sections are non-linear, and can exhibit shocks and discontinuities. The problem is similar to that of propagation of a flood wave caused by breaking of a dam, and it has been solved using special algorithms for problems involving shocks However, classical algorithms can provide enough accuracy with a smaller computational effort. Here we will use the two-step Taylor Galerkin algorithm introduced by Peraire, Zienkiewicz and Morgan and Donea [ZIE 86].

The method consists on performing a Taylor series expansion on time which allows substitution of time by space derivatives.

The first step, from t^n to $t^{n+1/2}$ gives

$$\phi^{n+1/2} = \phi^n + \frac{\Delta t}{2} \left. \frac{\partial \phi}{\partial t} \right|^n \tag{32}$$

from which, using the system of PDE's results in:

$$\bar{\phi}^{n+1/2} = \bar{\phi}^n + \frac{\Delta t}{2} \left(\bar{S} \text{-} di\bar{v} \, F \right)^n \tag{33}$$

where the overbar denotes values averaged over elements

$$\bar{\phi} = \frac{1}{\Omega^e} \int_\Omega \phi \, d\Omega \tag{34}$$

Once we have obtained the values of the vector of unknowns at $t^{n+1/2}$ we compute the fluxes and sources, which will be used in the second step:

$$\phi^{n+1} = \phi^n + \Delta t \left(\bar{S} \cdot \overline{div} \, F \right)^{n+1/2} \tag{35}$$

This equation is discretized using the classical Galerkin method:

$$\underline{M} \, \Delta \phi = \Delta t \left(\int_\Omega \underline{N} \, S^{n+1/2} d\Omega + \int_\Omega \underline{F}^{n+1/2} \, grad \, \underline{N} \, d\Omega - \int_{\Gamma_N} \underline{N} \, \underline{F}^{n+1/2} d\Omega \right) \tag{36}$$

6. Examples

6.1. *Tip No. 7 flowslide at Aberfan (1966)*

On October 21st 1966, a flow slide developed at a tip of loose colliery waste in Aberfan. It propagated downhill and into the village of Aberfan, causing 144 deaths. The material was loose coal mine waste, which was deposited by end-tipping. The height of the tip was 67 m. from the toe, and the natural slope of natural terrain was 12%. The failure mechanism and geotechnical properties have been described by Bishop et al. [BIS 69] and summarized in "The stability of tips and spoil heaps" [BIS 73] and "A sliding consolidation model for flow slides" [HUT 86], where the interested reader can find a detailed description. Here we will recall some aspects which will be of interest in the analysis.

First of all, failure was caused by artesian pore pressures at the toe, which probably saturated the lower part of the tip, while the upper part remained unsaturated. Bishop reports that "...the rescue work was complicated by water which flowed, after the slip, from the sandstone at the base of the rotational slip, where the boulder clay and head was stripped off...". Once failure was triggered, the flowslide propagated downhill 275 m., then divided into two, a north and a south lobe. It was the larger south lobe which ran into the village, while the northern lobe stopped after reaching an embankment.

The Aberfan flowslide had a runout of 600 m., although it is believed that it could have been larger if no obstructions were within its path. The velocity has been estimated from witnesses as 4.5-9 m/s.

One of the main difficulties when modelling the Aberfan flowslide is the role of pore pressures. Coal debris was a material composed of solid, fluid and gas phases, with a strong interaction between them. However, models cast in terms of effective stresses, pore pressures, and velocities of all constituents have not been used so far, because of the problem of moving interfaces mentioned above. Depth integrated models can provide important information about runout, propagation paths and velocities, which is frequently sufficient to design protection structures. The main shortcoming of depth integrated models comes from the fact that pore fluid and solid particles are modelled as a single phase material, with properties that do not change with time. This is why the Aberfan flowslide has been modelled assuming a Bingham fluid rheological law for the debris. For instance, Jeyapalan et al. [JEY 83] and Jin & Fread [JIN 97] obtained results which fitted well the observations choosing $\tau_y = 4794$ Pa., $\mu = 958$ Pa.s and $\rho = 1760$ kg/m^3. Even if the results are good, it is possible to argue that waste coal was not fully saturated, and the material was frictional. Of course, the apparent angle of friction introduced above will be much smaller than ϕ', but vertical consolidation could have made it to change during the propagation phase. Hutchinson [HUT 86] proposed a simple "sliding-consolidation" model in which it was clear that the combination of friction with basal pore pressures could provide accurate results of runout and velocities.

Here we have used the simplified frictional fluid law described in Section 2, in combination with the pore pressure dissipation mechanism therein described. The information provided in the literature does not give enough data to perform a realistic analysis in two dimensions. Therefore, we have used a simple 1D model with the terrain profiles given by Jeyapalan et al. [JEY 83] . The main purpose of this example is to show that a depth integrated model using pore pressure dissipation can reproduce the basic patterns observed.

Density of the mixture ρ and friction angle ϕ', have been taken as 1740 kg/m^3 and $36°$ respectively. Consolidation time T_v has been estimated from the range of values proposed by Hutchinson (1986) for the depth of the saturated liquefied layer $h_s = 0.1$ m and the coefficient of consolidation $c_v = 6.4\,10^{-5}$ m^2/s. Assuming $T_v = \frac{4h_s^2}{\pi^2 c_v}$, we arrive at $T_v = 64$ s. The initial value of the pore pressure coefficient r_{u0} has been taken as 0.78.

The results obtained in the simulation are given in Figure 8 where sections of the free surface of the flowslide are given at times 0., 5, 10, 20. and 60 s. The basic features of the flow obtained in the simulation (propagation distance ≈ 600 m, freezing time ≈ 40 s and average velocity 15 m/s) coincide reasonably well with those reported in the literature. Figure 9 shows the position of the front as a function of time.

6.2. Air Liquefaction Flow Slides

If the soil is partially saturated, its behaviour depends on the coupling between the solid skeleton and the pore air and water. In the limit case of a dry soil, the air has to flow out from the pores for the material to consolidate, but typical air consolidation times are much smaller than those of the water. Therefore, in practical cases, the role of air pressures is neglected, as the characteristic time of loading is much larger than consolidation time. However, it is possible to imagine situations with much smaller loading times, where coupling between pore air and soil skeleton plays a paramount

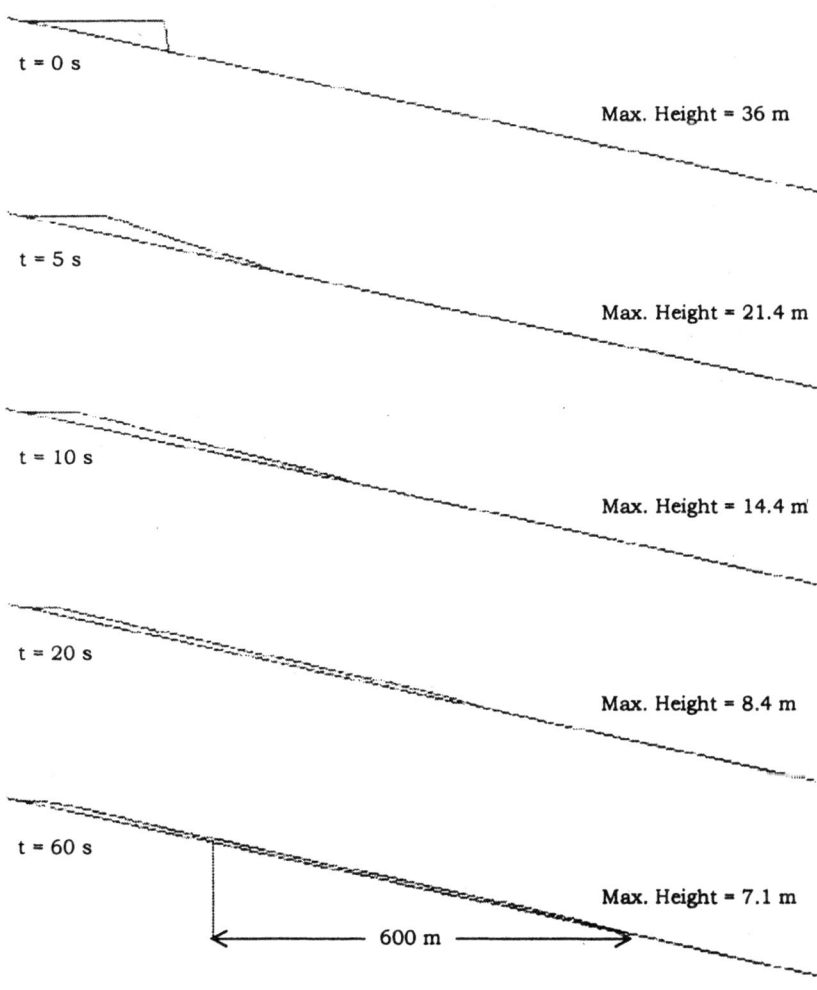

t = 0 s

Max. Height = 36 m

t = 5 s

Max. Height = 21.4 m

t = 10 s

Max. Height = 14.4 m

t = 20 s

Max. Height = 8.4 m

t = 60 s

Max. Height = 7.1 m

←————— 600 m —————→

Figure 8. *Free surface of Aberfan flowslide at different times.*

role. This is the case of fluidized granular beds, just to mention a particular example of industrial interest.

Bishop [BIS 73] describes the case of the Jupille flowslide, which happened in Belgium in February 1961. A tip of uncompacted fly ash located in the upper part of

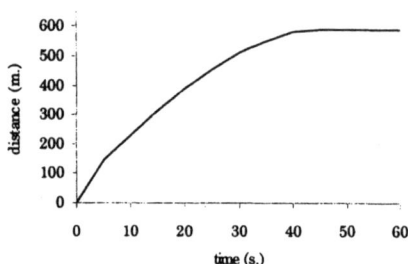

Figure 9. *Position of head of Aberfan flowslide as a function of time*

a narrow valley collapsed, and the subsequent flowslide travelled for about 600 m at very high speeds (130 km/h) until it stopped. Triggering mechanism was suggested to be "collapse due to undermining of a steep, partly saturated and slightly cohesive face". Bishop referred to Calembert and Dantinne [CAL 64], who pointed out the role of the entrapped air. The mechanism of pore air consolidation can explain the "sort of fog" which formed above the flowing material, as warmer air met the colder winter air in the exterior. The "honeycomb" like structure of soil in the tip was responsible of its sudden collapse when movement started.

As, unfortunately, there are not available data in the literature to model accurately the conditions of Jupille flowslide, we will propose a "thought numerical experiment" in which we will consider a flowslide with initial air pore pressures with a dissipation time characterized by $T_v = 10\ s$. This is a crude simplification of reality, where there will coexist a saturated layer in the bottom with high water pore pressures and a second layer of unsaturated soil.

The layout of the problem is given in Figure 10. Vertical height of the slope is 40 m, and the liquefied soil has a maximum depth of 10 m.

We have assumed that the mass of soil depicted at $t = 0$ has zero effective pore pressures due to its sudden collapse caused by an earthquake, which has also eliminated any initial cohesion or cementation. Density has been taken as $\rho = 1300\ Kg/m^3$ friction angle is $\phi' = 30°$. Once failure has been triggered, the flowslide propagates downhill as shown in Figure 10 until it stops at $t = 7.5\ s$.

7. Acknowledgements

The authors would like to gratefully acknowledge the kind help provided by many colleagues who, during these years, encouraged them to study this difficult problem and provided them with examples, information and field data. Special thanks to Profs.

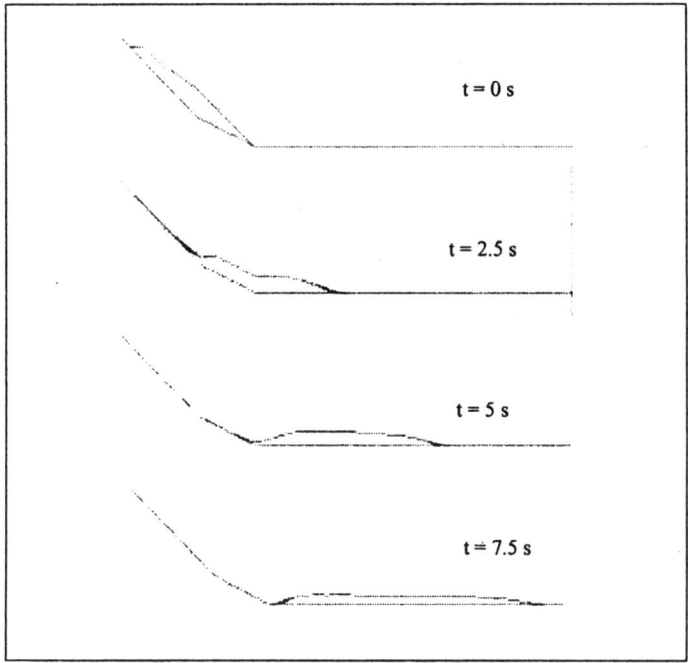

Figure 10. *Flowslide propagation*

Zhang Sucheng and Yu Bin, from the Institute of Mountain Hazards of the Chinese Academy of Sciences, and to Profs. Li and Liu (Hohai Univ.) for everything we have shared during several journeys in China, and to the members of the Spanish mission in El Salvador in 2001 (Belén Benito, Paco Lamas, Carlos López Casado, Ana Negredo and Antonio Pazos).

8. References

[BAG 54] Bagnold, R.A. (1954). "Experiments on a gravity-free dispersion of large solid spheres in a Newtonian fluid under shear", **Proc.Royal Society A**, 225, pp.49-63.

[BIS 69] Bishop, A.W., Hutchsinson, J.N., Penman, A.D.M. & Evans, H.E. (1969). "Geotechnical Investigations into the causes and circumstances of the disaster of 21st October 1966". **A selection of technical reports submitted to the Aberfan Tribunal**, pp. 1-80, Welsh Office, H.M.S.O. London, England.

[BIS 73] Bishop A.W. ,. The stability of tips and spoil heaps, **Quart.J.Eng.Geol.** 6, 335-376, 1973.

[CAL 64] Calembert, L. & Dantinne, R. (1964). "The avalanche of ash at Jupille (Liege) on February 3rd, 1961. From: **The commemorative volume dedicated to Professeur F.Campus**, pp. 41-57, Liége, Belgium.

[CHE 96] Chen, C.L. & Ling, C.H. (1996). "Granular-Flow Rheology: Role of Shear-Rate Number in Transition Regime", **J.Engng.Mech. ASCE**, 122, 5, 469-481.

[DUK 87] Du R., Kang Z. and Zhu P., **Debris Flow of Xiaojiang basin in photographs**, Sichuan Publishing House of Science and Technology, 1987.

[HUN 95] Hungr, O. (1995). "A model for the runout analysis of rapid flow slides, debris flows and avalanches", **Can.Geotech.J.** 32, pp 610-623.

[HUT 86] Hutchinson, J.N. (1986). "A sliding-consolidation model for flow slides", **Can.Geotech.J.**, 23, 115-126.

[HUT 91] Hutter, K. & Koch, T. (1991). "Motion of a granular avalanche in an exponentially curved chute: experiments and theoretical predictions",**Phil.Trans.R.Soc.London, A 334**, pp 93-138.

[JEY 83] Jeyapalan, J.K., Duncan, J.M. & Seed, H.B. (1983). "Investigation of flow failures of tailing dams", **J.Geotech.Engng.ASCE 109**, pp 172-189.

[JIN 97] Jin, M. & Fread, D.L. (1997). One-dimensional routing of mud/debris flows using NWS FLDWAV model, in C.L.Chen (Ed.), **Debris-Flow Hazards Mitigation: Mechanics, Prediction and Assessment, ASCE**, pp. 687-696.

[LAI 97] Laigle, D. & Coussot, P. (1997). "Numerical Modelling of Mudflows", **J.Hydr.Engng. ASCE, Vol.123**, pp. 617-623.

[LAI 97b] Laigle, D. (1997). "A two-dimensional model for the study of debris-flow spreading on a torrent debris fan", in C.L.Chen (Ed.), **Debris-Flow Hazards Mitigation: Mechanics, Prediction and Assessment, ASCE** , pp. 123-132.

[PER 86] Peraire, J "**A Finite Element Method for Convection Dominated Flows**", Ph.D. Thesis, University of Wales, Swansea (1986).

[ZIE 86] Peraire,J. , Zienkiewicz, O.C. and Morgan,K. "Shallow Water Problems. A General Explicit Formulation", **International Journal for Numerical Methods in Engineering.** 22, pp. 547-574 (1986).

[RIC 97] Rickenmann, D. & Koch, T. (1997). "Comparison of debris flow modelling approaches", in **Debris-Flow Hazards Mitigation:Mechanics, Prediction and Assessment,** Proc.1st Int.Conference, ASCE, C-l.Chen (Ed.), pp.576-585.

[SAV 91] Savage, S.B. and Hutter, K. (1991). "The dynamics of avalanches of granular materials from initiation to runout. Part I: Analysis", **Acta Mechanica 86**, pp 201-223.

[VUL 88] Vulliet, L. & Hutter, K. (1988). "Continuum model for natural slopes in slow movement", **Géotechnique 38**, pp. 199-217.

[ZHA 93] S.Zhang, "A comprehensive Approach to the Observation and prevention of Debris Flows in China", **Natural Hazards**, 7, 1-23, 1993.

[ZIE 91] O.C.Zienkiewicz and R.L.Taylor, **The Finite Element Method** (4th. edition), Vol.2, McGraw-Hill, 1991.

Index